Advanced X-by-Wire Technologies in Design, Control and Measurement for Vehicular Electrified Chassis

Advanced X-by-Wire Technologies in Design, Control and Measurement for Vehicular Electrified Chassis

Editors

Yong Li
Xing Xu
Lin Zhang
Yechen Qin
Yang Lu

MDPI • Basel • Beijing • Wuhan • Barcelona • Belgrade • Manchester • Tokyo • Cluj • Tianjin

Editors
Yong Li
Jiangsu University
Zhenjiang
China

Xing Xu
Jiangsu University
Zhenjiang
China

Lin Zhang
Tongji University
Shanghai
China

Yechen Qin
Beijing Institute of
Technology
Beijing
China

Yang Lu
Tsinghua University
Beijing
China

Editorial Office
MDPI
St. Alban-Anlage 66
4052 Basel, Switzerland

This is a reprint of articles from the Special Issue published online in the open access journal *World Electric Vehicle Journal* (ISSN 2032-6653) (available at: https://www.mdpi.com/journal/wevj/special_issues/Vehicular_Electrified_Chassis).

For citation purposes, cite each article independently as indicated on the article page online and as indicated below:

LastName, A.A.; LastName, B.B.; LastName, C.C. Article Title. *Journal Name* **Year**, *Volume Number*, Page Range.

ISBN 978-3-0365-8056-2 (Hbk)
ISBN 978-3-0365-8057-9 (PDF)

Contents

World Electric Vehicle Journal

MDPI

Editorial

Advanced X-by-Wire Technologies in Design, Control and Measurement for Vehicular Electrified Chassis

Yong Li

Automotive Engineering Research Institute, Jiangsu University, 301 Xuefu Road, Zhenjiang 212013, China; liyongthinkpad@outlook.com

Citation: Li, Y. Advanced X-by-Wire Technologies in Design, Control and Measurement for Vehicular Electrified Chassis. *World Electr. Veh. J.* **2023**, *14*, 136. https://doi.org/10.3390/wevj14060136

Received: 15 May 2023
Accepted: 15 May 2023
Published: 25 May 2023

1. Introduction

Advanced X-by-wire technologies for vehicular electrified chassis play an essential role in developing new energy-intelligent vehicles, which is the inevitable choice for intelligent vehicles in the future. This technology is involved in mechanical engineering, electronic and electrical engineering, computer technology, control engineering, signal processing, and artificial intelligence. Advanced electrified chassis control technology transmits control signals through cables and acts directly on the actuator to implement its corresponding actions. The application of X-by-wire technologies for vehicular electrified chassis has changed complex mechanical connections among actuators and hydraulic and pneumatic equipment in the past, significantly promoting energy efficiency, integration, and intelligence.

This Special Issue focuses on advanced X-by-wire technologies in durable reliability design, modeling, integration control, thermal management, energy management, fault diagnosis, and fault-tolerant control with the vehicular electrified chassis. Therefore, this Special Issue aimed to solicit recent advanced X-by-wire technologies for vehicular electrified chassis.

The topics of interest included but were not limited to:

➢ Modeling, analysis, control, and management of electrified chassis;
➢ Coordinated control of integrated chassis;
➢ Highly integrated design technology of electronic control suspension, steering by wire, braking by wire;
➢ High-efficiency motor drive control, thermal management, electric drive system design;
➢ Autonomous driving and intelligent linearization control technology;
➢ Testing and signal analysis technology of electrified chassis;
➢ Vibration and noise suppression;
➢ Reliability design and evaluation;
➢ System operation condition monitoring and fault diagnosis technology;
➢ Highly reliable fault-tolerant control technology.

A total of 15 papers (from 17 submitted) were published. These papers can be loosely categorized into four sections: (1) Suspension System; (2) Trajectory Planning and Control; (3) Vehicle Torque Distribution; (4) Motor Control Review. In this article, we provide a brief overview of the published papers.

2. Overview of Contribution

2.1. Suspension System

Reference [1] studies using a mechatronic inerter to enhance vibration isolation in vehicle seat suspensions by introducing it into a half-vehicle model and optimizing the seat suspension layout parameters using the particle swarm optimization algorithm. Numerical simulations show that the mechatronic inerter improved the vibration isolation performance of the suspension compared to a passive suspension and increased the transfer function-order of the external electrical network further by reducing seat acceleration and pitch acceleration RMS values. Reference [2] proposes a design method for vehicle ISD (Inerter Spring Damper) suspension systems that utilize the fractional-order electrical network structure of a mechatronic inerter. This method used the inerter's external electrical network to simulate the corresponding mechanical network structure equivalently. A 1/4 dynamic model of the suspension was constructed, and the improved oustaloup filtering algorithm was used to simulate fractional-order components. The optimal fractional electrical network structure and parameters were obtained through the structure-immittance approach and an optimization algorithm. Reference [3] proposes a new active suspension control strategy that combines a fuzzy neural network and a proportional-integral-derivative (PID) controller to improve vehicle comfort and smoothness and reduce the vibrations caused by uneven road surfaces. The main optimization target was body acceleration, and the PID controller parameters were adjusted in real-time. An offline optimization and online fine-tuning method for fuzzy neural network parameters were proposed using a particle swarm optimization algorithm and gradient descent method. Reference [4] focuses on developing a mechatronic inerter consisting of a ball-screw inerter and permanent magnet electric machinery. The study proves the feasibility of using electrical element impedances to simulate corresponding mechanical elements. The impedance characteristics of the bridge and series-parallel electrical networks were introduced, and their effectiveness in improving the vibration isolation performance of the mechatronic inertial suspension was compared. The advantages of the bridge network were demonstrated, and a real vehicle test shows that the mechatronic inertial suspension based on the bridge network is superior to the passive suspension. Reference [5] proposes a fault diagnosis design method for the solenoid valve in the electronically controlled air suspension (ECAS) system based on multiple extended Kalman filter banks (EKFs). The fault model of the solenoid valve was built by the fault tree analysis of the ECAS system and considered the correlation between the duty cycle and flow rate of the air spring solenoid valve. An adaptive threshold was used for fault diagnosis, and an active fault-tolerant control was carried out based on an analytical model. The real controller based on d2p rapid prototyping technology and the vehicle model based on AMESim were tested on the hardware-in-the-loop (HiL) simulation platform.

2.2. Trajectory Planning and Control

Reference [6] proposes an automatic driving trajectory-planning method using a variable Gaussian safety field to improve planning efficiency and safety. This method used a time series bird's-eye view as the input to extract features of the surrounding traffic environment and the policy gradient algorithm to generate the planned trajectory. The variable Gaussian safety field improved the safety of the reinforcement learning vehicle tracking algorithm. The simulation results demonstrated the method's excellent trajectory planning ability in highway scenes with high safety and precision tracking control. Reference [7] presents a trajectory-tracking control algorithm for X-by-wire electric vehicles based on a hierarchical control architecture. The algorithm consists of three layers: trajectory tracking, tire force distribution, and actuator control. The trajectory tracking layer used the model predictive control algorithm to control the vehicle and follow the desired trajectory. The tire force distribution layer solves the tire force distribution problem using quadratic programming with constraints. The actuator control layer obtained longitudinal and lateral forces of each tire and calculated the vehicle's steer angle and driving torque. Simulation results show that the proposed algorithm could accurately track the desired trajectory under different driving conditions. Reference [8] proposes a path-tracking controller for distributed drive electric vehicles to achieve safe obstacle avoidance. The path planning was based on a sixth-degree polynomial with anti-collision and anti-rollover conditions. The Model Predictive Control (MPC) controller outputs the front wheel steering angle and additional yaw moment, while the torque distribution controller distributes the wheel torque. The obstacle avoidance path-tracking control was achieved through the additional yaw moment and the vertical force ratio of the wheel. Reference [9] proposes an adaptive robust control framework for the path-tracking control of X-by-wire autonomous vehicles. The non-singular fast terminal sliding mode control algorithm was used to formulate the control law, and the radial basis forward neural network was introduced to estimate system uncertainty in real time. The dynamic model of an active front steering system was established, and the model reference control algorithm was applied to the steering torque control. Reference [10] proposes a multi-agent coordinated control system for active collision avoidance in intelligent vehicles. This system uses hierarchical control and blackboard model methods to handle conflicts between different agent decisions and achieve multi-decisions and planning simultaneously. The fuzzy sliding mode control theory was used to ensure accurate path tracking in lateral collision avoidance.

2.3. Vehicle Torque Distribution

Reference [11] proposes an anti-roll and anti-rollover control strategy to improve the roll stability of distributed drive electric vehicles (DDEV). The control strategy used the active control of wheel torque adjustment to achieve an effect similar to active suspension. This strategy decoupled roll motion and yaw motion and used the LQR algorithm and sliding mode variable structure to calculate the direct yaw moment and additional rolling moment, respectively. Reference [12] proposed a regenerative-braking torque optimization method for dual-motor electric vehicles (EVs) that integrated energy recovery and braking stability. This method used the genetic algorithm theory and considered the state of charge (SOC), vehicle speed, and braking intensity to design an energy recovery-dominated regenerative braking torque distribution rule. Reference [13] proposes a torque distribution method for four in-wheel motor drive (4IWMD) electric vehicles that aim to optimize torque distribution and energy efficiency. The dynamic programming (DP) control algorithm was used to distribute the torque between the front and rear in-wheel motors for optimal torque distribution and energy efficiency. Reference [14] proposed an acceleration slip regulation (ASR) control strategy based on nonlinear model predictive control (NMPC) for front and rear dual-motor four-wheel drive electric vehicles (4WD EVs). The ASR controller tracks reference speed or optimal slip rate, including intervention and exit mechanisms. The motor output torque was determined according to the wheel with the increased slip rate to enhance the passibility of split road surfaces. Simulation experiments on different road conditions demonstrated that the proposed controller exhibited better dynamic performance and stability than the PID control, particularly under low speed and low adhesion road conditions, and met robustness requirements.

2.4. Motor Control Review

Reference [15] reviews the position's sensorless compound control technology for permanent magnet synchronous motors (PMSMs). This technology improved motor reliability, reduced costs, and expanded the speed range of PMSMs. The article elaborated on the compound control technology of PMSMs without a position sensor and summarized the existing problems and development trends of sensorless compound control technology.

3. Final Thoughts

The Special Issue covers advances in the design, control and measurement of X-by-Wire technologies. The Guest Editorial Board hopes that this Special Issue provides state-of-the-art research in the field of vehicular electrified chassis. This Special Issue also presents various innovative approaches with promising results to address the challenges in vehicular electrified chassis. The Guest Editorial Board also hopes that more researchers will enter into this interesting field to promote respective research in the near future.

Acknowledgments: We appreciate the contributions of all authors and the efforts of all reviewers in the peer-reviewing of submitted papers. We also would like to thank the Editor-in-Chief of World Electric Vehicle Journal, and the Editorial office for their great support and help during the publication of this Special Issue.

Conflicts of Interest: The author declares no conflict of interest.

References

1. Qiu, C.; Liu, X.; Shen, Y. Improvement of the Vehicle Seat Suspension System Incorporating the Mechatronic Inerter Element. *World Electr. Veh. J.* **2023**, *14*, 29. [CrossRef]
2. Hua, J.; Shen, Y.; Yang, X.; Zhang, Y.; Liu, Y. Optimal Design of Fractional-Order Electrical Network for Vehicle Mechatronic ISD Suspension Using the Structure-Immittance Approach. *World Electr. Veh. J.* **2023**, *14*, 12. [CrossRef]
3. Li, M.; Li, J.; Li, G.; Xu, J. Analysis of Active Suspension Control Based on Improved Fuzzy Neural Network PID. *World Electr. Veh. J.* **2022**, *13*, 226. [CrossRef]
4. Zhang, T.; Yang, X.; Shen, Y.; Liu, X.; He, T. Performance Enhancement of Vehicle Mechatronic Inertial Suspension, Employing a Bridge Electrical Network. *World Electr. Veh. J.* **2022**, *13*, 229. [CrossRef]
5. Jiang, X.; Xu, X.; Shan, H. Model-Based Fault Diagnosis of Actuators in Electronically Controlled Air Suspension System. *World Electr. Veh. J.* **2022**, *13*, 219. [CrossRef]
6. Zhu, Z.; Teng, C.; Cai, Y.; Chen, L.; Lian, Y.; Wang, H. Vehicle Safety Planning Control Method Based on Variable Gauss Safety Field. *World Electr. Veh. J.* **2022**, *13*, 203. [CrossRef]
7. Wang, Z.; Li, Y.; Kaku, C.; Zheng, H. Trajectory Tracking Control of Intelligent X-by-Wire Vehicles. *World Electr. Veh. J.* **2022**, *13*, 205. [CrossRef]
8. Wu, H.; Zhang, H.; Feng, Y. MPC-Based Obstacle Avoidance Path Tracking Control for Distributed Drive Electric Vehicles. *World Electr. Veh. J.* **2022**, *13*, 221. [CrossRef]
9. Chen, M.; Ren, Y.; Ou, M. Adaptive Robust Path Tracking Control for Autonomous Vehicles Considering Multi-Dimensional System Uncertainty. *World Electr. Veh. J.* **2023**, *14*, 11. [CrossRef]
10. Yuan, C.; Lin, Y.; Shen, J.; Chen, L.; Cai, Y.; He, Y.; Weng, S.; Wu, X.; Yuan, Y.; Gong, Y.; et al. Research on Active Collision Avoidance and Hysteresis Reduction of Intelligent Vehicle Based on Multi-Agent Coordinated Control System. *World Electr. Veh. J.* **2023**, *14*, 16. [CrossRef]
11. Chang, X.; Zhang, H.; Yan, S.; Hu, S.; Meng, Y. Analysis and Roll Prevention Control for Distributed Drive Electric Vehicles. *World Electr. Veh. J.* **2022**, *13*, 210. [CrossRef]
12. Wu, T.; Wang, F.; Ye, P. Regenerative Braking Strategy of Dual-Motor EV Considering Energy Recovery and Brake Stability. *World Electr. Veh. J.* **2023**, *14*, 19. [CrossRef]
13. Adeleke, O.P.; Li, Y.; Chen, Q.; Zhou, W.; Xu, X.; Cui, X. Torque Distribution Based on Dynamic Programming Algorithm for Four In-Wheel Motor Drive Electric Vehicle Considering Energy Efficiency Optimization. *World Electr. Veh. J.* **2022**, *13*, 181. [CrossRef]
14. Shi, W.; Jiang, Y.; Shen, Z.; Yu, Z.; Chu, H.; Liu, D. Nonlinear MPC-Based Acceleration Slip Regulation for Distributed Electric Vehicles. *World Electr. Veh. J.* **2022**, *13*, 200. [CrossRef]
15. Li, Y.; Hu, H.; Shi, P. A Review of Position Sensorless Compound Control for PMSM Drives. *World Electr. Veh. J.* **2023**, *14*, 34. [CrossRef]

World Electric Vehicle Journal

MDPI

Article

Improvement of the Vehicle Seat Suspension System Incorporating the Mechatronic Inerter Element

Chengqun Qiu [1,*], Xiaofu Liu [2] and Yujie Shen [3]

[1] Jiangsu Province Intelligent Optoelectronic Devices and Measurement-Control Engineering Research Center, Yancheng Teachers University, Yancheng 224007, China
[2] College of Engineering, China Agricultural University, Beijing 100083, China
[3] Automotive Engineering Research Institute, Jiangsu University, Zhenjiang 212013, China
* Correspondence: yctcqcq@126.com; Tel.: +86-15805108661

Abstract: A mechatronic inerter can simulate the equivalent mechanical network through the external electrical network and can be used in a wide range of mechanical device design applications. In this paper, we study the use of a mechatronic inerter to enhance vibration isolation in vehicle seat suspensions. Firstly, the vertical and pitch movements of the vehicle's sprung mass and the vertical vibration of the seat are considered in a half vehicle model. Then, the mechatronic inerter is introduced and the external electrical network is presented. The particle swarm optimization algorithm was used to optimize the seat suspension layout parameters with different transfer function-orders. Numerical simulations under different speeds were performed, and the results show that the application of the used mechatronic inerter's seat suspension vibration isolation performance outperforms passive suspension. In addition, with an increase in the external electrical network transfer function-order, the seat acceleration and pitch acceleration RMS values will be further reduced. The results of the study will contribute to a new approach to vehicle seat suspension design.

Keywords: vehicle; seat suspension; inerter; mechatronic system

Citation: Qiu, C.; Liu, X.; Shen, Y. Improvement of the Vehicle Seat Suspension System Incorporating the Mechatronic Inerter Element. *World Electr. Veh. J.* **2023**, *14*, 29. https://doi.org/10.3390/wevj14020029

Academic Editors: Yong Li, Xing Xu, Lin Zhang, Yechen Qin and Yang Lu

Received: 12 December 2022
Revised: 15 January 2023
Accepted: 18 January 2023
Published: 23 January 2023

1. Introduction

Seat suspension is used in vehicles in order to enhance the driver's comfort and protect the health of the driver from vibrations caused by uneven roads. Improved driver ride comfort can be achieved by designing the ride quality of the main and cockpit suspensions [1–4]. However, for commercial vehicles, the driver is primarily exposed to high amplitude and low frequency vibrations, which are a major factor in health disorders [5,6]. Therefore, using the seat's suspension to reduce unwanted vibration is a simple and effective method. In particular, Deng proposed, in 2019, a novel seat suspension capable of variable stiffness and damping (VSVD) that improved ride comfort with magnetorheological fluid dampers [7]. In [8], Ning proposed an electrical variable stiffness device (EVSD) and applied it to suspensions in 2019. The introduction of a negative stiffness structure into a cab seat suspension structure improved the cab's working environment and the seating comfort in [9]. In [10], Liu proposed a semi-active electromagnetic device capable of varying inertance and damping (VIVD), using an energy storage priority control (ESPC) strategy to reduce the vibrations in vehicle seat suspensions in 2021. Until now, several scholars have proposed a number of different structures for seat suspension, and the structural aspect of improving the performance of seat suspension has reached a bottleneck. The question of how to design a new seat suspension structure with excellent performance has becomes a heated problem.

Inerters, such as springs and dampers, are strictly mechanical components with two endpoints [11–13]; the use of such results in significant improvements in vehicle suspension system performance [14,15]. Professor Smith gave the physical definition and dynamic equations of the inerter device, and designed the rack-pinion inerter and ball-screw inerter. The application of an inerter element was further studied in a vehicle suspension system [16–18] and in civil engineering [19,20]. In [21], Liu proposed a new semi-active suspension system based on a hydro-pneumatic inerter, and an MPC control strategy was designed to suppress vehicle vibration in 2021. In [22], use of electrical network elements for equivalent fractional-order mechanical networks in vehicle suspension design showed that the results of numerical simulations confirm the performance advantages of vehicle mechatronic ISD suspension with a fractional-order electrical network. In [23], Li effectively used the advantages of an inertial suspension and GH control strategy to reduce dynamic tire loads in HMDVs over a wider frequency range in 2022. In [24–26], the effect of inerter nonlinearity on vehicle suspension vibration isolation performance was studied. In [27], inerter development led to a new theory of electromechanical similarity, where an inerter corresponds to a capacitor, a damper corresponds to a resistor, and a spring corresponds to an inductor. Shen proposed an optimal design method for vehicle mechatronic ISD suspension based on the structure-immittance approach in 2021 [28]. However, the application of an inerter element to a vehicle seat suspension system lacks research. Based on the new electromechanical similarity theory, the ability to use an external electrical network can increase the order of the seat suspension's transfer function and simplify the mechanic. Therefore, in order to improve vibration isolation performance and the ride comfort of seat suspension, this paper integrates a mechatronic inerter element into vehicle seat suspension. The article is structured as follows.

Firstly, in Section 2, a half vehicle model is established with a seven-degree-of-freedom model that takes into account the vertical and pitch movements of the vehicle's sprung mass and the vertical vibration of the seat. Then, in Section 3, the seat suspension, employing a mechatronic inerter device, is introduced, and the different external electrical networks are presented. In Section 4, the particle swarm optimization algorithm is used to optimize the designed seat suspension parameters. Then, Section 5 provides an analysis of the seat suspension dynamics according to the half vehicle model. Section 6 is the concluding section of this paper.

2. Half Vehicle Model

We refer to the half vehicle model built by Shen in 2021 [28]. Figure 1 shows the half vehicle model built for this study; this seven-degree-of-freedom model takes into account the vertical and pitch movements of the vehicle's sprung mass and the vertical vibration of the seat. This study was carried out with the vehicle unloaded, therefore, the mass and inertia of the driver are not taken into account.

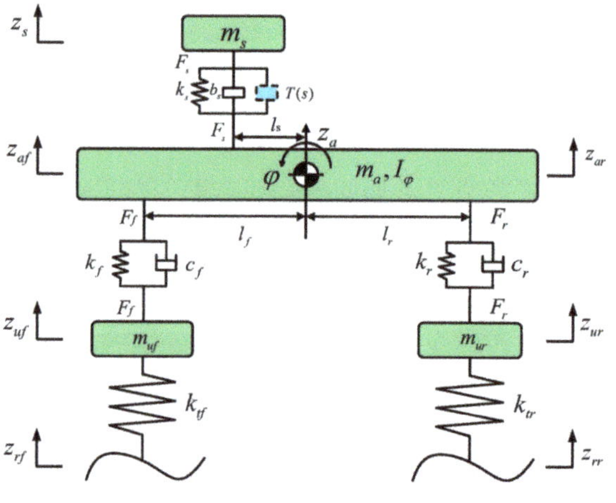

Figure 1. Seven-degree-of-freedom vehicle model.

The vertical motion equation of the seat mass is

$$m_s \ddot{z}_s = F_s \tag{1}$$

The equation representing the sprung mass vertical motion is

$$m_a \ddot{z}_a = k_f(z_{uf} - z_{af}) + c_f(\dot{z}_{uf} - \dot{z}_{af}) + k_r(z_{ur} - z_{ar}) + c_r(\dot{z}_{ur} - \dot{z}_{ar}) - F_s \tag{2}$$

The equation representing the sprung mass pitch motion is

$$I_\varphi \ddot{\varphi} = l_r F_r + l_s F_s - l_f F_f \tag{3}$$

The equations of the front and rear unsprung masses are

$$\begin{cases} m_{uf}\ddot{z}_{uf} = k_{tf}(z_{rf} - z_{uf}) - k_f(z_{uf} - z_{af}) - c_f(\dot{z}_{uf} - \dot{z}_{af}) \\ m_{ur}\ddot{z}_{ur} = k_{tr}(z_{rr} - z_{ur}) - k_r(z_{ur} - z_{ar}) - c_r(\dot{z}_{ur} - \dot{z}_{ar}) \end{cases} \tag{4}$$

The pitch angle can be approximately equal to the following equation when the angle is relatively small.

$$\begin{cases} z_{as} = z_a - l_s\varphi \\ z_{sf} = z_a - l_f\varphi \\ z_{sr} = z_a + l_r\varphi \end{cases} \tag{5}$$

where m_s is the vehicle seat mass, z_s is the seat's vertical displacement, m_a is the vehicle's sprung mass, z_a is the vertical displacement of the body centroid, F_s is the seat's suspension force, F_f and F_r are the forces of the front and rear suspensions, l_s is the horizontal distance from the seat to the centroid, l_f and l_r are the distances from the front and rear axles to the body centroid, φ is the body pitch angle, I_φ is the body pitch moment of inertia, k_f and c_f are the spring's stiffness and the damping coefficient of the front suspension, k_r and c_r are the spring's stiffness and the damping coefficient of the rear suspension, m_{uf} and m_{ur} are the front and rear unsprung mass, z_{uf} and z_{ur} are the vertical displacements of the front and rear unsprung mass, k_{tf} and k_{tr} are the equivalent stiffness of the front and rear tires, z_{rf} and z_{rr} are the displacement inputs of the front and rear wheels, z_{af} and z_{ar} are the vertical displacements of the front corner and rear corner of the vehicle's body. This study was carried out on the basis of a passenger car model in order to achieve an effective increase in ride comfort in passenger cars; the model for this study was built on a mature, commercially available model. Table 1 shows the parameters of the half vehicle model.

Table 1. Main parameters of the vehicle model.

Name	Value
Seat mass m_s (kg)	48
Body centroid mass m_a (kg)	928.2
Unsprung mass of front wheels m_{uf} (kg)	26.5
Unsprung mass of rear wheels m_{ur} (kg)	24.4
Distance from seat to centroid l_s (m)	0.324
Distance from front axle to centroid l_f (m)	0.968
Distance from rear axle to centroid l_r (m)	1.392
Moment of inertia aound the Y axis I_φ (kg·m^2)	1058
Front suspension stiffness k_f (kN·m^{-1})	25
Rear suspension stiffness k_r (kN·m^{-1})	22
Front suspension damping c_f (Ns/m)	1500
Rear suspension damping c_r (Ns/m)	1300
Tire stiffness k_t (kN·m^{-1})	192

3. Seat Suspension Layout

3.1. The Ball-Screw Mechatronic Inerter

The inerter is a mass element at both ends, and its output force is proportional to the relative acceleration at both ends. Inerters are available in ball-screw, rack and pinion, and fluid types. In this study, we have designed a ball-screw mechatronic inerter. The ball-screw mechatronic inerter comprises a rotary motor device and a ball-screw inerter. Among them, the ball-screw converts the linear reciprocating motion into a rotary motion and transmits it to the rotary motor. Figure 2 shows the working principle of the ball-screw mechatronic inerter.

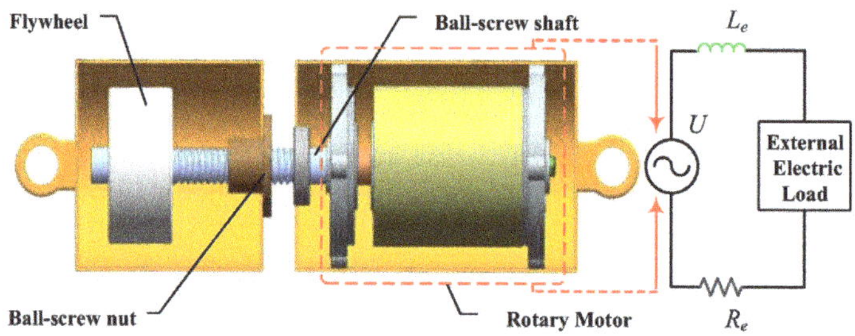

Figure 2. Ball-screw mechatronic inerter.

According to Figure 2, the inertance of the ball-screw mechatronic inerter can be changed by modifying the rotational inertia of the flywheel mounted on the ball-screw shaft. When both endpoints of the ball-screw mechatronic inerter move in a straight line, the rotating motor rotor is driven by the ball-screw shaft, producing a voltage U that flows through the external electrical load. R_e and L_e are the coil resistance and inductance. In this paper, the coil factor is not considered in the optimization. The external electrical load can be adopted to simulate the corresponding mechanical network in the optimization process.

3.2. The Seat Suspension Layout

The designed seat suspension, using the mechatronic inerter, includes a mechanical part and an electrical network part. For the mechanical section, we needed to provide load-bearing capacity for the seat suspension by means of a parallel spring, which includes a spring and an inerter, and provides a fundamental seat suspension layout to protect the system in the event of electrical network failure. The electrical section involves resistors, inductors, and capacitors to simulate the dampers, springs, and inerters. Figure 3 shows the layout of the mechatronic seat suspension, where the mechatronic inerter is connected, in parallel, with the spring.

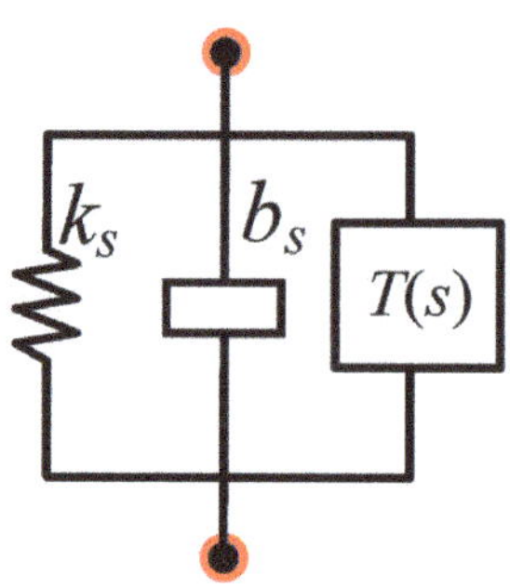

Figure 3. General layout of the seat suspension.

Where k_s and b_s are mechanical structures and $T(s)$ is the impedance expression to be solved; the double primary impedance transfer function is as follows:

$$T(s) = \frac{\alpha_1 s + \alpha_0}{\beta_1 s + \beta_0} \tag{6}$$

where $\alpha_i \geq 0$, $\beta_i \geq 0$ (β_i are not all 0). The positive real constraints of the double primary impedance transfer function are as follows:

$$\alpha_1 \beta_1 \geq 0 \tag{7}$$

The biquadratic impedance transfer function is as follows:

$$T(s) = \frac{\alpha_2 s^2 + \alpha_1 s + \alpha_0}{\beta_2 s^2 + \beta_1 s + \beta_0} \tag{8}$$

where $\alpha_i \geq 0$, $\beta_i \geq 0$ (β_i are not all 0). The positive realness constraints of the biquadratic impedance transfer function are as follows:

$$\left(\sqrt{\alpha_2 \beta_0} - \sqrt{\alpha_0 \beta_2}\right)^2 \leq \alpha_1 \beta_1 \tag{9}$$

The bicubic impedance transfer function is as follows:

$$T(s) = \frac{\alpha_3 s^3 + \alpha_2 s^2 + \alpha_1 s + \alpha_0}{\beta_3 s^3 + \beta_2 s^2 + \beta_3 s + \beta_0} \tag{10}$$

where $\alpha_i \geq 0$, $\beta_i \geq 0$ (β_i are not all 0). The positive realness constraints of the bicubic impedance transfer function are as follows:

$$\begin{cases} (\alpha_1 + \beta_1)(\alpha_2 + \beta_2) \geq (\alpha_0 + \beta_0)(\alpha_3 + \beta_3); \\ a_3 = 0, a_2 \geq 0, a_0 \geq 0, -a_1 \leq 2\sqrt{a_0 a_2}; \\ a_3 > 0, a_0 \geq 0, a_1 \geq 0, -a_2 \leq \sqrt{3a_1 a_3} or a_2^2 > 3a_1 a_3, 2a_2^3 - 9a_1 a_2 a_3 + 27a_0 a_3^2 \geq 2(a_2^2 - 3a_1 a_3)^{3/2} \end{cases} \tag{11}$$

where $a_0 = \alpha_0 \beta_0$, $a_1 = \alpha_1 \beta_1 - \alpha_0 \beta_2 - \alpha_2 \beta_0$, $a_2 = \alpha_2 \beta_2 - \alpha_1 \beta_3 - \alpha_3 \beta_1$, $a_3 = \alpha_3 \beta_3$.

4. Optimal Design of the Mechatronic Seat Suspension

For optimum performance of the mechatronic seat suspension, the parameters of the designed seat suspension systems are optimized via particle swarm optimization. To begin with, the particle is initialized, and the fit value of the particle is then compared with the best location it passes through, and the speed and position of the particle are updated. This ends when the termination condition is met. As the model for this study was built on a mature, commercially available model, before optimization, the main spring coefficient for protection remained the same as the conventional suspension, and only the parameters in $T(s)$ and b were optimized to give the optimization results practical significance.

Assuming the vehicle is driving at 30 km/h on a C grade road [29], and taking into account the vertical acceleration of the seat and the acceleration of the pitch motion; then, the objective function is defined as

$$f = \frac{J_1}{J_{1\text{pas}}} + \frac{J_2}{J_{2\text{pas}}} \tag{12}$$

where $J_{1\text{pas}}$ and $J_{2\text{pas}}$ are the root-mean-square (RMS) values of the seat vertical acceleration and the pitch motion acceleration of traditional passive suspension. Here, $J_{1\text{pas}} = 0.9913 \text{ m/s}^2$ and $J_{2\text{pas}} = 1.2852 \text{ rad/s}^2$. J_1 and J_2 are the RMS values of the seat vertical acceleration and the pitch motion acceleration of the designed suspension system. The mechanical inertance b and the $T(s)$ transfer function as the optimization variables. Particle swarm optimization was used to find the global optimum by following the current search. This algorithm has attracted academic attention for its ease of implementation, high accuracy and fast convergence. Equations (13) and (14) are the updated formulas for the particle velocity and position properties.

$$V^{n+1} = \lambda V^n + d_1 r_1 (P_{id}^n - X^n) + d_2 r_2 \left(P_{gd}^n - X^n \right) \quad (13)$$

$$X^{n+1} = X^n + V^{n+1} \quad (14)$$

where λ is the inertia factor, V is the velocity of the particle, X is the particle's position, n is the iterations number, and d_1 and d_2 are non-negative constants. The random numbers r_1 and r_2 usually have a value between 0 and 1, while P_{id} and P_{gd} are the individual extremum and global extremum. Figure 4 shows the optimization process of the algorithm.

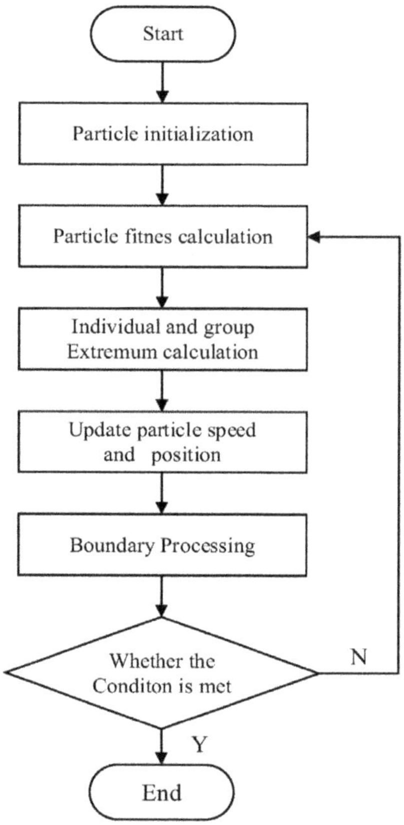

Figure 4. The optimization process of particle swarm optimization.

The optimized results of the transfer functions can be passively realized by the electrical network involving resistors, capacitors, and inductors. Figure 5 illustrates the corresponding network; Table 2 shows the detailed parameters.

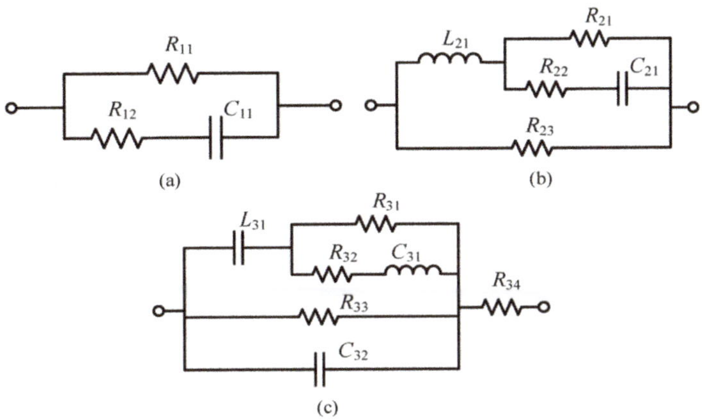

Figure 5. Electrical network. (**a**) Bivariate transfer function electrical network; (**b**) Biquadratic transfer function electrical network; (**c**) Bicubic transfer function electrical network.

Table 2. Electrical network parameters.

Name	Value
Resistor R_{11} (Ω)	4877
Resistor R_{12} (Ω)	654
Capacitor C_{11} (F)	0.0047
Resistor R_{21} (Ω)	98,754
Resistor R_{22} (Ω)	35,789
Resistor R_{23} (Ω)	5711
Capacitor C_{21} (F)	0.0079
Inductor L_{21} (H)	0.34
Resistor R_{31} (Ω)	81
Resistor R_{32} (Ω)	6998
Resistor R_{33} (Ω)	74,223
Resistor R_{34} (Ω)	158
Capacitor C_{31} (F)	0.0024
Capacitor C_{32} (F)	0.0017
Inductor L_{31} (H)	0.76

5. Performance Evaluation

The structural optimization of the designed seat suspension was required for the application of the mechatronic inerter. Table 3 shows the RMS values of the seat acceleration and pitch acceleration among the different suspension systems at a speed of 30 km/h.

Table 3. Comparison of the different seat suspension systems.

	RMS of Seat Acceleration	Improvement	RMS of Pitch Acceleration	Improvement
Passive suspension	0.9913	/	1.2852	/
Layout S1	0.9664	2.51%	1.2638	1.67%
Layout S2	0.9459	4.58%	1.2414	3.41%
Layout S3	0.8866	10.56%	1.1879	7.57%

Table 3 shows that, for the S1 seat suspension layout, the RMS values of the seat acceleration and the pitch acceleration decreased by 2.51% and 1.67%, respectively. The improvements are not obvious. Therefore, as the transfer function of the external electrical network increases in order, the RMS values of the seat acceleration and the pitch acceleration of the S2 seat suspension layout are further reduced by 4.58% and 3.41%, respectively. Then, for the S3 seat suspension layout, which used a bicubic transfer function electrical network, the RMS values of the seat acceleration and the pitch acceleration decreased by 10.56% and 7.57%, respectively; this resulted in a significant improvement in the ride comfort of the vehicle seat suspension. A comparison of the time domain characteristics of the seat acceleration and the pitch acceleration are shown in Figures 6 and 7. Figures 8 and 9 show the comparisons of the RMS values of the seat acceleration and the pitch acceleration at different speeds.

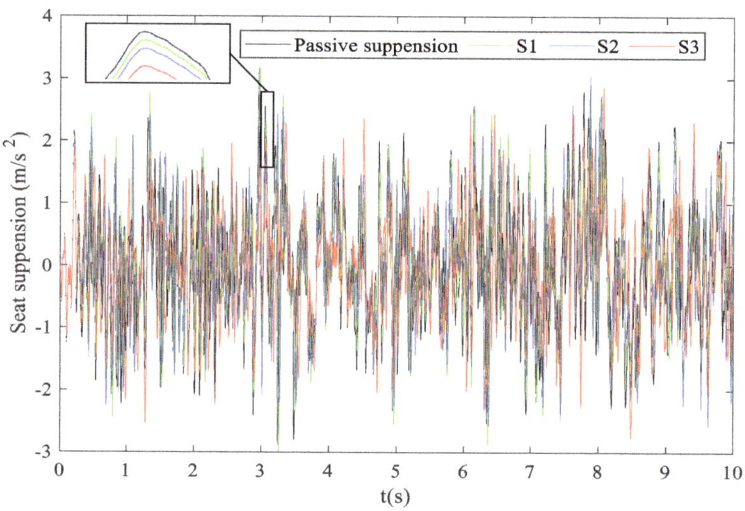

Figure 6. Comparison of the seat accelerations in the time domain.

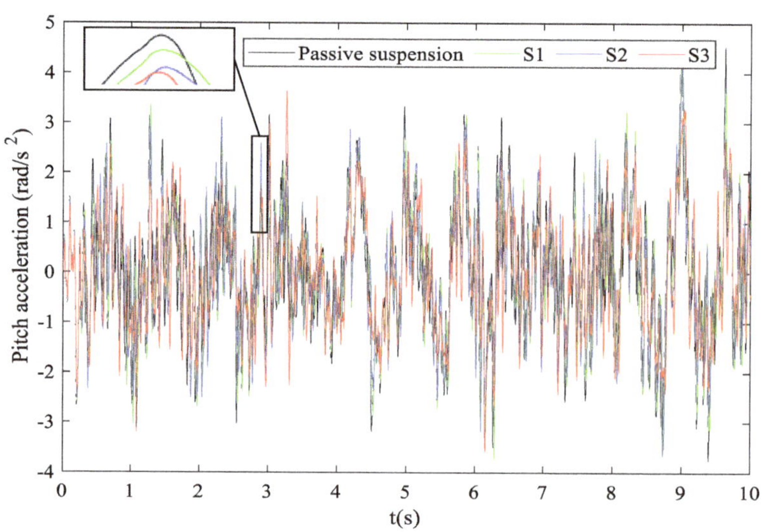

Figure 7. Comparison of the pitch accelerations in the time domain.

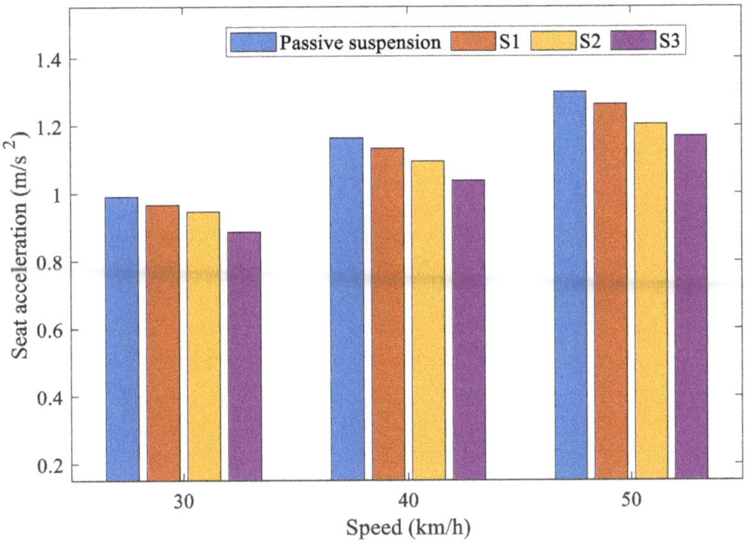

Figure 8. Comparison of the RMS of seat acceleration under different speeds.

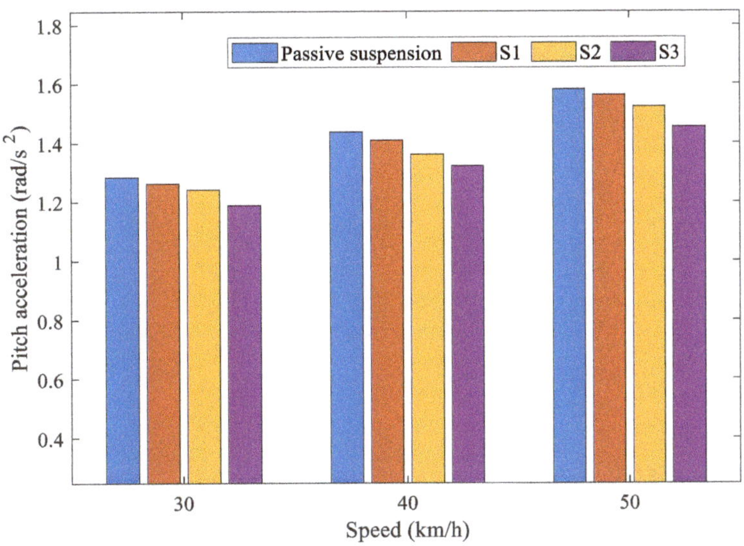

Figure 9. Comparison of the RMS of pitch acceleration under different speeds.

It is noted that, as the vehicle speed increases, the RMS values of the seat acceleration under the same suspension structure also increase. Under the same vehicle speed, the higher the transfer function order is, the smaller the RMS values of the seat acceleration are. For the pitch acceleration, as can be seen in the bar chart, when the speed of the vehicle increased, the pitch acceleration under the same type of suspension also increased. As the order of transfer function of the external electrical network increased, the pitch acceleration was reduced. In conclusion, the RMS values of the seat acceleration and pitch acceleration of the S3 layout are significantly lower than those of the S1 and S2 layouts, and passive suspension.

6. Conclusions

In this study, a mechatronic inerter element was introduced into the structural design of vehicle seat suspensions, and the problem of optimizing the design of the vehicle seat suspension, to integrate the mechatronic inerter element, was investigated. In addition, the vertical and pitch movements of the vehicle's sprung mass and the vertical vibration of the seat were considered in a half vehicle model. Based on a ball-screw mechatronic inerter, the external electrical networks, using different transfer function-orders, were optimized via the particle swarm optimization algorithm. The results show that, as the external electrical network transfer function-order is increased, the RMS values of the seat acceleration and pitch acceleration will be further reduced. The RMS values of the seat acceleration and pitch acceleration can be simultaneously reduced by 10.56% and 7.57%, respectively, at most. The performance of vehicle seat suspensions with an integrated mechatronic inerter element can be improved by increasing the order of the external electrical network transfer function.

Author Contributions: Conceptualization, C.Q.; methodology, Y.S.; software, X.L.; validation, X.L.; formal analysis, C.Q.; investigation, Y.S.; writing—original draft preparation, C.Q.; writing—review and editing, Y.S.; supervision, X.L. All authors have read and agreed to the published version of the manuscript.

Funding: This work was supported by the Natural Science Foundation of Beijing Municipality (No. 3214045), and the Natural Science Foundation of Jiangsu Province (No. BK20211364).

Data Availability Statement: The data used to support the findings of the study are available from the corresponding author upon request.

Conflicts of Interest: The authors declare no conflict of interest.

References

1. Choi, S.; Choi, Y.T.; Chang, E.G.; Han, S.J.; Kim, C.S. Control characteristics of a continuously variable ER damper. *Mechatronics* **1998**, *8*, 143–161. [CrossRef]
2. Rakheja, S.; Afework, Y.; Sankar, S. An analytical and experimental investigation of the driver-seat-suspension system. *Veh. Syst. Dyn.* **1994**, *23*, 501–524. [CrossRef]
3. Tong, R.T.; Amirouche, F. Ride control-a two state suspension design for cabs and seats. *Veh. Syst. Dyn.* **1999**, *33*, 578–589. [CrossRef]
4. Wereley, N.M.; Pang, L. Nondimensional analysis of semi-active electrorheological and magnetorheological dampers using approximate parallel plate models. *Smart Mater. Struct.* **1997**, *7*, 732–743. [CrossRef]
5. Amirouche, F.; Palkovics, L.; Woodrooffe, J. Optimal driver seat suspension design for heavy trucks. *Transp. Syst. ASME* **1994**, *2*, 277–291.
6. Andersson, R. The low back pain of bus drivers in an urban area of California. *Spine* **1982**, *17*, 1481–1488. [CrossRef]
7. Deng, H.; Deng, J.; Yue, R.; Han, G.; Zhang, J.; Ma, M.; Zhong, X. Design and verification of a seat suspension with variable stiffness and damping. *Smart Mater. Struct.* **2019**, *28*, 65015. [CrossRef]
8. Ning, D.; Du, H.; Sun, S.; Li, W.; Zhang, N.; Dong, M. A novel electrical variable stiffness device for vehicle seat suspension control with mismatched disturbance compensation. *IEEE ASME Trans. Mechatron.* **2019**, *24*, 2019–2030. [CrossRef]
9. Liao, X.; Du, X.; Li, S. Design of cab seat suspension system for construction machinery based on negative stiffness structure. *Adv. Mech. Eng.* **2021**, *13*, 1–15. [CrossRef]
10. Liu, P.; Ning, D.; Luo, L.; Zhang, N.; Du, H. An electromagnetic variable inertance and damping seat suspension with controllable circuits. *IEEE Trans. Ind. Electron.* **2021**, *69*, 2811–2821. [CrossRef]
11. Smith, M.C.; Wang, F. Performance benefits in passive vehicle suspensions employing inerters. *Veh. Syst. Dyn.* **2004**, *42*, 235–257. [CrossRef]
12. Zhang, C.; Zhang, Z.; Zhao, H. Analysis of dynamic characteristics and ride performance of automobile active suspension. *Chin. Agric. Mech.* **2015**, *36*, 176–179. [CrossRef]
13. Huang, C.; Chen, L.; Yuan, Z.; Jiang, H.; Niu, L. Hybrid fuzzy control of semi-active suspension system. *Automot. Eng.* **2014**, *36*, 999–1018.
14. Smith Malcolm, C. Synthesis of mechanical networks: The inerter. *IEEE Trans. Autom. Control* **2002**, *47*, 1648–1662. [CrossRef]
15. Zhao, Z.-P.; Chen, Q.-J.; Zhang, R.-F.; Pan, P.; Jian, Y.-Y. Energy dissipation mechanism of inerter systems. *Int. J. Mech. Sci.* **2020**, *184*, 105845. [CrossRef]
16. Chen, M.Z.Q.; Hu, Y.L.; Li, C.Y.; Chen, G. Application of semi-active inerter in semi-active suspensions via force tracking. *J. Vib. Acoust.* **2016**, *138*, 041014. [CrossRef]

17. Hu, Y.L.; Wang, K.; Chen, Y.H.; Chen, M.Z.Q. Inerter-based semi-active suspensions with low-order mechanical admittance via network synthesis. *Trans. Inst. Meas. Control* **2018**, *40*, 4233–4245. [CrossRef]
18. Liu, Y.L.; Zhao, W.T.; Yang, X.F.; Shen, Y.J. Predictive control of vehicle ISD suspension based on a hydraulic electric inerter. *Shock Vib.* **2019**, *2019*, 9230736. [CrossRef]
19. Zhang, S.Y.; Jiang, J.Z.; Neild, S.A. Optimal configurations for a linear vibration suppression device in a multi-storey building. *Struct. Control Health Monit.* **2016**, *24*, e1887. [CrossRef]
20. Giaralis, A.; Petrini, F. Wind-induced vibration mitigation in tall buildings using the tuned mass-damper-inerter. *J. Struct. Eng.* **2017**, *142*, 04017127. [CrossRef]
21. Yang, L.; Wang, R.; Ding, R.; Liu, W.; Zhu, Z. Investigation on the dynamic performance of a new semi-active hydro-pneumatic inerter-based suspension system with MPC control strategy. *Mech. Syst. Signal Process.* **2021**, *154*, 107569. [CrossRef]
22. Shen, Y.; Hua, J.; Fan, W.; Liu, Y.; Yang, X.; Chen, L. Optimal design and dynamic performance analysis of a fractional-order electrical network-based vehicle mechatronic ISD suspension. *Mech. Syst. Signal Process.* **2023**, *184*, 109718. [CrossRef]
23. Li, Y.; Yang, X.; Shen, Y.; Liu, Y.; Wang, W. Optimal design and dynamic control of the HMDV inertial suspension based on the ground-hook positive real network. *Adv. Eng. Softw.* **2022**, *171*, 103171. [CrossRef]
24. Shen, Y.; Chen, L.; Liu, Y.; Zhang, X. Influence of fluid inerter nonlinearities on vehicle suspension performance. *Adv. Mech. Eng.* **2017**, *9*, 1687814017737257. [CrossRef]
25. Sun, X.Q.; Chen, L.; Wang, S.H.; Zhang, X.L.; Yang, X.F. Performance investigation of vehicle suspension system with nonlinear ball-screw inerter. *Int. J. Automot. Technol.* **2016**, *17*, 399–408. [CrossRef]
26. Liu, C.; Chen, L.; Zhang, X.; Yang, Y.; Nie, J. Design and tests of a controllable inerter with fluid-air mixture condition. *IEEE Access* **2020**, *8*, 125620–125629. [CrossRef]
27. Yang, X.; He, T.; Shen, Y.; Liu, Y.; Yan, L. Research on predictive coordinated control of ride comfort and road friendliness for heavy vehicle ISD suspension based on the hybrid-hook damping strategy. *Proc. Inst. Mech. Eng. D J. Automob. Eng.* **2022**. [CrossRef]
28. Shen, Y.; Hua, J.; Wu, B.; Chen, Z.; Xiong, X.; Chen, L. Optimal design of the vehicle mechatronic ISD suspension system using the structure-immittance approach. *Proc. Inst. Mech. Eng. D J. Automob. Eng.* **2022**, *236*, 512–521. [CrossRef]
29. *ISO 8608:2016*; Mechanical Vibration—Road Surface Profiles—Reporting of Measured Data. International Organization for Standardization: Geneva, Switzerland, 2016. Available online: https://www.iso.org/standard/71202.html. (accessed on 22 November 2022).

World Electric Vehicle Journal

MDPI

Article

Optimal Design of Fractional-Order Electrical Network for Vehicle Mechatronic ISD Suspension Using the Structure-Immittance Approach

Jie Hua [1], Yujie Shen [1,*], Xiaofeng Yang [2], Ying Zhang [3] and Yanling Liu [2]

1 Automotive Engineering Research Institute, Jiangsu University, 301 Xuefu Road, Zhenjiang 212013, China
2 School of Automotive and Traffic Engineering, Jiangsu University, 301 Xuefu Road, Zhenjiang 212013, China
3 School of Electromechanical Engineering, Guangdong University of Technology, Guangzhou 510006, China
* Correspondence: shenyujie@ujs.edu.cn

Abstract: In order to more effectively design the structure of vehicle ISD (Inerter Spring Damper) suspension system using the inerter, this paper proposed a design method using a fractional-order electrical network structure of a mechatronic inerter for fractional-order electrical network components, according to the characteristics that the external electrical network of a mechatronic inerter can simulate the corresponding mechanical network structure equivalently. First, the 1/4 dynamic model of the suspension is constructed. The improved Oustaloup filtering algorithm is used to simulate fractional calculus, and the fractional order components are simulated. Then, the simulation model of the vehicle mechatronic ISD suspension is established. In order to simplify the electrical network, one resistance, one fractional inductance and one fractional capacitance are limited in the design of the fractional electrical network at the outer end of the mechatronic inerter. The structure-immittance approach is used to obtain two general layouts of all possible structures of three elements. At the same time, the optimal fractional electrical network structure and parameters are obtained by combining the optimization algorithm. The simulation results verify the performance of the fractional ISD suspension with the optimized structure, which can provide a new idea for the structural design of a fractional-order electrical network applied in vehicle mechatronic ISD suspension.

Keywords: vehicle; suspension; mechatronic inerter; fractional-order electrical network; structure-immittance approach; optimal design

Citation: Hua, J.; Shen, Y.; Yang, X.; Zhang, Y.; Liu, Y. Optimal Design of Fractional-Order Electrical Network for Vehicle Mechatronic ISD Suspension Using the Structure-Immittance Approach. *World Electr. Veh. J.* **2023**, *14*, 12. https://doi.org/10.3390/wevj14010012

Academic Editors: Yong Li, Xing Xu, Lin Zhang, Yechen Qin and Yang Lu

Received: 30 November 2022
Revised: 30 December 2022
Accepted: 3 January 2023
Published: 4 January 2023

1. Introduction

The proposition of the inerter [1] breaks through the inherent structure of the existing suspension system "spring damper" parallel connection, and forms a new suspension structure system. This suspension, consisting of spring, damper and inerter elements, is called ISD suspension. Scholars all over the world have adopted many methods to realize the inerter [2–9], and after the application of the inerter, the performance potential of the vibration isolation system has also been expanded to include aircraft [10], trains [11], buildings [12], bridges [13], etc. The structural design of ISD suspension plays an important role in meeting various performance indicators of vehicles. The question of how to design the structure of ISD suspension has attracted the attention of scholars at home and abroad.

Common ISD suspension structure design methods include the structure approach, the immitance approach and the structure-immitance approach. The structure approach [14] limits the number of components in the suspension, and integrates them into parameter optimization according to the feasible range of component parameters. The disadvantage is that the arrangement and combination method has a huge workload, which makes it difficult to cover a wide range of mechanical networks, and it is easy to omit structures with excellent performance. The immitance approach [15] replaces the suspension structure with a fixed form of impedance or admittance expression, uses the parameter optimization method to optimize the solution and finally realizes it passively through network synthesis. The suspension component parameters obtained by the immitance approach often do not conform to the routine, which is not conducive to engineering realization. The structure-immitance approach [16] can be used to express the structure with a predetermined number of elements with a general impedance expression. However, the structural complexity of a pure mechanical network is high, which has affected the engineering design of ISD suspension. The mechatronic inerter [17] is a device coupled by a mechanical inerter and a rotating motor. The external circuit impedance of the mechatronic inerter can be used to simulate the target mechanical impedance to achieve the passive structure design of complex mechanical network, overcome the space limitation of pure mechanical network suspension structure and expand the design idea of suspension system structure. However, for the electrical network at the outer end of the mechatronic inerter, the increase of the order of its impedance transfer function will bring higher performance improvement [18], and at the same time, the difficulty of network integration will also increase greatly. For example, the bicubic impedance transfer function requires no more than 13 elements to realize passively [19], and its structure is complex.

In the structural design of the suspension system, fractional calculus theory has also been commonly used [20–23], and its feasibility has also been verified [24,25], indicating that the fractional-order function can more accurately describe the dynamic characteristics of complex systems than the integral-order function. Using fractional-order electrical network elements to replace the original integer-order electrical network at the outer end of the mechatronic inerter can effectively avoid the high complexity of pure integer order network structures. However, the structural design of fractional-order electrical networks in vehicle mechatronic ISD suspension has not been reported yet, and in the design of a fractional-order vehicle mechatronic ISD suspension, a simple and clear fractional-order electrical network structure is essential. Therefore, this paper will use the structure-immitance approach to study the optimal design of the fractional-order electrical network structure for a vehicle mechatronic ISD suspension. The content layout of the rest of this paper is as follows.

First, in Section 2, the definition and algorithm realization of fractional calculus are introduced, and the equivalent realization relationship between fractional electrical components and fractional mechanical components are analyzed. In Section 3, a quarter suspension dynamic simulation model is established, and fractional-order electrical components are used in the design of the electrical network, and the structure-immitance approach is used to design the electrical network structure. Then, in Section 4, the electrical network structure and parameters of the suspension system are obtained through optimization. Finally, in Section 5, the dynamic performance of the optimized fractional-order ISD suspension is evaluated by comparison, and some conclusions are made in Section 6.

2. Equivalent Realization of Fractional Passive Network Elements

Fractional calculus has the basic operator ${}_{t_0}D_t^{\alpha}$, among which, α is limited to real numbers, t and t_0 are the upper and lower bounds of the operator. The unified definition of fractional calculus operator [26] is:

$$ {}_{t_0}D_t^{\alpha} f(t) = \begin{cases} \frac{\mathrm{d}^{\alpha}}{\mathrm{d}t^{\alpha}} f(t), & \alpha > 0 \\ f(t), & \alpha = 0 \\ \int_{t_0}^{t} f(\tau) d\tau^{-\alpha}, & \alpha < 0 \end{cases} \tag{1} $$

There are many definitions of fractional calculus. This paper adopts the Grünwald–Letnikow [26] fractional calculus definition. The Grünwald–Letnikow of the α derivative of a given function $f(t)$ is defined as:

$$ {}_{t_0}^{GL}D_t^{\alpha} f(t) = \lim_{h \to 0} \frac{1}{h^{\alpha}} \sum_{j=0}^{[(t-t_0)/h]} (-1)^j \binom{\alpha}{j} f(t - jh) \tag{2} $$

where [·] means taking the nearest integer. At the same time, in order to ensure the approximation effect at the frequency band boundary and ensure that the transfer function is regular, the improved Oustaloup filtering algorithm [27] is considered to approximate fractional calculus. The mathematical model of the improved Oustaloup filter is:

$$s^{\gamma} \approx \left(\frac{d\omega_{\rm h}}{b}\right)^{\gamma}\left(\frac{ds^2 + b\omega_{\rm h}s}{d(1-\gamma)s^2 + b\omega_{\rm h}s + d\gamma}\right)\prod_{k=1}^{N}\frac{s+\omega'_k}{s+\omega_k} \tag{3}$$

$$\omega'_k = \omega_{\rm b}\omega_{\rm u}{}^{(2k-1-\gamma)/N}, \omega_k = \omega_{\rm b}\omega_{\rm u}{}^{(2k-1+\gamma)/N}, \omega_{\rm u} = \sqrt{\omega_{\rm h}/\omega_{\rm b}} \tag{4}$$

where N is the filter order, γ is the fractional order, ω'_k and ω_k are zero point and the pole, respectively. $\omega_{\rm h}$ and $\omega_{\rm b}$ are the upper and lower limits of frequency bands, respectively. In general, the weighting parameters $b = 10$, $d = 9$. In this paper, the filter frequency band is $(10^{-3}, 10^{3})$ rad/s. The larger the filter order, the higher the approximation accuracy. In this frequency band, the fifth order Oustaloup filtering effect has met the accuracy requirements, so the selection order is five.

In the new mechanical and electrical analogy, the spring and the inductance, the damper and the resistance, and the inerter and capacitance are similar, respectively [28]. According to the above fractional definition and approximation method, the impedance expression of fractional network elements (including mechanical network and electrical network) is obtained by using the form of pull transform, with excitation force as the input and corresponding speed as the output, as shown in Table 1, where s is the Laplace variable, α and β are fractional orders.

Table 1. Impedance expression of fractional network elements.

Mechanical Network Elements	Impedance	Electrical Network Elements	Impedance
Spring	k/s^{α}	Inductor	$1/Ls^{\alpha}$
Damper	c	Resistor	$1/R$
Inerter	bs^{β}	Capacitor	Cs^{β}

3. Model Construction of Vehicle Mechatronic ISD Suspension System

3.1. The Ball-Screw Mechatronic Inerter

A mechatronic inerter is considered in this paper, which is formed by coupling a ball-screw inerter with a rotary motor. Its structural diagram is shown in Figure 1. The relative linear motion of the two ends of the mechanical inerter can be converted into the rotary motion of the motor. The inductor, resistor and capacitor in the electrical network at the outer end of the rotary motor can equivalently simulate the spring, damper, and inerter in the mechanical network structure.

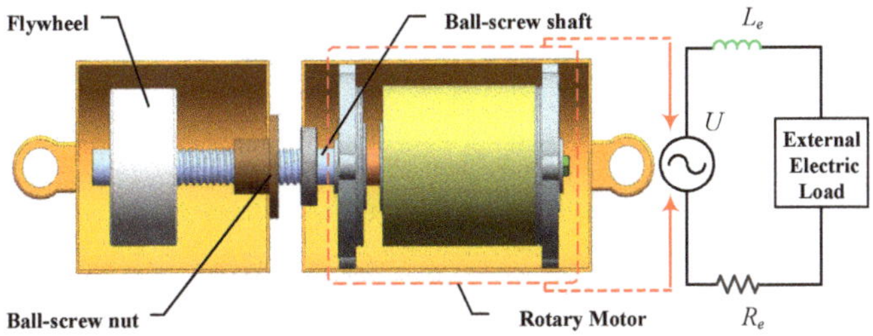

Figure 1. The schematic of ball-screw mechatronic inerter.

3.2. Mechatronic ISD Suspension Structure Layout

The quarter suspension model is a typical vibration model of vehicle suspension system, which is a basic dynamic model for studying its vertical performance. In this paper, a quarter vehicle mechatronic ISD suspension dynamics model is established, as shown in

Figure 2. Based on a mature vehicle model, Table 2 illustrates the parameters for the model.

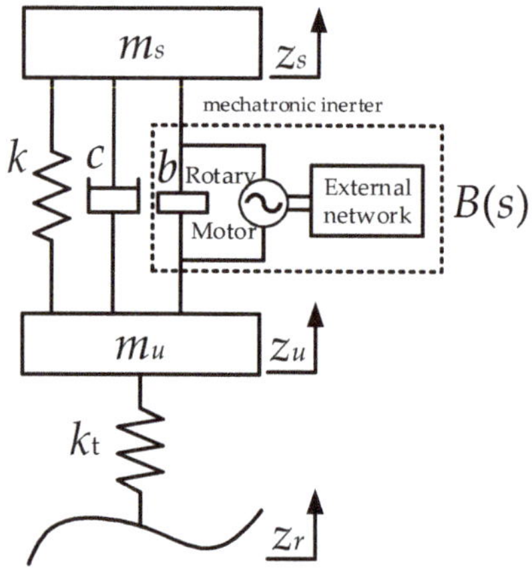

Figure 2. A quarter vehicle suspension model.

Table 2. Parameters of quarter vehicle suspension model.

Parameters	Values
Sprung Mass m_s/kg	320
Unsprung Mass m_u/kg	45
Spring Stiffness k/N m^{-1}	22,000
Tire Stiffness k_t/N m^{-1}	190,000

The dynamic Laplace equation of the suspension model is shown in Equation (5):

$$\begin{cases} m_s s^2 Z_s + [k + cs + sB(s)](Z_s - Z_u) = 0 \\ m_u s^2 Z_u - [k + cs + sB(s)](Z_s - Z_u) + k_t(Z_u - Z_r) = 0 \end{cases} \tag{5}$$

where k, k_t, and c are spring stiffness, tire stiffness, and the damping coefficient, respectively, m_s and m_u are the sprung mass and the unsprung mass, respectively. z_s, z_u and z_r are the vertical displacements of the sprung mass, the unsprung mass, and road roughness, respectively, and Z_s, Z_u and Z_r are their Laplace transforms, respectively. $B(s)$ is the impedance expression of the mechatronic inerter, which is shown as follows [17]:

$$\begin{cases} B(s) = bs + \frac{K_{\mathrm{m}}}{Z_e(s)} \\ K_{\mathrm{m}} = \left(\frac{2\pi}{P}\right)^2 k_t k_e \end{cases} \tag{6}$$

where b is the inertance of the ball-screw mechatronic inerter, P is the pitch of the ball-screw mechanism, k_e is the induced electromotive force coefficient of the rotary motor, k_t is the thrust coefficient of the rotary motor. K_{m} is the electromechanical parameter conversion coefficient of the ball-screw mechatronic inerter, which is taken as 7056 HN/m in this paper. $Z_e(s)$ is the impedance expression of the external electrical network of the rotary motor. The fractional-order external electrical network of the mechatronic inerter includes resistor(s), fractional-order capacitor(s) and fractional-order inductor(s). In order to simplify the electrical network, the number of resistors, fractional capacitors and fractional inductors is limited to one in the optimal design. Eight structures of the three element arrangement are summarized using the structure-immittance approach, and two general structures are used for general expression, as shown in Figures 3 and 4.

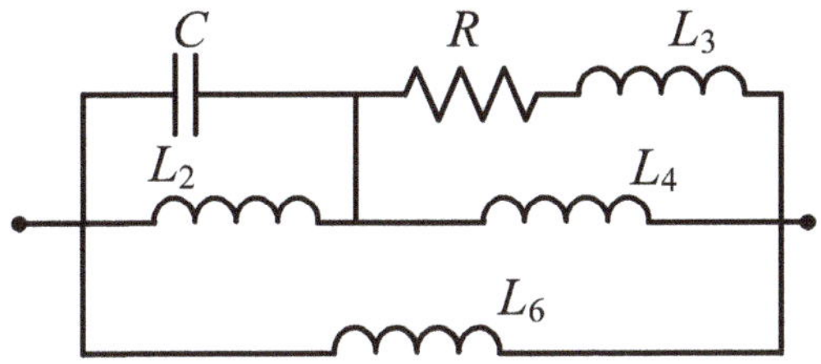

Figure 3. General structure of Y_1(s).

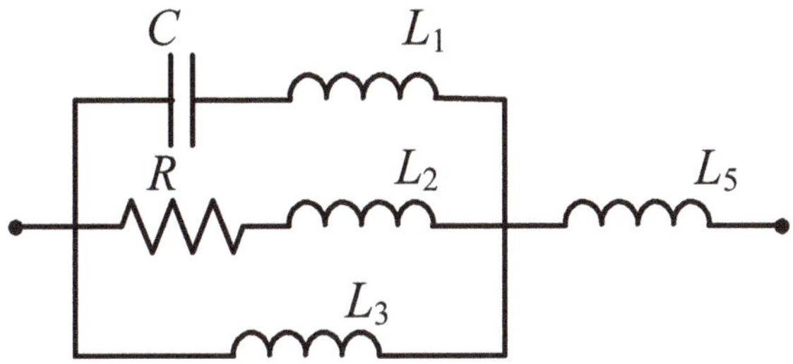

Figure 4. General structure of Y_2(s).

The impedance transfer function expressions of the two general structures in Figures 3 and 4 are, respectively, as follows:

$$Y_1(s) = \frac{C\frac{1}{R}s^{\alpha+\beta} + C(\frac{1}{L_4} + \frac{1}{L_6})s^{\beta} + \frac{1}{R}(\frac{1}{L_2} + \frac{1}{L_6})}{C\frac{1}{R}L_3 s^{2\alpha+\beta} + Cs^{\alpha+\beta} + \frac{1}{R}s^{\alpha} + \frac{1}{L_2} + \frac{1}{L_4}} \tag{7}$$

$$Y_2(s) = \frac{C\frac{1}{R}(L_1 + L_2)s^{2\alpha+\beta} + Cs^{\alpha+\beta} + \frac{1}{R}s^{\alpha} + \frac{1}{L_3}}{C(L_1 + L_5)s^{2\alpha+\beta} + \frac{1}{R}(L_2 + L_5)s^{\alpha+\beta} + s^{\alpha}} \tag{8}$$

where L_1, L_2, L_3, L_4, L_5, and L_6 are fractional-order inductors, R and C are the resistor and the fractional-order capacitor, respectively. α and β are the fractional-order inductance order and the fractional-order capacitance order, respectively. In the Y_1 (s) structure, at least three of L_2, $1/L_3$, L_4 and L_6 are zero, and in the Y_2 (s) structure, at least three of $1/L_1$, $1/L_2$, L_3 and $1/L_5$ are zero. For example, in Figure 3, when L_3, L_4 and L_6 are zero, it is a structure in which a fractional-order inductor is connected in parallel with a fractional-order capacitor, and then connected in series with a resistor. In Figure 4, when $1/L_2$, $1/L_3$ and $1/L_5$ are zero, it is a fractional-order capacitor in series with a fractional-order inductor, and then in parallel with a resistor.

4. Parameter Optimization Design

4.1. Pattern Search Optimization Algorithm

In this paper, the pattern search method [29] is used for the optimization design of the suspension system. As a general algorithm for solving the optimal value of a function, the greatest advantage of pattern search method is that it does not need to use the derivative of the objective optimization function in the algorithm program of pattern search method. Therefore, pattern search method can effectively solve the optimization problems of non-derivative functions and complex derivative functions. The specific steps of pattern search method are shown in Figure 5.

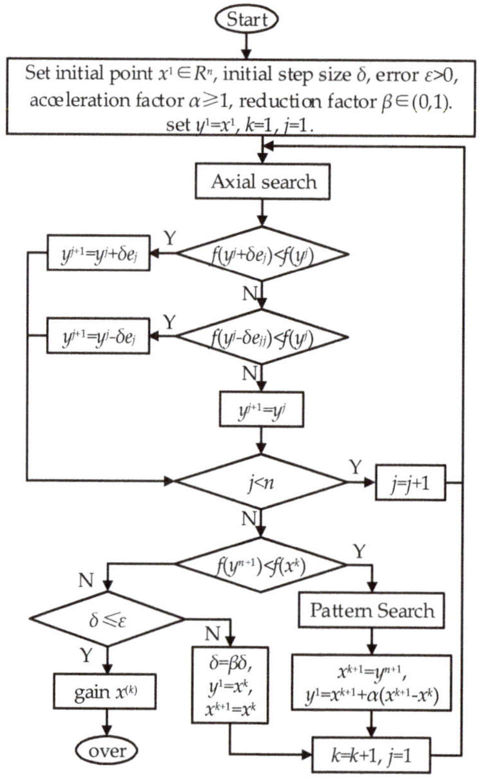

Figure 5. Pattern search optimization algorithm.

4.2. Optimization Results

To ensure vehicle ride comfort, the RMS (root-mean-square) values of the vehicle body acceleration, the suspension working space and the dynamic tire load are selected as evaluation indicators, and the traditional passive suspension is chosen as evaluation benchmark to establish the optimization objective function, as shown below:

$$f = \frac{BA(P)}{BA_{\text{pas}}} + \frac{SWS(P)}{SWS_{\text{pas}}} + \frac{DTL(P)}{DTL_{\text{pas}}} \tag{9}$$

$$P = [b\ c\ L_e\ C_e\ R_e\ \alpha\ \beta] \tag{10}$$

where BA and BA_{pas} are the RMS values of the vehicle body acceleration of the suspension to be optimized and the traditional passive suspension, respectively, SWS and SWS_{pas} are the RMS values of the suspension working space of the suspension to be optimized and the traditional passive suspension, respectively, and DTL and DTL_{pas} are the RMS values of the dynamic tire load of the suspension to be optimized and the traditional passive suspension, respectively. BA_{pas}, SWS_{pas} and DTL_{pas} are calculated by a mature traditional passive suspension [30], and their performances have reached a high level, which are 1.3096 m·s^{-2}, 0.0130 m and 900.4704 N, respectively. P represents the set of parameters to be optimized for the suspension system, including inertance b, damping coefficient c, fractional-order inductance coefficient L_e, fractional-order capacitance coefficient C_e, resistance coefficient

R_e, fractional-order inductance order α, and fractional-order capacitance order β. Their constraints are as follows:

$$\begin{cases} b, c \geq 0 \\ L_e, R_e, C_e \geq 0 \\ 1 \geq \alpha, \beta \geq 0 \end{cases} \tag{11}$$

The optimized fractional-order electrical network structure is shown in Figure 6. This structure is the case when L_2, $1/L_3$ and L_6 are zero in $Y_1(s)$ structure. Set the fractional-order inductance order α and the fractional-order capacitance order β to 1 for optimization, and get the integer-order ISD suspension system parameters. The optimization parameters of fractional-order ISD suspension and integral-order ISD suspension are shown in Table 3.

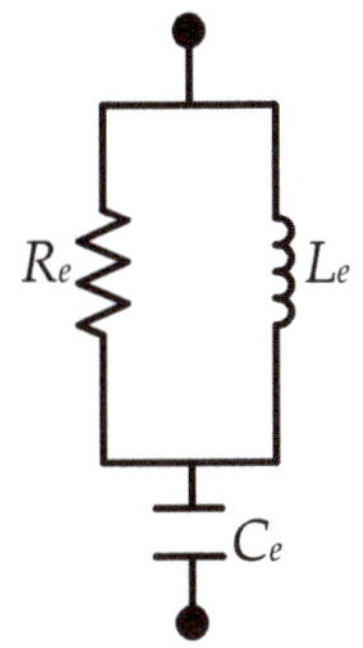

Figure 6. The optimized fractional-order electrical network structure.

Table 3. Optimization parameters.

Fractional-Order ISD Suspension Parameters	Values	Integer-Order ISD Suspension Parameters	Values
Inertance b/kg	5	Inertance b/kg	13
Damping coefficient c/N·s·m^{-1}	1074	Damping coefficient c/N·s·m^{-1}	232
Fractional-order inductance coefficient L_e/H	1.05	Inductance coefficient L_e/H	1.34
Fractional-order capacitance coefficient C_e/F	0.06	Capacitance coefficient C_e/F	0.03
Resistance coefficient R_e/Ω	320.73	Resistance coefficient R_e/Ω	5.56
Fractional-order inductance order α	0.28	-	-
Fractional-order capacitance order β	0.81	-	-

5. Simulation Analysis

5.1. The Characteristics of Bode Diagram

Compared with the Bode diagram of the traditional passive suspension, Figure 7 shows the Bode diagram of vehicle mechatronic ISD suspension applying the optimized fractional-order electrical network structure.

It can be seen that for the fractional-order ISD suspension, in the low frequency range [10^{-2}, 2] Hz, the optimized structure shape is similar to the spring. In the range [2,4] Hz, the structure shape is similar to the damper, and above 4 Hz, the optimized structure is similar to the inerter, which is the difference between the traditional passive suspension system and the optimized structure. The traditional suspension system composed of "spring damper" mechanical components cannot show inertia characteristics, which is the main factor limiting the performance improvement of the traditional suspension structure, and also the reason why the ISD suspension of vehicles containing the inerter has better vibration isolation performance.

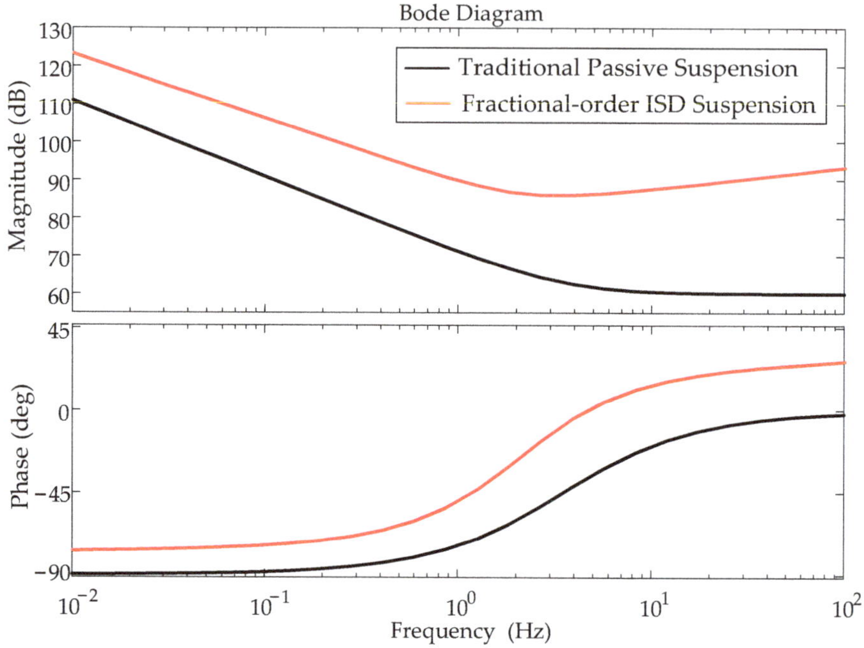

Figure 7. Bode diagram comparison of two suspension systems.

5.2. Random Road Input

The random road input is selected as the road input model to study the advantages of the optimized fractional-order ISD suspension compared with the integral-order ISD suspension and the traditional passive suspension. The random road input model is as follows [31]:

$$\dot{z}_r(t) = -0.111[uz_r(t) + 40\sqrt{G_q(n_0)uw(t)}] \tag{12}$$

where u, $z_r(t)$, $w(t)$ and $G_q(n_0)$ are the vehicle speed, the vertical input displacement, the white noise with the mean value of 0, and the road roughness coefficient, respectively. Class C pavement is selected in this paper, and the pavement roughness coefficient is 2.56×10^{-4} m^3. Figures 8–10 and Table 4 show the comparison of the RMS values of the vehicle body acceleration, the suspension working space and the dynamic tire load of the three suspension systems at a speed of 20 m/s.

The optimization adopts multi-objective optimization, and the final optimization result is the best case of comprehensive improvement. From the data point of view, the RMS value of suspension working space has the best effect, and the other two indexes have also been improved. It can be seen that, compared with the traditional passive suspension, the RMS values of the vehicle body acceleration of the integral-order ISD suspension and the fractional-order ISD suspension are reduced by 3.44% and 4.12%, respectively. The RMS values of the suspension working space of the two suspensions are reduced by 22.31% and 23.08%, respectively. Furthermore, the RMS values of the dynamic tire load of the two suspensions are reduced by 2.73% and 5.31%, respectively, showing the advantages of the designed fractional-order electric network structure. It shows that the vehicle mechatronic ISD suspension with optimized fractional-order electrical network structure can further improve the vibration isolation performance of the suspension system.

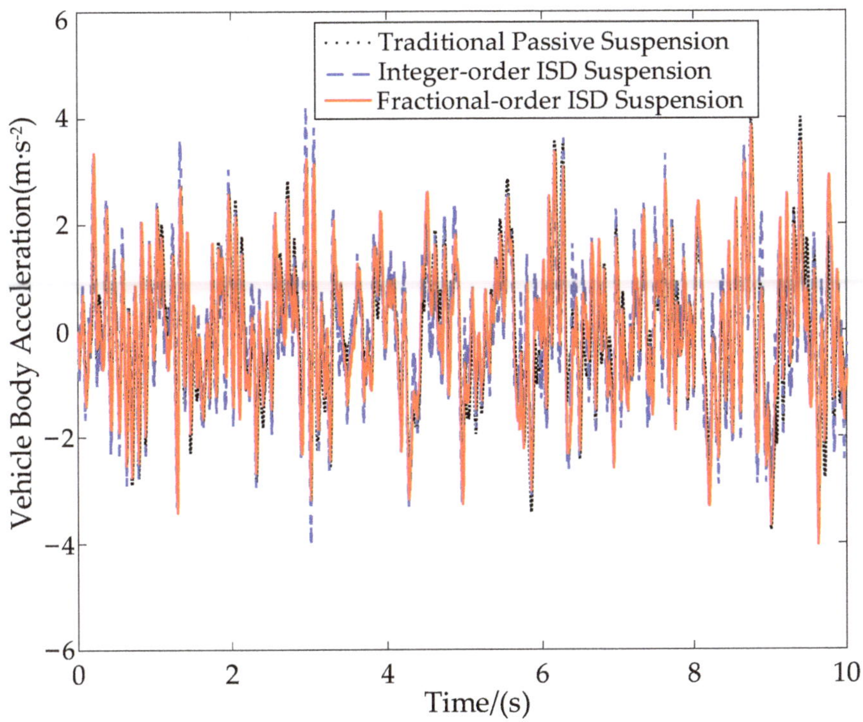

Figure 8. Comparison of the vehicle body acceleration.

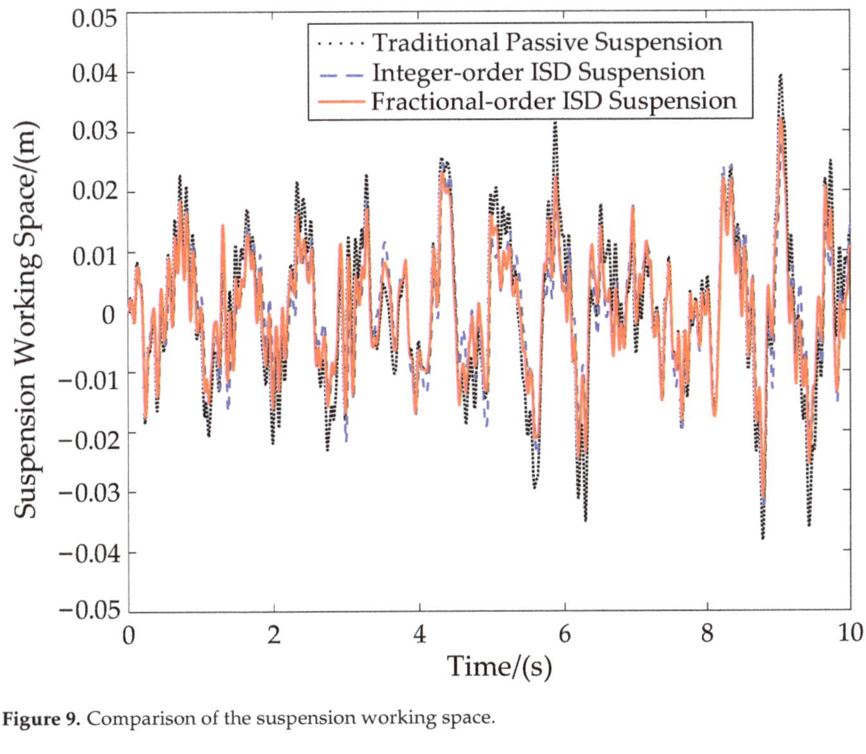

Figure 9. Comparison of the suspension working space.

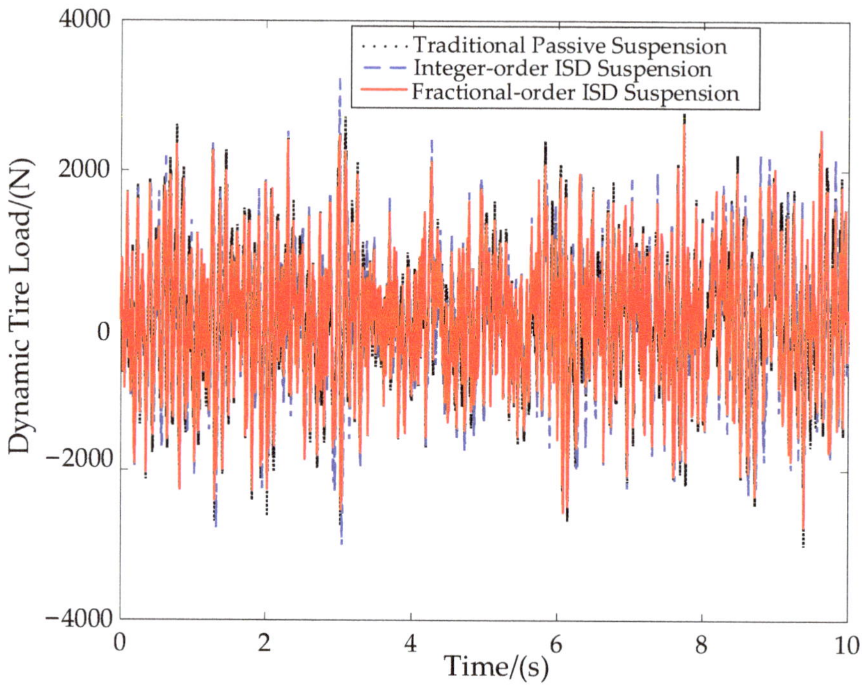

Figure 10. Comparison of the dynamic tire load.

Table 4. Comparison of performance indexes of three suspensions.

Performance Index	Traditional Passive Suspension	Integer-Order Isd Suspension	Improvement	Fractional-Order Isd Suspension	Improvement
RMS of vehicle body acceleration/($m{\cdot}s^{-2}$)	1.3096	1.3051	3.44%	1.3042	4.12%
RMS of suspension working space/(m)	0.0130	0.0101	22.31%	0.0100	23.08%
RMS of dynamic tire load/(N)	900.4704	875.8558	2.73%	852.6704	5.31%

6. Conclusions

In this paper, the optimal design of fractional-order electrical network for vehicle mechatronic ISD suspension is studied. An optimization design method of fractional-order electrical network for vehicle mechatronic ISD suspension is proposed by using the structure-immittance approach. The structural parameters of the fractional-order vehicle mechatronic ISD suspension are optimized by establishing a 1/4 dynamic model of the suspension. Through simulation comparison, the results show that the performance of the vehicle mechatronic ISD suspension system applying the fractional-order electrical network structure obtained by optimization design is further improved, which provides a reference for the structural design of fractional-order electrical network components based vehicle mechatronic ISD suspension.

Author Contributions: Conceptualization, J.H. and Y.S.; methodology, Y.Z.; software, J.H.; validation, X.Y.; formal analysis, Y.L.; investigation, Y.S.; writing—original draft preparation, J.H.; writing—review and editing, X.Y.; supervision, Y.Z. All authors have read and agreed to the published version of the manuscript.

Funding: This research was funded by the National Natural Science Foundation of China under Grant 52002156, 52072157 and 52008259, the Natural Science Foundation of Jiangsu Province under Grant BK20200911.

Data Availability Statement: Not applicable.

Conflicts of Interest: The authors declare no conflict of interest.

References

1. Smith, M.C. Synthesis of mechanical networks: The inerter. *IEEE Trans. Autom. Control.* **2002**, *47*, 1648–1662. [CrossRef]
2. Wen, H.; Guo, J.; Li, Y.; Zhang, K. The transmissibility of a vibration isolation system with ball-screw inerter based on complex mass. *J. Low Freq. Noise Vib. Act. Control.* **2018**, *37*, 1097–1108. [CrossRef]
3. Li, Y.; Cheng, Z.; Hu, N.; Yang, Y.; Xiao, Z. Modeling, design and experiments of a ball-screw inerter with mechanical diodes. *J. Sound Vib.* **2021**, *504*, 116121. [CrossRef]
4. Wang, R.; Meng, X.; Shi, D.; Zhang, X.; Chen, Y.; Chen, L. Design and test of vehicle suspension system with inerters. *Proc. Inst. Mech. Eng. Part C J. Mech. Eng. Sci.* **2014**, *228*, 2684–2689. [CrossRef]
5. Papageorgiou, C.; Houghton, N.E.; Smith, M.C. Experimental Testing and Analysis of Inerter Devices. *J. Dyn. Syst. Meas. Control.* **2009**, *131*, 101–116. [CrossRef]
6. Wang, F.; Hong, M.; Lin, T. Design and testing a hydraulic inerter. *Proc. Inst. Mech. Eng. Part C J. Mech. Eng. Sci.* **2011**, *225*, 66–72. [CrossRef]
7. Shen, Y.; Shi, D.; Chen, L.; Liu, Y.; Yang, X. Modeling and experimental tests of hydraulic electric inerter. *Sci. China Technol. Sci.* **2019**, *62*, 2161–2169. [CrossRef]
8. Domenico, D.D.; Ricciardi, G.; Zhang, R. Optimal design and seismic performance of tuned fluid inerter applied to structures with friction pendulum isolators. *Soil Dyn. Earthq. Eng.* **2020**, *132*, 106099. [CrossRef]
9. Liu, X.; Jiang, J.Z.; Titurus, B.; Harrison, A. Model identification methodology for fluid-based inerters. *Mech. Syst. Signal Process.* **2018**, *106*, 479–494. [CrossRef]
10. Li, Y.; Howcroft, C.; Neild, S.A.; Jiang, J.Z. Using continuation analysis to identify shimmy-suppression devices for an aircraft main landing gear. *J. Sound Vib.* **2017**, *408*, 234–251. [CrossRef]
11. Wang, F.; Liao, M.; Liao, B.; Su, W.; Chan, H. The performance improvements of train suspension systems with mechanical networks employing inerters. *Veh. Syst. Dyn.* **2009**, *47*, 805–830. [CrossRef]
12. De Domenico, D.; Ricciardi, G. An enhanced base isolation system equipped with optimal tuned mass damper inerter (TMDI). *Earthq. Eng. Struct. Dyn.* **2018**, *47*, 1169–1192. [CrossRef]
13. Wang, Q.; Zheng, Z.; Qiao, H.; De Domenico, D. Seismic protection of reinforced concrete continuous girder bridges with inerter-based vibration absorbers. *Soil Dyn. Earthq. Eng.* **2023**, *164*, 107526. [CrossRef]
14. Smith, M.C.; Wang, F. Performance benefits in passive vehicle suspensions employing inerters. *Veh. Syst. Dyn.* **2004**, *42*, 235–257. [CrossRef]
15. Shen, Y.; Jiang, J.Z.; Neild, S.A.; Chen, L. Vehicle vibration suppression using an inerter-based mechatronic device. *Proc. Inst. Mech. Eng. Part D J. Automob. Eng.* **2020**, *234*, 2592–2601. [CrossRef]
16. Zhang, S.Y.; Sheng, X.; Jiang, J.Z.; Zhou, H.; Ren, W.; Zhang, Z. Vibration suppression of bridges under moving loads using the structure-immittance approach. *Int. J. Mech. Sci.* **2021**, *211*, 106792. [CrossRef]
17. Wang, F.; Chan, H. Vehicle suspensions with a mechatronic network strut. *Veh. Syst. Dyn.* **2011**, *49*, 811–830. [CrossRef]
18. Wang, F.; Hsieh, M.; Chen, H. Stability and performance analysis of a full-train system with inerters. *Veh. Syst. Dyn.* **2012**, *50*, 545–571. [CrossRef]
19. Zhang, S.Y.; Jiang, J.Z.; Wang, H.; Neild, S. Synthesis of essential-regular bicubic impedances. *Int. J. Circuit Theory Appl.* **2017**, *45*, 1482–1496. [CrossRef]
20. You, H.; Shen, Y.; Xing, H.; Yang, S. Optimal control and parameters design for the fractional-order vehicle suspension system. *J. Low Freq. Noise Vib. Act. Control.* **2018**, *37*, 456–467. [CrossRef]
21. Sun, H.; Li, R.; Xu, J.; Xu, F.; Zhang, B.; Dong, X. Fractional Modeling and Characteristic Analysis of Hydro-Pneumatic Suspension for Construction Vehicles. *Processes* **2021**, *9*, 1414. [CrossRef]
22. Chen, E.; Xing, W.; Wang, M.; Ma, W.; Chang, Y. Study on chaos of nonlinear suspension system with fractional-order derivative under random excitation. *Chaos Solitons Fractals* **2021**, *152*, 111300. [CrossRef]
23. Chen, Y.; Xu, J.; Tai, Y.; Xu, X.; Chen, N. Critical damping design method of vibration isolation system with both fractional-order inerter and damper. *Mech. Adv. Mater. Struct.* **2022**, *9*, 1348–1359. [CrossRef]
24. Shah, Z.M.; Khanday, F.A. Analysis of Disordered Dynamics in Polymer Nanocomposite Dielectrics for the Realization of Fractional-Order Capacitor. *IEEE Trans. Dielectr. Electr. Insul.* **2021**, *28*, 266–273. [CrossRef]
25. Redman-White, W.; Kennedy, H.; Bodnar, R.; Lee, T. Adaptive Tuning of Large-Signal Resonant Circuits Using Phase-Switched Fractional Capacitance. *IEEE Trans. Circuits Syst. II Express Briefs* **2017**, *64*, 1072–1076. [CrossRef]
26. Xue, D. Definition and Calculation of Fractional Calculus. In *Fractional Calculus and Fractional-Order Control*; Zhang, Z., Jiang, H., Eds.; Science Press: Beijing, China, 2018; pp. 31–35, ISBN 978-7-03-054398-1.

27. Xue, D. Fractional Calculus Operators and Approximation of Systems. In *Fractional Calculus and Fractional-Order Control*; Zhang, Z., Jiang, H., Eds.; Science Press: Beijing, China, 2018; pp. 115–121, ISBN 978-7-03-054398-1.
28. Shen, Y.; Hua, J.; Fan, W.; Liu, Y.; Yang, X.; Chen, L. Optimal design and dynamic performance analysis of a fractional-order electrical network-based vehicle mechatronic ISD suspension. *Mech. Syst. Signal Process.* **2023**, *184*, 109718. [CrossRef]
29. Gasparo, M.G.; Papini, A.; Pasquali, A. Nonmonotone algorithms for pattern search methods. *Numer. Algorithms* **2001**, *28*, 171–186. [CrossRef]
30. Shen, Y.; Chen, L.; Yang, X.; Shi, D.; Yang, J. Improved design of dynamic vibration absorber by using the inerter and its application in vehicle suspension. *J. Sound Vib.* **2016**, *361*, 148–158. [CrossRef]
31. Sun, X.; Cai, Y.; Chen, L.; Liu, Y.; Wang, S. Vehicle height and posture control of the electronic air suspension system using the hybrid system approach. *Veh. Syst. Dyn.* **2016**, *54*, 328–352. [CrossRef]

World Electric Vehicle Journal

Article

Analysis of Active Suspension Control Based on Improved Fuzzy Neural Network PID

Mei Li *, Jiapeng Li, Guisheng Li and Jie Xu

Mechanical and Electrical Engineering College, Hainan University, Haikou 570100, China
* Correspondence: meili@hainanu.edu.cn

Abstract: To improve the comfort and smoothness of vehicle driving and reduce the vehicle vibration caused by uneven road surface. In this paper, a new active suspension control strategy is pro-posed by combining a fuzzy neural network and a proportional-integral-derivative (PID) controller, taking body acceleration as the main optimization target and adjusting the parameters of the PID controller in real time. Meanwhile, a fuzzy neural network parameter optimization algorithm combining a particle swarm optimization algorithm and gradient descent method is proposed to realize offline optimization and online fine-tuning of fuzzy neural network parameters. Finally, the active suspension model of a 2-degree-of-freedom 1/4 vehicle is established using MATLAB/Simulink, and the proposed control scheme is verified through simulation studies. The results show that the active suspension system with a particle swarm-optimized fuzzy neural network control method improves the spring mass acceleration, dynamic deflection of suspension, and dynamic tire deformation by 30.4%, 17.8%, and 15.5%, respectively, compared with the passive suspension. In addition, there are also 14.6%, 12.1%, and 11.2% performance improvements, respectively, compared to the PID-controlled active suspension system. These results indicate that the control strategy proposed in this paper can improve the vehicle driving performance and can support the design and development of active suspension systems.

Keywords: fuzzy neural network; particle swarm algorithm; PID control; active suspension; MATLAB/Simulink simulation

Citation: Li, M.; Li, J.; Li, G.; Xu, J. Analysis of Active Suspension Control Based on Improved Fuzzy Neural Network PID. *World Electr. Veh. J.* **2022**, *13*, 226. https://doi.org/10.3390/wevj13120226

Academic Editors: Yong Li, Xing Xu, Lin Zhang, Yechen Qin and Yang Lu

Received: 6 October 2022
Accepted: 2 November 2022
Published: 24 November 2022

1. Introduction

The suspension system's qualities determine the vehicle's smoothness and handling stability [1]. Traditional passive suspension systems have fixed parameters such as stiffness and damping, so they cannot effectively suppress vehicle vibration in the face of complex driving conditions. In order to reduce body vibration brought on by outside disturbances and give passengers a comfortable ride experience under various driving conditions, the active suspension can adjust the vehicle suspension in real time through active control force according to the road condition information [2–4].

For the controller design and optimization of active suspension systems, many researchers have proposed some simple and feasible control methods, such as linear quadratic regulator (LQR) control, PID control, optimal control, adaptive control, and sliding mode control [5–8]. Among them, PID control is favored by many researchers due to its relatively mature technology and wide application market. The traditional proportional-integral-derivative (PID) controller has the advantages of simple structure, good real-time performance, and low cost [9]. However, in today's practical engineering applications, many new control strategies have good improvements compared to PID control [10]. The parameters of the traditional PID controller are difficult to set accurately and are fixed after setting, and they cannot adapt to all suspension conditions as the vehicle working conditions change in real time [11].

In order to be able to improve the control effect of the PID controller, much in-depth research has been carried out. For example, fuzzy control is used to correct the PID parameters, allowing the PID parameters to vary in real time within a specific range, improving the accuracy of the controller [12]. In the literature [13], a type of PID transverse interconnected electronically controlled air suspension system controller based on an optimization algorithm has been designed, and the optimal solution of PID controller parameters obtained. In addition, the highly parallel structure and powerful learning capability of the neural network system can also be well used to achieve online real-time adjustment of PID parameters [14–16]. It can be seen that, by combining different optimization strategies with PID control and real-time optimization of PID controller parameters, better control results can be achieved. However, a single optimization strategy also has certain defects; for example, fuzzy control has a strong subjectivity and uncertainty, which can be considered as a subjective means of expressing domain expert knowledge, and as the number of fuzzy rules increases, the parameters and structure of this control system will become increasingly difficult to establish. Neural network control is also not perfect, and there are many drawbacks in practical applications, such as the uncontrollability of network behavior, convergence and stability being difficult to guarantee, and multiple instances of trial and error being needed for the network.

It was found that hybrid control has better results for PID controllers compared to a single optimization strategy. Based on the 14-degree-of-freedom whole vehicle model, in [17], two control systems, the fuzzy PID controller and the neural network controller, have been used to substantially improve lateral stability and vehicle handling. In addition, in [18], the fuzzy road information is collected in real time and used to adjust the control performance of the fuzzy PID, so as to develop a new road condition-based fuzzy PID control strategy that meets the control performance requirements under different road conditions. A vibration-controlled active suspension based on an adaptive fuzzy fractional-order PID controller is proposed in [19], which was very effective in reducing driver body vibrations, thus improving the ride quality of the driver. In order to optimize the control system, an attempt was made in [20] to optimize the PID controller and Fuzzy PID controller using a particle swarm optimization (PSO) algorithm with excellent efficiency in reducing the vertical displacement of the body and obtaining a suitable control signal.

In addition to achieving active suspension control through a combination of control methods, a new control method with better performance has been proposed by combining fuzzy control theory and neural network control theory. In a sense, this design approach does not have the same heavy reliance on expert experience as fuzzy control, while retaining the adaptive performance and learning capability of neural network control. For example, the Adaptive Neuro-Fuzzy Inference System (ANFIS) is an artificial neural network integrated fuzzy logic control system whose control rules are obtained by implementing the Sugeno first-order fuzzy inference system in the form of a network. In [21], a comparison of the control of a semi-vehicle suspension system using PID, LQR, FUZZY, and ANFIS controllers was analyzed, and it was found that the ANFIS controller provided the best performance in terms of "stabilization time" and "peak overshoot". Meanwhile, in [22], a fuzzy control and neuro-fuzzy inference system was tested; the solenoid valve was controlled by an ANFIS, and the proposed method improved the ride comfort while maintaining road safety. The latter was a fuzzy neural network composed of an RBF neural network model, whose biggest advantage is that the fuzzy inference process and the RBF function have functional equivalence and are suitable for real-time control of the system. In [23], a Takagi–Sugeno fuzzy controller was designed to control the contraction–expansion factor to satisfy the control input current of the MR damper by introducing a fuzzy neural network controller with PSO and BP learning and training algorithms, and the results showed that the system approach was effective. For the regulation of the PID controller, an effective real-time control strategy is needed. Therefore, this paper proposes an active suspension control strategy based on particle swarm optimization with fuzzy neural network PID control. The real-time control performance of the fuzzy neural network is used to achieve the rectification of the PID controller parameters for the purpose of real-time control of the suspension system. However, the fuzzy neural network has many parameters such as center, width, and weights, and it is difficult to obtain a reliable set of parameters. For this reason, the particle swarm optimization method is used to calculate a set of optimal parameters offline based on the objective function. In addition, it can avoid the problem that the neural network using gradient descent method may lead to gradient explosion, or the network not being able to converge for a long time due to too large or too small optimization weights [24]. Through MATLAB/Simulink simulation, it is shown that the FNN-PID control strategy of particle swarm optimization has a certain control effect on the active suspension system.

The rest of this paper is organized as follows. In Section 2, the mathematical model of the active suspension system for a 2-degree-of-freedom 1/4 vehicle and the road excitation model are presented, and the principles of PID control and fuzzy neural network PID control, as well as the combined optimization algorithm of particle swarm optimization algorithm and gradient descent method, are introduced. In Section 3, the simulation results of the suspension system controller design are shown. Finally, the conclusion and summary are presented in Section 4.

2. Materials and Methods

2.1. Active Suspension Simulation Model

The 2-degree-of-freedom 1/4 suspension model is shown in Figure 1, and the following assumptions are made regarding the model.

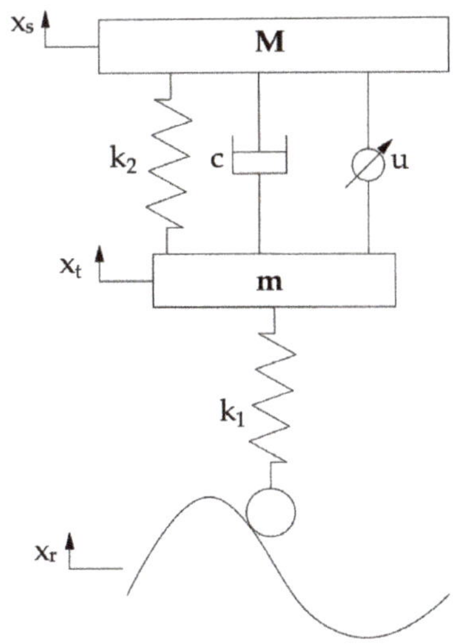

Figure 1. 2-degree-of-freedom 1/4 active suspension model.

1. The elastic center of the vehicle body coincides with the center of mass;
2. The vehicle body is rigid, and the occupants move in the same way as the vehicle body;
3. There is no sliding between the tires and the road, and the wheels are always in contact with the ground;
4. The vertical vibration characteristics of the wheel are reduced by a spring that does not take into account the damping effect.

As shown in Figure 1, M denotes the mass on the spring and m denotes the mass under the spring; x_r denotes the road disturbance excitation, x_t denotes the vertical displacement of the mass under the spring, and x_s denotes the vertical displacement of the mass on the spring; c denotes the suspension equivalent damping; k_1 is the tire equivalent stiffness and k_2 is the suspension stiffness; and u is the actuator active control force.

According to Newton's second law, combined with the suspension system dynamics model, the 1/4 active suspension dynamics equation is established as follows:

$$\mathrm{M}\ddot{\mathrm{x}}_{\mathrm{s}} + \mathrm{k}_2(\mathrm{x}_{\mathrm{s}} - \mathrm{x}_{\mathrm{t}}) + \mathrm{c}(\dot{\mathrm{x}}_{\mathrm{s}} - \dot{\mathrm{x}}_{\mathrm{t}}) - \mathrm{u} = 0 \tag{1}$$

$$\mathrm{m}\ddot{\mathrm{x}}_{\mathrm{t}} - \mathrm{k}_1(\mathrm{x}_{\mathrm{r}} - \mathrm{x}_{\mathrm{t}}) - \mathrm{k}_2(\mathrm{x}_{\mathrm{s}} - \mathrm{x}_{\mathrm{t}}) - \mathrm{c}(\dot{\mathrm{x}}_{\mathrm{s}} - \dot{\mathrm{x}}_{\mathrm{t}}) + \mathrm{u} = 0 \tag{2}$$

Meanwhile, the state variable is selected as: $x_1 = \mathrm{x_s} - \mathrm{x_t}$, $x_2 = \dot{\mathrm{x}}_{\mathrm{s}}$, $x_3 = \mathrm{x_r} - \mathrm{x_t}$, $x_4 = \dot{\mathrm{x}}_{\mathrm{t}}$. The state vector is $X = [x_1\, x_2\, x_3\, x_4]^T$. The output variable is $y_1 = \dot{x}_2 = \ddot{\mathrm{x}}_{\mathrm{s}}$, $y_2 = x_1 = \mathrm{x_s} - \mathrm{x_t}$, $y_3 = x_3 = \mathrm{x_r} - \mathrm{x_t}$. The output vector is represented as $Y = [y_1\, y_2\, y_3]^T$. The input vector is $U = [u\, x_r]^T$.

Then, the state equation of the model is shown in (3).

$$\begin{cases} \dot{X} = AX + BU \\ Y = CX + DU \end{cases} \tag{3}$$

Among them, $A = \begin{bmatrix} 0 & 1 & 0 & -1 \\ -\frac{\mathrm{k}_2}{\mathrm{M}} & -\frac{c}{\mathrm{M}} & 0 & \frac{\mathrm{c}}{\mathrm{M}} \\ 0 & 0 & 0 & -1 \\ \frac{\mathrm{k}_2}{\mathrm{m}} & \frac{\mathrm{c}}{\mathrm{M}} & \frac{\mathrm{k}_1}{\mathrm{m}} & -\frac{\mathrm{c}}{\mathrm{M}} \end{bmatrix}$, $B = \begin{bmatrix} 0 & 0 \\ \frac{1}{\mathrm{M}} & 0 \\ 0 & 1 \\ \frac{1}{\mathrm{M}} & 0 \end{bmatrix}$, $C = \begin{bmatrix} -\frac{\mathrm{k}_2}{\mathrm{M}} & -\frac{\mathrm{c}}{\mathrm{M}} & 0 & \frac{\mathrm{k}_2}{\mathrm{M}} \\ 1 & 0 & 0 & 0 \\ 0 & 0 & 1 & 0 \end{bmatrix}$, $D = \begin{bmatrix} \frac{1}{\mathrm{M}} & 0 \\ 0 & 0 \\ 0 & 0 \end{bmatrix}$.

2.2. *Road Excitation Model*

2.2.1. White Noise Road Excitation

The difference between different grades of road mainly lies in the difference in road roughness, which is generally expressed by the road unevenness coefficient, $\mathrm{G_q}$. According to the "Draft Method for Representation of Road Unevenness" presented by the International Organization for Standardization in ISO/TC108/SC2N67, the power spectrum density of a road can be expressed as follows [25]:

$$\mathrm{G_q(n)} = \mathrm{G_q(n_0)}\left(\frac{\mathrm{n}}{\mathrm{n_0}}\right)^{-\mathrm{w}} \tag{4}$$

In the formula, n is the spatial frequency, $\mathrm{n_0}$ is the reference spatial frequency, $\mathrm{n_0} = 0.1\ \mathrm{m^{-1}}$, $\mathrm{G_q(n_0)}$ is the reference spatial frequency of the road power spectral density, and w is the frequency index, often taken as $\mathrm{w} = 2$.

For the analysis of vehicle suspension system dynamics, the vehicle travel speed is also a factor to be considered [26]. Converting the spatial frequency power spectral density, $\mathrm{G_q(n)}$, to the temporal frequency power spectral density, $\mathrm{G_q(f)}$, the variable of vehicle speed can be introduced. When a vehicle travels at a certain speed on a road surface with spatial frequency n, its equivalent time frequency can be expressed as:

$$\mathrm{f} = \mathrm{vn} \tag{5}$$

In the formula, v is the speed of the vehicle in $\mathrm{m{\cdot}s^{-1}}$ and f is the time frequency in $\mathrm{s^{-1}}$.

As a result, the following may be deduced about the road excitation model created using the filtered white noise method:

$$\dot{\mathrm{x}}(\mathrm{t}) + 2\pi\mathrm{f_0}\mathrm{x}(\mathrm{t}) = 2\pi\mathrm{n_0}\sqrt{\mathrm{G_q(n_0)vW(t)}} \tag{6}$$

In the formula, $\mathrm{x(t)}$ is the road displacement, $\mathrm{W(t)}$ is the mean value of 0 Gaussian white noise, $\mathrm{f_0}$ is the lower cutoff frequency, $\mathrm{n_0} = 0.1\ \mathrm{m^{-1}}$, and $\mathrm{f_0} = 0.1$ Hz.

The basic idea of the model is to abstract the random fluctuations of the road process as white noise satisfying certain conditions, and then fit the time domain model of the random unevenness of the road by a hypothetical system with appropriate transformation. The pavement unevenness refers to the deviation of the road surface from the ideal plane. The rougher the pavement and the worse the pavement grade, the higher the geometric mean of power spectral density. In this paper, we simulate and analyze the A–D pavements, respectively, and the specific parameters are shown in Table 1.

Table 1. Parameters for each grade of pavement.

Road Grade	Geometric Mean of Power Spectral Density $G_q(n_0)/10^{-6}$ m^3
A	16
B	64
C	256
D	1024

2.2.2. Step Noise Road Excitation

The white noise pavement excitation is mainly used to simulate continuously uneven pavement, such as asphalt pavement, gravel road surface, etc. However, it is usually necessary to consider the response to an encountered shock in addition to the continuous vibration. Here, step pavement excitation is used for simulation. The specific mathematical expression is as follows:

$$x(t) = \begin{cases} 0, & t < t_1 \\ x, & t \geq t_1 \end{cases} \tag{7}$$

From the equation, x denotes the displacement of the step and t_1 denotes the time when the step occurs.

2.3. *Controller Design Principle*

2.3.1. FNN-PID Controller

The structure of the FNN-PID controller is shown in Figure 2.

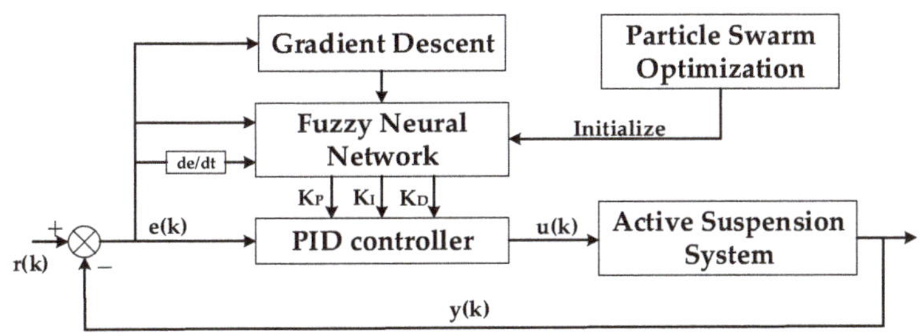

Figure 2. FNN-PID controller structure model.

In Figure 2, the input of the FNN is the deviation of Sprung Mass Acceleration, e(k), and the rate of change of deviation, de/dt; the input of the PID controller is e(k) [27] and u(k) is the control quantity. The anticipated value of the system is denoted by r(k), and the actual output value is denoted by y(k). After the fuzzy neural network algorithm has been trained, the best control parameters for the PID controller are obtained. According to the optimal control parameters, the PID controller enables real-time control of the suspension system by adjusting the magnitude of the control quantity, u(k).

2.3.2. PID Control

In the field of industrial automation control, the PID algorithm is a common control algorithm [28]. The discrete control rate of a commonly used PID algorithm is shown in Equation (8).

$$u(k) = K_p e(k) + K_i \sum_{i=0}^{k} e(i) + K_d[e(k) - e(k-1)] \tag{8}$$

In the formula, the error between the system's input and output is denoted as e(k); $K_i \sum_{i=0}^{k} e(i)$ is the cumulative sum of the error, and the error's rate of change is $e(k) - e(k-1)$.

In PID control, the proportional link is used to quickly eliminate the error between input and output; the larger the proportional coefficient, K_p, the faster the system response. The integral link is used to lower the system's static error; the larger the integral coefficient, K_i, the more accurate the system response. The differential link is used to eliminate the oscillation in the control process; the larger the differential coefficient, K_d, the more robust the system response process.

The control of the active suspension system often uses incremental PID control. According to Formulas (9) and (10), it can be seen that, when the three coefficients, K_p, K_i, and K_d, in the PID control are determined after only using the deviation measured before and after the moment to derive the control increment by the formula, the control amount corresponds to the increment of the last few errors. There is no accumulation of errors; only those related to the last three sampling values belong to the recursive algorithm.

$$\Delta u(k) = K_p[e(k) - e(k-1)] + K_i e(k) + K_d[e(k) - 2e(k-1) + e(k-2)] \tag{9}$$

$$u(k) = u(k-1) + \Delta u(k) \tag{10}$$

2.3.3. FNN Control

The structure of the FNN is displayed in Figure 3, and it is split into five layers, including input layer, fuzzification layer, fuzzy inference layer, normalization layer, and output layer.

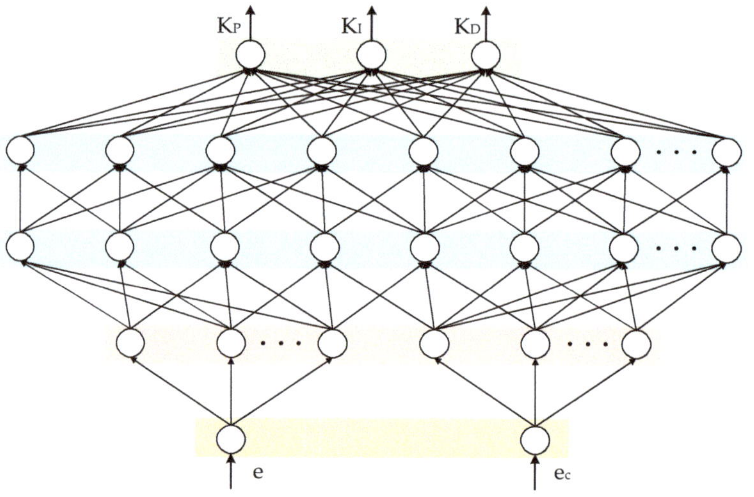

Figure 3. FNN structure model.

Input layer: the vector of input is $x = [x_1, x_2, \cdots, x_n]$. The nodes of this layer of the network are directly connected to each component, x_i, and the error, e(k), and the error rate of change, $e_c(k)$, are the inputs to the general FNN. The result of the first layer is y_i^1.

$$y_i^1 = x_i, i = 1, 2, \cdots, n \tag{11}$$

Fuzzification layer: Each neuron in this layer represents 1 Gaussian subordinate function, whose role is to divide the input values into fuzzy intervals and fuzzy them [29]. The output is represented as follows:

$$y_{ij}^2 = \exp\,[\frac{(u_{ij} - c_{ij})^2}{b_{ij}{}^2}] \tag{12}$$

In the formula, $i = 1, 2, \cdots, n$, $j = 1, 2, \cdots, m_i$, the quantity of input vectors is n, and the number of fuzzy rules is m_i. The Gaussian function's center and width, respectively, are denoted by c_{ij} and b_{ij}. The output quantity is expressed as y_{ij}^2.

Fuzzy inference layer: A fuzzy rule from the fuzzy rule base is represented by each neuron. The aim is to determine each rule's fitness, and the common calculation methods are the minimum value method and the product method. The product method is chosen here for calculation.

$$y_h^3 = x_1^{i_1} \cdot x_2^{i_2} \cdots x_m^{i_n} \tag{13}$$

In the formula, $i_j = 1, 2, \cdots, m_i$, $h = 1, 2, \cdots, m$, $m = \prod_{j=1}^{n} m_j$.

Normalization layer: The output of the fuzzy inference layer is normalized. The quantity of nodes in this layer is equal to the amount of nodes in the fuzzy inference layer, and the output of the normalization layer is as stated below.

$$y_h^4 = \frac{y_h^3}{\sum_{h=1}^{m} y_h^3} \tag{14}$$

Output layer: This layer uses the center of gravity approach to realize the clarification and defuzzification process, and the output obtained after calculation is the result of PID parameter adjustment.

$$y_k = \sum_{h=1}^{m} y_h^4 \omega_s^j = \sum_{h=1}^{m} \frac{y_h^3}{\sum_{h=1}^{m} y_h^3} \omega_s^j \tag{15}$$

In the formula, in the fuzzy inference layer, there are m nodes, $k = 1, 2, \cdots, r$; the output layer's node count is denoted by the symbol r, and ω_s^j is the jth weight corresponding to the sth output.

2.3.4. FNN Optimization Algorithm

The learning optimization strategy of FNN is usually to continuously adjust the parameters of the network by the gradient descent method to obtain the ideal control parameters. Determining the best learning parameters and avoiding local optimality can be accomplished using a particle swarm algorithm (PSO) to solve continuous and discrete optimization problems [30,31]. However, the optimization speed of PSO is not very quick, its local search performance is subpar, and it is simple to fall into a local extremum. In general, the PSO method is very slow in searching around the global optimum. To address the shortcomings of existing optimization algorithms, the gradient descent algorithm and PSO algorithm are combined to optimize the network parameters. The approximate optimal solutions of the network weight parameters are found by using the particle swarm algorithm's global search capability, after which they are adjusted and optimized using the gradient descent approach, thus improving the training accuracy of the fuzzy neural network.

1. Gradient Descent;

The network weight learning error metric is defined as:

$$E(k) = \frac{1}{2} e(k)^2 = \frac{1}{2} [r(k) - y(k)]^2 \tag{16}$$

From the gradient descent method, the learning algorithm of the network is expressed as [32]:

$$\Delta\omega_s^j(k) = -\eta\frac{\partial E(k)}{\partial\omega_s^j} \tag{17}$$

$$\omega_s^j(k) = \omega_s^j(k-1) + \Delta\omega_s^j(k) + \alpha\left[\omega_s^j(k-1) - \omega_s^j(k-2)\right] \tag{18}$$

where η is the learning rate, α is the momentum factor, and $\eta \in [0,1]$, $\alpha \in [0,1]$.

The same can be obtained:

$$c_{ij}(k) = c_{ij}(k-1) + \Delta c_{ij}(k) + \alpha\left[c_{ij}(k-1) - c_{ij}(k-2)\right] \tag{19}$$

$$b_{ij}(k) = b_{ij}(k-1) + \Delta b_{ij}(k) + \alpha\left[b_{ij}(k-1) - b_{ij}(k-2)\right] \tag{20}$$

2. Particle swarm algorithm

The population intelligence optimization approach designated as particle swarm optimization (PSO) is frequently employed in multi-objective optimization situations [33,34]. The specific working process is as follows: Suppose a particle swarm with M particles searches for the optimal position in a space of N dimensions. Assuming that the position of the ith particle (i =1, 2, ... , M) is x_i and the velocity is v_i, the individual extreme value, p_{ibest}, and the group extreme value, p_{gbest}, of the particle are determined according to the particle fitness value, and the particle is continuously updated according to p_{ibest} and p_{gbest}. Its own position and velocity are updated to find the global optimal solution. The particle velocity and position update formulas are expressed as:

$$v_{id}^{t+1} = \omega v_{id}^t + c_1 r_1\left(p_{ibestd}^t - x_{id}^t\right) + c_2 r_2\left(p_{gbestd}^t - x_{id}^t\right) \tag{21}$$

$$x_{id}^{t+1} = x_{id}^t + v_{id}^{t+1} \tag{22}$$

In the formula, ω is the inertia weight, t is the number of current iteration steps, c_1 and c_2 are learning factors, r_1 and r_2 are random numbers between 0 and 1, and $d = 1, 2, \cdots, D$; $i = 1, 2, \cdots, M$.

2.3.5. Hybrid Algorithm Optimization Process

The hybrid algorithm of the particle swarm algorithm and gradient descent method to optimize the fuzzy neural network PID controller is denoted as PSO-FNN-PID. The specific steps of the hybrid algorithm to optimize the fuzzy neural network are shown in Figure 4.

1. The fuzzy neural network parameters, c_{ij}, b_{ij}, ω_s^j, are initialized;
2. Particle swarm initialization. Parameters such as those of population size, particle dimensions, and initial inertia weight, as well as learning factor, are set first, after which a set of particle positions is generated at random and the particle's maximum and minimum velocities are determined; between the extremes of highest and minimum velocity, each particle's velocity is determined randomly;
3. After updating the velocity and position of the particle, the fitness value of the particle at each iteration step is calculated, and the individual optimal extremum, p_{ibest}, and the population optimal extremum, p_{gbest}, are updated;
4. If the termination condition is satisfied, the corresponding network parameters are passed to the FNN;
5. The FNN acquires the initial values of the parameters and then calculates them and updates the network parameters online by back-propagation through the gradient descent method. The final optimal solutions are output.

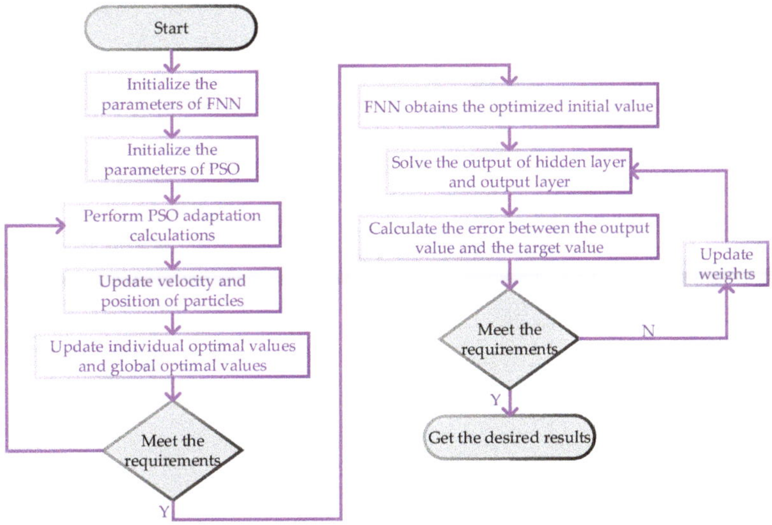

Figure 4. Hybrid algorithm optimization flow chart.

3. Results

Simulation of the models was performed using MATLAB/Simulink. According to the differential equations of motion and the 1/4 suspension system's stochastic road input model, the simulation models of the 1/4 active and passive suspension systems were created in Simulink. Table 2 illustrates the fundamental parameters for the 1/4 suspension model.

Table 2. Fundamental data of suspension system.

Variable	Value
Sprung mass M/kg	240
Unsprung mass m/kg	30
Tire stiffness K_1/(N/m)	160,000
Spring rate K_2/(N/m)	16,000
Suspension damping c/(N•s)/m	980

For the sake of highlighting the optimization effect of the FNN-PID control strategy and to verify its effectiveness, the passive suspension, PID, and FNN-PID control active suspension were emulated and analyzed, respectively. The PID controller's parameters were $K_P = 5$, $K_I = 430$, and $K_D = 0.1$. The structure of the FNN was designed as 2-14-49-49-3, and the number of network parameters to be adjusted was $14 \times 2 + 49 \times 3 = 175$. Therefore, the dimension of the particles is set to 175, and then the following other pertinent settings are made to the particle swarm algorithm: the overall population size is 150, the learning factor is c1 = c2 = 2, and the inertia weight is 0.8. The particle velocity interval for the width of the affiliation function, b_{ij}, and the center value, c_{ij}, is set to [−3, 3], and the particle velocity interval for the connection weight, ω_s^j, of the FNN is [−1, 1]. The learning rate of the FNN is $\eta = 0.5$ and the momentum factor is $\alpha = 0.2$.

When vehicles are driven on actual highways and dirt roads, they are often subjected to impact-type road surfaces, such as gravel and speed bumps, which affect all vehicle driving performance. In order to study the control effect of the FNN-PID-controlled active suspension under such operating conditions, a stepped road model was established to examine the vibration response properties of the suspension under such conditions. The step excitation with a step amplitude of 0.01 m was selected, as well as the suspension system's vibration response curve, which is displayed in Figures 5–7.

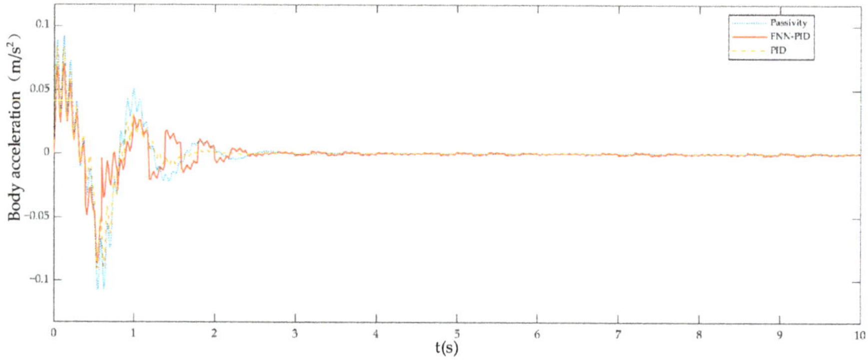

Figure 5. Step response diagram of Sprung Mass Acceleration.

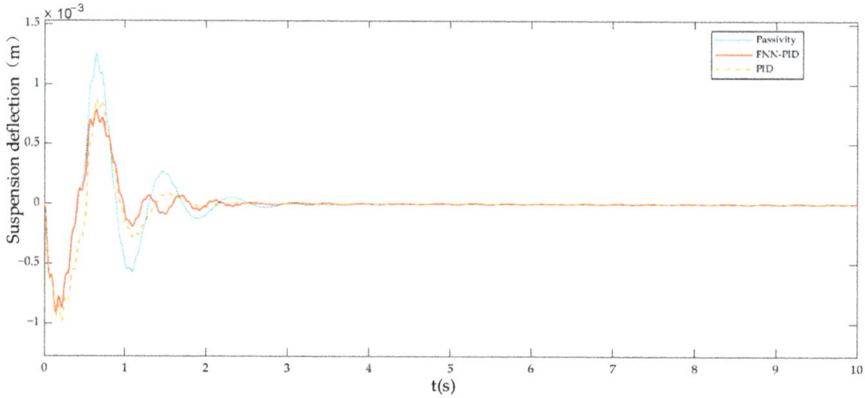

Figure 6. Step response diagram of Dynamic Deflection of Suspension.

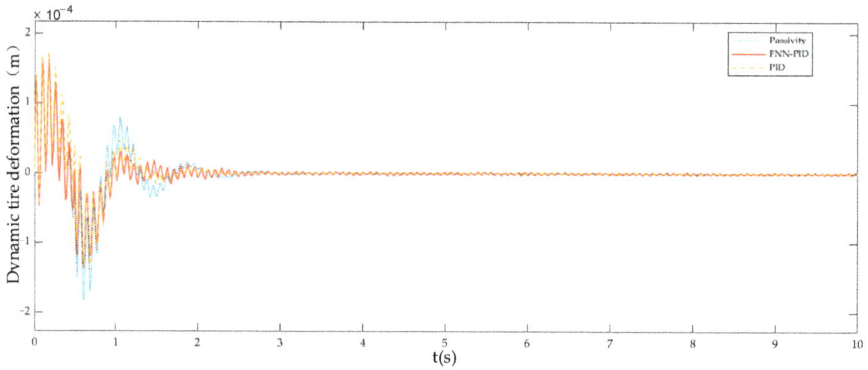

Figure 7. Step response diagram of Dynamic Tire Deformation.

Figure 5 demonstrates that the vehicle suspension system with the FNN-PID controller has better Sprung Mass Acceleration (SMA) than the conventional suspension system, which can make the body more stable with good control effect and can make the vehicle amplitude stable in a short time and quickly converge to 0. Additionally, Figures 6 and 7 show that the Dynamic Deflection of Suspension (DDS) and Dynamic Tire Deformation (DTD) can also reduce the amplitude under FNN-PID control and cause it to quickly converge to 0. Therefore, the active suspension controlled by FNN-PID can effectively reduce vibration and recover quickly, which greatly improves the passenger's ride experience.

Meanwhile, Table 3 shows the root mean square of each suspension index, demonstrating that the SMA, DDS, and DTD of the active suspension system with the FNN-PID controller are improved to some extent. The SMA, DDS, and DTD are decreased by 30.7%, 23.4%, and 16.3%, respectively, when compared to passive suspension. Compared to the PID-controlled active suspension system, the three performance indicators are reduced by 14.6%, 11.3%, and 8.2% respectively. The FNN-PID controller clearly has the potential to significantly lower the suspension's performance indices and improve the vehicle's passenger comfort.

Table 3. Comparison of root-mean-square suspension performance under step excitation.

Index	Passive	PID Controller	FNN-PID Controller
SMA (m/s^2)	2.265×10^{-2}	1.838×10^{-2}	1.570×10^{-2}
DDS (m)	2.896×10^{-4}	2.502×10^{-4}	2.219×10^{-4}
DTD (m)	3.671×10^{-5}	3.350×10^{-5}	3.074×10^{-5}

In addition, the issue of the time between the change in road conditions and the response achieved by the suspension system is taken into account. A set of control tests was set up with the objective of achieving a steady state of vehicle vertical displacement under step response. As shown in Figure 8, the suspension is given a step signal of 0.01 m, 0.05 m, and 0.08 m at 1 s, and a reasonable steady-state error, Δ, is set. When the step signal is 0.01 m, $\Delta = 0.0002$, and when the step signal is 0.05 m and 0.08 m, $\Delta = 0.001$. When the step signal is 0.01 m and 0.05 m, the time for the FNN-PID controller to reach steady state is approximately 2.1 s, and the overall response time is 1.1 s. The time for the PID controller to reach steady state is approximately 2.4 s, with an overall response time of 1.4 s, and the passive system reaches steady state in approximately 3.1 s, with an overall response time of 2.1 s. When the step signal is 0.08, the time for the FNN-PID controller to reach steady-state is approximately 2.6 s, and the overall response time is 1.6 s. The time for the PID controller to reach steady-state is approximately 2.7 s, the overall response time is 1.7 s, and the passive system reaches steady state in approximately 3.4 s, with an overall response time of 2.4 s. The results show that the response time of the FNN-PID controller is reduced by 21.4% compared to the PID controller and 47.6% compared to the passive suspension when the road conditions are less variable. When the road conditions vary widely, the response time of the FNN-PID controller is reduced by only 5.9% compared to the PID controller and 33.3% compared to the passive suspension.

In order to obtain each suspension performance index under normal vehicle driving, the proposed random road excitation is used for simulation analysis. It is assumed that the vehicle is driven in a straight line at 30 km/h on the Class B road, with a simulation time of 20 s. The simulation is performed under the control of the PID controller and FNN-PID controller, respectively. The SMA, DDS, and DTD were still chosen as the main indexes to evaluate the performance of suspension, and the simulation result curves are shown in Figures 9–11.

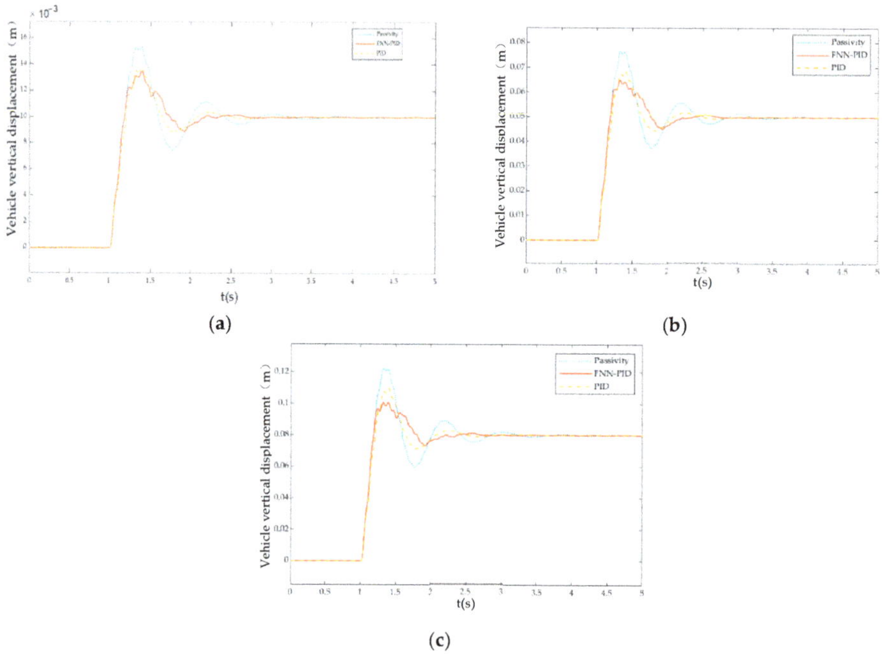

Figure 8. Plot of vehicle vertical displacement change under step signal. (**a**) The step signal is 0.01 m. (**b**) The step signal is 0.05 m. (**c**) The step signal is 0.08 m.

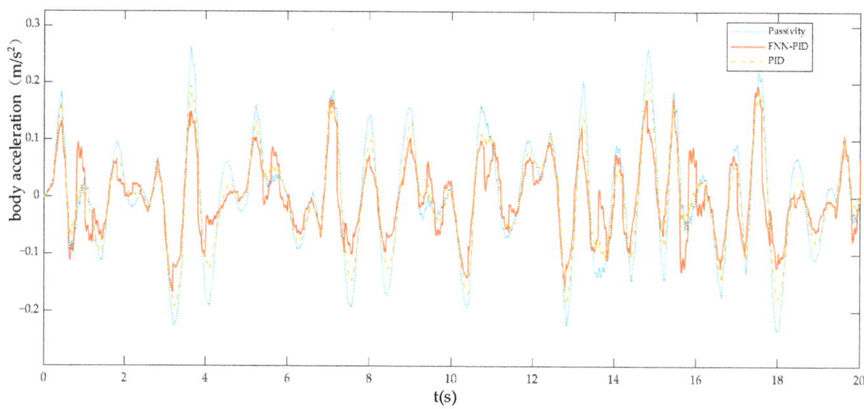

Figure 9. Comparison of Sprung Mass Acceleration simulation results.

From Figures 9–11, it is clear that, in comparison with the passive suspension system, the SMA, DDS, and DTD of the active suspension system with the PID controller and the FNN-PID controller are reduced to a certain extent, indicating that both designed active suspension control systems are able to curb the overall vehicle vibration.

To make the analysis of the control effect of different controllers on the suspension system more intuitive, the above graphs were data processed to obtain the root mean square values of each curve, as demonstrated in Table 4.

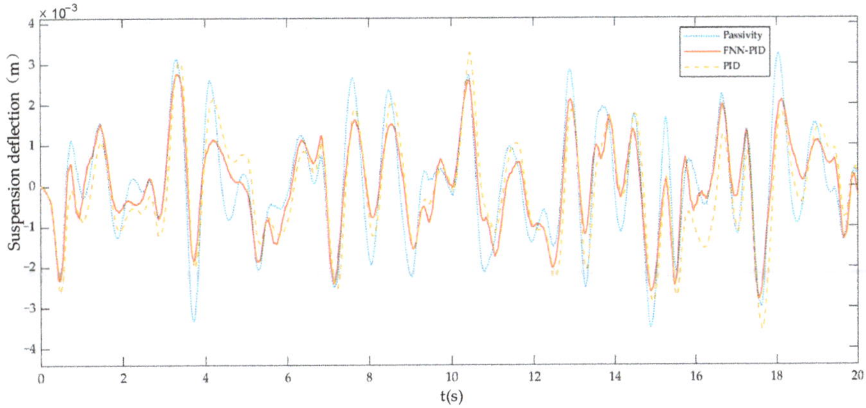

Figure 10. Comparison of Dynamic Deflection of Suspension simulation results.

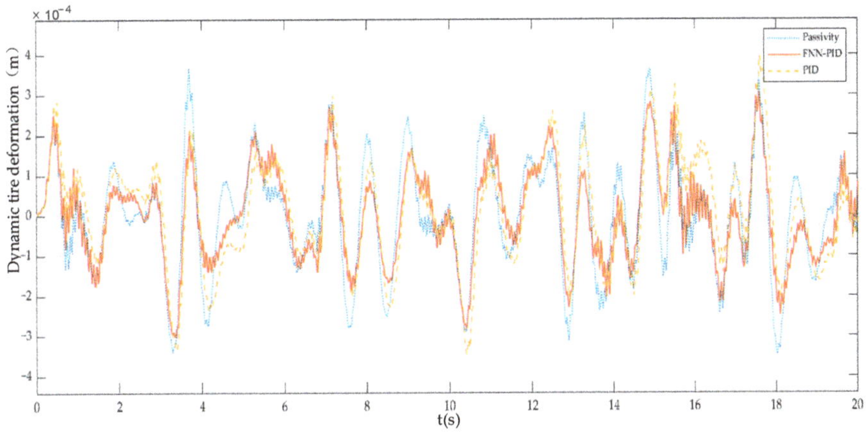

Figure 11. Comparison of Dynamic Tire Deformation simulation results.

Table 4. Root mean square comparison of suspension performance indexes under random.

Class	Index	Passive	PID Controller	FNN-PID Controller
A	SMA (m/s^2)	0.0490	0.0391	0.0334
	DDS (m)	6.735×10^{-4}	6.286×10^{-4}	5.525×10^{-4}
	DTD (m)	7.017×10^{-5}	6.701×10^{-5}	5.890×10^{-5}
B	SMA (m/s^2)	0.0979	0.0782	0.0681
	DDS (m)	1.346×10^{-3}	1.254×10^{-3}	1.107×10^{-3}
	DTD (m)	1.402×10^{-4}	1.336×10^{-4}	1.185×10^{-4}
C	SMA (m/s^2)	0.1820	0.1440	0.1257
	DDS (m)	2.515×10^{-3}	2.319×10^{-3}	2.012×10^{-3}
	DTD (m)	2.576×10^{-4}	2.508×10^{-4}	2.151×10^{-4}
D	SMA (m/s^2)	0.3476	0.2673	0.2390
	DDS (m)	4.747×10^{-3}	4.354×10^{-3}	3.820×10^{-3}
	DTD (m)	4.933×10^{-4}	4.602×10^{-4}	4.003×10^{-4}

Table 4 illustrates that, in comparison with the passive suspension, the active suspension with FNN-PID control has 30.4%, 17.8%, and 15.5% reduction in SMA, DDS, and DTD, respectively. On the other hand, the SMA, DDS, and DTD of the active suspension with FNN-PID control were reduced by 14.6%, 12.1%, and 11.2%, respectively, compared to the active suspension with PID control. From these data, it is possible to draw the conclusion that, when compared with the other two suspension systems, the FNN-PID controller is able to suppress the variation of SMA, so that the wheels can closely follow the road, ensuring good maneuverability while giving passengers a more comfortable ride.

4. Discussion

A 1/4 active suspension simulation model was established in MATLAB/Simulink, and a PSO-FNN-PID control algorithm was designed. By combining the PSO algorithm and the gradient descent method, the initial parameters of the FNN were optimized offline and then finetuned online, so as to obtain the optimal control rules and realize the real-time online adjustment of the PID control parameters. The combination of the two methods allowed the learning process to avoid falling into local minima while reaching the exact value at a later stage of learning. The simulation analysis indicated that the active suspension system with PSO-FNN-PID control had a preferable control effect compared with the passive suspension system and active suspension system with PID control, and could effectively attenuate the body vibration caused by external disturbance. Therefore, an examination of the graphs and data reveals that the PSO-optimized FNN-PID controller has significantly superior performance, and is capable of enhancing the vehicle's road adhesion and ensuring a smooth and comfortable ride. In the future, we will continue to explore the possibility of implementing this control system in the whole vehicle system and consider the development and implementation of related hardware and software. In addition, we also hope to make some changes to the basic PID control, such as including an integral anti-saturation control strategy, in order to obtain better results.

Author Contributions: Conceptualization, M.L. and J.L.; methodology, M.L.; software, J.L.; validation, J.L. and J.X.; formal analysis, J.L.; investigation, G.L.; resources, M.L.; data curation, J.L.; writing—original draft preparation, G.L.; writing—review and editing, M.L.; visualization, J.X.; supervision, M.L.; funding acquisition, M.L. All authors have read and agreed to the published version of the manuscript.

Funding: This research was funded by Natural Science Foundation of Hainan Province, grant number 2019RC144, and supported by China Scholarship Council, grant number 202007565009.

Data Availability Statement: Not applicable.

Conflicts of Interest: The authors declare no conflict of interest.

References

1. Manolache-Rusu, I.-C.; Suciu, C.; Mihai, I. Analysis of Passive vs. Semi-Active Quarter Car Suspension Models. In *Advanced Topics in Optoelectronics, Microelectronics and Nanotechnologies X*; SPIE: Bellingham, WA, USA, 2020; Volume 11718, p. 117181Q. [CrossRef]
2. Kumar, S.; Medhavi, A.; Kumar, R. Active and Passive Suspension System Performance under Random Road Profile Excitations. *Int. J. Acoust. Vib.* **2020**, *25*, 532–541. [CrossRef]
3. Babawuro, A.Y.; Tahir, N.M.; Muhammed, M.; Sambo, A.U. Optimized State Feedback Control of Quarter Car Active Suspension System Based on LMI Algorithm. *J. Phys. Conf. Ser.* **2020**, *1502*, 012019. [CrossRef]
4. Haemers, M.; Derammelaere, S.; Ionescu, C.-M.; Stockman, K.; De Viaene, J.; Verbelen, F. Proportional-Integral State-Feedback Controller Optimization for a Full-Car Active Suspension Setup Using a Genetic Algorithm. *IFAC-PapersOnLine* **2018**, *51*, 1–6. [CrossRef]
5. Youness, S.F.; Lobusov, E.C. Networked Control for Active Suspension System. *Procedia Comput. Sci.* **2019**, *150*, 123–130. [CrossRef]
6. Nguyen, T.A. Control an Active Suspension System by Using PID and LQR Controller. *Int. J. Mech. Prod. Eng. Res. Dev.* **2020**, *10*, 7003–7012. [CrossRef]
7. Bai, R.; Wang, H.-B. Robust Optimal Control for the Vehicle Suspension System with Uncertainties. *IEEE Trans. Cybern.* **2021**, *52*, 9263–9273. [CrossRef] [PubMed]
8. Jibril, M.; Alluvada, P. Quarter Car Active Suspension System Design Using Optimal and Robust Control Method. *Ind. Eng. Lett.* **2020**, *10*, 43.

9. Li, Z. Evaluating Car's Ride Comfort and Controlling Vibration of Suspension System Based on Adaptive PID Control. *Tech. J. Daukeyev Univ.* **2021**, *1*, 1–9. [CrossRef]
10. Li, Z.; Chen, L.; Zheng, Q.; Dou, X.; Yang, L. Control of a Path Following Caterpillar Robot Based on a Sliding Mode Variable Structure Algorithm. *Biosyst. Eng.* **2019**, *186*, 293–306. [CrossRef]
11. Saifi, D.; Kumar, P. Modelling of Active Suspension System for Quarter Car (PID Control, MATLAB). *Int. J. Eng. Appl. Sci. Technol.* **2021**, *5*, 155–160. [CrossRef]
12. Khodadadi, H.; Ghadiri, H. Self-Tuning PID Controller Design Using Fuzzy Logic for Half Car Active Suspension System. *Int. J. Dyn. Control* **2018**, *6*, 224–232. [CrossRef]
13. Cao, K.; Li, Z.; Gu, Y.; Zhang, L.; Chen, L. The Control Design of Transverse Interconnected Electronic Control Air Suspension Based on Seeker Optimization Algorithm. *Proc. Inst. Mech. Eng. Part D J. Automob. Eng.* **2021**, *235*, 2200–2211. [CrossRef]
14. Muderrisoğlu, K.; Arisoy, D.O.; Ahan, A.O.; Akdogan, E. PID Parameters Prediction Using Neural Network for A Linear Quarter Car Suspension Control. *Int. J. Intell. Syst. Appl. Eng.* **2016**, *4*, 20–24. [CrossRef]
15. Wang, M.; Pang, H.; Wang, P.; Luo, J. BP Neural Network and PID Combined Control Applied to Vehicle Active Suspension System. In Proceedings of the 2021 40th Chinese Control Conference (CCC), Shanghai, China, 26–28 July 2021; pp. 8187–8192.
16. Qiu, R. Adaptive Control of Vehicle Active Suspension Based on Neural Network Optimization. *E3S Web Conf.* **2021**, *261*, 03046. [CrossRef]
17. Ahmed, A.A.; Saleh Alshandoli, A.F. Using Of Neural Network Controller and Fuzzy PID Control To Improve Electric Vehicle Stability Based On A14-DOF Model. In Proceedings of the 2020 International Conference on Electrical Engineering (ICEE), Istanbul, Turkey, 28–27 September 2020; pp. 1–6.
18. Han, S.-Y.; Dong, J.-F.; Zhou, J.; Chen, Y.-H. Adaptive Fuzzy PID Control Strategy for Vehicle Active Suspension Based on Road Evaluation. *Electronics* **2022**, *11*, 921. [CrossRef]
19. Swethamarai, P.; Lakshmi, P. Adaptive-Fuzzy Fractional Order PID Controller-Based Active Suspension for Vibration Control. *IETE J. Res.* **2022**, *68*, 3487–3502. [CrossRef]
20. Sadeghi, M.S.; Varzandian, S.; Barzegar, A. Optimization of Classical PID and Fuzzy PID Controllers of a Nonlinear Quarter Car Suspension System Using PSO Algorithm. In Proceedings of the 2011 1st International Conference on Computer and Knowledge Engineering (ICCKE), Mashhad, Iran, 13–14 October 2011; pp. 172–176.
21. Gandhi, P.; Adarsh, S.; Ramachandran, K.I. Performance Analysis of Half Car Suspension Model with 4 DOF Using PID, LQR, FUZZY and ANFIS Controllers. *Procedia Comput. Sci.* **2017**, *115*, 2–13. [CrossRef]
22. Shalabi, M.E.; Fath Elbab, A.M.R.; El-Hussieny, H.; Abouelsoud, A.A. Neuro-Fuzzy Volume Control for Quarter Car Air-Spring Suspension System. *IEEE Access* **2021**, *9*, 77611–77623. [CrossRef]
23. Pang, H.; Liu, F.; Xu, Z. Variable Universe Fuzzy Control for Vehicle Semi-Active Suspension System with MR Damper Combining Fuzzy Neural Network and Particle Swarm Optimization. *Neurocomputing* **2018**, *306*, 130–140. [CrossRef]
24. Ghosh, G.; Mandal, P.; Mondal, S.C. Modeling and Optimization of Surface Roughness in Keyway Milling Using ANN, Genetic Algorithm, and Particle Swarm Optimization. *Int. J. Adv. Manuf. Technol.* **2019**, *100*, 1223–1242. [CrossRef]
25. Zhao, Q.; Zhu, B. Multi-Objective Optimization of Active Suspension Predictive Control Based on Improved PSO Algorithm. *J. Vibroeng.* **2019**, *21*, 1388–1404. [CrossRef]
26. Yuan-chun, K. Fatigue Analysis of Suspension Control Arm Based on Road Spectrum. *IOP Conf. Ser. Mater. Sci. Eng.* **2019**, *538*, 012062. [CrossRef]
27. Zhou, Z.C.; Chen, R. Design on Fuzzy Neural Network PID Control System of Diesel Engine. *Adv. Mater. Res.* **2013**, *756–759*, 425–429. [CrossRef]
28. Borase, R.P.; Maghade, D.K.; Sondkar, S.Y.; Pawar, S.N. A Review of PID Control, Tuning Methods and Applications. *Int. J. Dyn. Control* **2021**, *9*, 818–827. [CrossRef]
29. Cao, P.; Zhao, W.; Liu, S.; Shi, L.; Gao, H. Using a Digital Camera Combined with Fitting Algorithm and T-S Fuzzy Neural Network to Determine the Turbidity in Water. *IEEE Access* **2019**, *7*, 83589–83599. [CrossRef]
30. Jain, N.K.; Nangia, U.; Jain, J. A Review of Particle Swarm Optimization. *J. Inst. Eng. India Ser. B* **2018**, *99*, 407–411. [CrossRef]
31. Bansal, J.C. Particle Swarm Optimization. In *Evolutionary and Swarm Intelligence Algorithms*; Bansal, J.C., Singh, P.K., Pal, N.R., Eds.; Studies in Computational Intelligence; Springer International Publishing: Cham, Switzerland, 2019; pp. 11–23, ISBN 978-3-319-91341-4.
32. Dogo, E.M.; Afolabi, O.J.; Nwulu, N.I.; Twala, B.; Aigbavboa, C.O. A Comparative Analysis of Gradient Descent-Based Optimization Algorithms on Convolutional Neural Networks. In Proceedings of the 2018 International Conference on Computational Techniques, Electronics and Mechanical Systems (CTEMS), Belgaum, India, 21–22 December 2018; pp. 92–99.
33. Lv, S.; Chen, G.; Dai, J. Active Suspension Control Based on Particle Swarm Optimization. *Recent Pat. Mech. Eng.* **2020**, *13*, 60–78. [CrossRef]
34. Bejarbaneh, E.Y.; Bagheri, A.; Bejarbaneh, B.Y.; Buyamin, S.; Chegini, S.N. A New Adjusting Technique for PID Type Fuzzy Logic Controller Using PSOSCALF Optimization Algorithm. *Appl. Soft Comput.* **2019**, *85*, 105822. [CrossRef]

World Electric Vehicle Journal

MDPI

Article

Performance Enhancement of Vehicle Mechatronic Inertial Suspension, Employing a Bridge Electrical Network

Tianyi Zhang [1], Xiaofeng Yang [1,*], Yujie Shen [2], Xiaofu Liu [3] and Tao He [1]

1 School of Automotive and Traffic Engineering, Jiangsu University, Zhenjiang 212013, China
2 Automotive Engineering Research Institute, Jiangsu University, Zhenjiang 212013, China
3 College of Engineering, China Agricultural University, Beijing 100083, China
* Correspondence: yangxf18@ujs.edu.cn; Tel.: +86-139-5285-3832

Abstract: Inerters, a new type of mass element, have been successfully applied in various fields, such as in automotive and civil engineering. The development of a new element, named a mechatronic inerter, which consists of a ball-screw inerter and permanent magnet electric machinery, proves the feasibility of adopting electrical element impedances to simulate corresponding mechanical elements. In this paper, the structures of the bridge electrical network and series-parallel electrical network and their impedance characteristics are first introduced. Then, a seven-degree-of-freedom vehicle model is established. In addition, by comparison with passive suspension, a bridge network and a series-parallel network with various basic topologies are used to improve the vibration isolation performance of mechatronic inertial suspension, and the advantages of the bridge network (a) are demonstrated. Finally, a bridge electrical network (a) was designed and a real vehicle test was carried out. The test results showed that the mechatronic inertial suspension based on the bridge network (a) was superior to the passive suspension; the RMS (root-mean-square) values of the suspension working space and dynamic tire load of the left rear wheel suspension were reduced by 21.1% and 6.3%, respectively; and the RMS value of the centroid acceleration was improved by 1.8%.

Keywords: suspension; mechatronic inerter; bridge network; high-order impedance; real vehicle test

Citation: Zhang, T.; Yang, X.; Shen, Y.; Liu, X.; He, T. Performance Enhancement of Vehicle Mechatronic Inertial Suspension, Employing a Bridge Electrical Network. *World Electr. Veh. J.* **2022**, *13*, 229. https://doi.org/10.3390/wevj13120229

Academic Editors: Yong Li, Xing Xu, Lin Zhang, Yechen Qin and Yang Lu

Received: 29 October 2022
Accepted: 23 November 2022
Published: 1 December 2022

1. Introduction

Vehicle suspension has gradually developed from passive suspension with fixed parameters, to semi-active suspension with variable parameters [1], and active suspension with active control [2]. However, the suspension structure still utilizes a "spring–damper" system, which are connected in parallel, and the improvement of suspension performance has encountered a bottleneck. Since Smith proposed the inerter in 2002 [3], the inerter-based vibration suppression system has been a popular direction in mechanical fields and has successfully made up for a lack of inertial element and promoted the structural development of vehicle suspension. This article refers to suspension with an inerter as inertial suspension. With the rapid development of inertial suspension systems, the inerter has developed various implementation forms. Papageorgiou et al. introduced a rack-and-pinion inerter in 2009 [4], Faraj proposed a ball-screw inerter in 2019 [5], and Liu designed a hydraulic inerter in 2018 [6]. Other inerters have been proposed by scholars, such as a fluid inerter [7,8] and hydraulic electric inerter [9]. Inertial suspension performs better than passive suspension [10–12]. Moreover, an inerter can effectively improve the performance of mechanical vibration isolation systems [13–16]. To date, inertial suspension has been widely applied in train suspension [17,18], bridges [19,20], buildings [21,22], and robots [23].

With the appearance of inerters, electromechanical similarity theory has achieved a complete correspondence, where an inerter corresponds to a capacitor, a damper corresponds to a resistor, and a spring corresponds to an inductor [24]. Smith et al. improved suspension performance by optimizing suspension layouts with an inerter in 2004 [25]. The optimization was further carried out using linear matrix inequalities, and a synthesized passive network was realized by Bott-Duffin, which indicated that the system performance could be further improved by allowing higher order passive impedance [26]. However, network synthesis for a high-order impedance can be hard to realize mechanically. Therefore, Wang et al. proposed a mechatronic inerter, consisting of a ball-screw inerter and permanent magnet electric machinery in 2011 [27], and the system impedance was realized with the combination of mechanical and electrical networks.

The research on mechatronic inerters mainly focuses in two directions. One is the mechatronic inerter [28–30] and its external circuit [31,32]. Ning et al. introduced a new controllable electrically interconnected suspension based on a mechatronic inerter in 2020, which was composed of a controllable electrical network and two independent electromagnetic suspensions [33], and the results showed that the vertical vibration and roll performance of the vehicle could be improved by controlling the resistance of the electrical network. The other direction is the simplification of higher-order impedance and synthesis of passive networks [34,35]. Shen et al. used the structure-immittance approach to implement an optimal design methodology for mechatronic ISD (inerter–spring–damper) suspension in 2022, which ensured the simplicity of the suspension structure [36]. In summary, although a mechatronic inerter can utilize an external electrical network to simulate a target mechanical impedance and improve the performance of the suspension system, the external electrical network is complex. The problem of the simplest realization of high-order impedances remains to be studied.

A bridge network is a typical electrical circuit that is widely used in electrical theory. Unlike a series-parallel network, the bridge network is an non-series-parallel network that has a special connection method; to solve the impedance expression, the equivalent conversion of the Δ structure and the Y structure is required, and the bridge network is converted into a series and parallel network and calculated [37]. Then, the simplest realization of a biquad impedance of the bridge network is realized [38]. Moreover, the impedance expressions of different bridge networks have been discussed [39], and the results indicated that a bridge network can achieve a high-order impedance, with fewer components compared with Brune synthesis. Considering the large number of high-order impedance elements and the few studies on the simplification of high-order impedance, in this paper, a bridge network is used to simplify high-order impedance and apply it to mechatronic inertial suspension. Based on electromechanical similarity theory, the electrical network of mechatronic inertial suspension is studied. In summary, the bridge network is advantageous for improving the vibration isolation performance. This paper will concentrate on the performance enhancement of a vehicle mechatronic inertial suspension system using a bridge electrical network, to make full use of the advantages of mechatronic inertial suspension.

The following parts of this paper are organized as follows: Section 2 establishes a seven-degree-of-freedom vehicle model and analyzes three bridge networks and three series-parallel networks. The parameters of the proposed bridge networks and series-parallel networks are optimized in Section 3. In Section 4, the characteristics of the bridge network (a) are simulated and analyzed, as well as compared with a series-parallel network (d). Section 5 conducts reports test of a real vehicle. Finally, the summary and conclusions are presented in Section 6.

2. Model Building

2.1. Seven-Degree-of-Freedom Vehicle Model

In order to evaluate the influence of a bridge network on the performance of a vehicle, a seven-degree-of-freedom vehicle model is established in Figure 1. The front axle suspension adopts passive suspension. As the comfort of the rear passengers is very important, this paper applies mechatronic inertial suspension to the rear suspension system, and the mechatronic inerter is connected in series with a damper and then connected in parallel with a spring [40], as shown in Figure 1. The mechatronic inertial suspension is presented in Figure 2. The mechatronic inerter is connected to the damper in series, and the spring is connected to the mechatronic inerter and damper in parallel, when the mechatronic inerter is stimulated by vibration, the linear motion is converted into the rotating motion of the flywheel and motor through a ball screw. As shown in Figure 1.

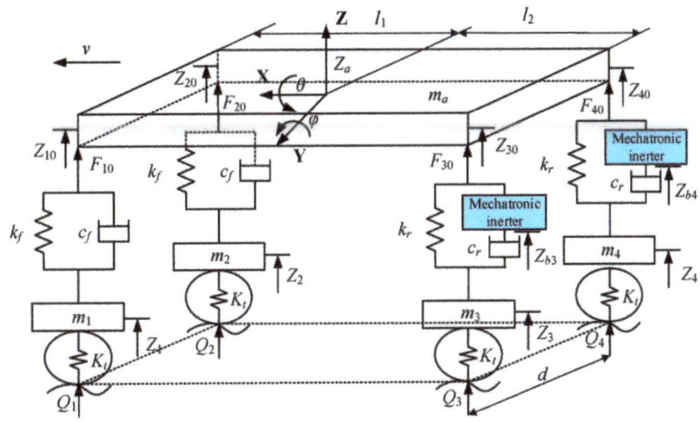

Figure 1. Seven-degree-of-freedom vehicle model.

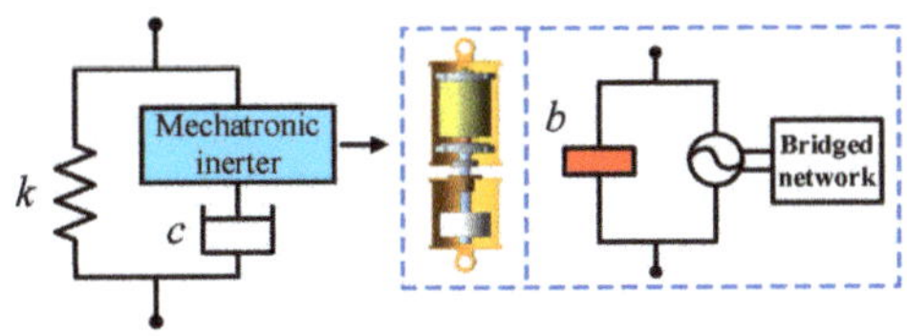

Figure 2. Mechatronic inertial suspension.

The vertical motion equation at the center of mass of the vehicle body is:

$$m_a \ddot{Z}_a = F_{10} + F_{20} + F_{30} + F_{40} \tag{1}$$

The roll motion equation of the vehicle body is:

$$I_x \ddot{\theta} = (F_{20} + F_{40} - F_{10} - F_{30}) \frac{d}{2} \tag{2}$$

The pitching motion equation of the vehicle body is:

$$I_y \ddot{\varphi} = l_2(F_{30} + F_{40}) - l_1(F_{10} + F_{20}) \tag{3}$$

When the pitch angle and roll angle are small, the following relationship holds:

$$\begin{cases} Z_{10} = Z_a - l_1\varphi - \frac{d\theta}{2} \\ Z_{20} = Z_a - l_1\varphi + \frac{d\theta}{2} \\ Z_{30} = Z_a + l_2\varphi - \frac{d\theta}{2} \\ Z_{40} = Z_a + l_2\varphi + \frac{d\theta}{2} \end{cases} \tag{4}$$

The vertical motion equation of the unsprung mass is:

$$\begin{cases} m_1\ddot{Z}_1 = K_t(Q_1 - Z_1) - F_{10} \\ m_2\ddot{Z}_2 = K_t(Q_2 - Z_2) - F_{20} \\ m_3\ddot{Z}_3 = K_t(Q_3 - Z_3) - F_{30} \\ m_4\ddot{Z}_4 = K_t(Q_4 - Z_4) - F_{40} \end{cases} \tag{5}$$

The forces of the four suspensions are:

$$\begin{cases} F_{10} = k_f(Z_1 - Z_{10}) + c_f(\dot{Z}_1 - \dot{Z}_{10}) \\ F_{20} = k_f(Z_2 - Z_{20}) + c_f(\dot{Z}_2 - \dot{Z}_{20}) \\ F_{30} = k_r(Z_3 - Z_{30}) + u_3 \\ u_3 = b(\ddot{Z}_{b3} - \ddot{Z}_{30}) + (\frac{2\pi}{P})^2\frac{k_tk_e}{Z}(\dot{Z}_{b3} - \dot{Z}_{30}) = c_r(\dot{Z}_3 - \dot{Z}_{b3}) \\ F_{40} = k_r(Z_4 - Z_{40}) + u_4 \\ u_4 = b(\ddot{Z}_{b4} - \ddot{Z}_{40}) + (\frac{2\pi}{P})^2\frac{k_tk_e}{Z}(\dot{Z}_{b4} - \dot{Z}_{40}) = c_r(\dot{Z}_4 - \dot{Z}_{b4}) \end{cases} \tag{6}$$

where m_a is the sprung mass. m_1, m_2, m_3, and m_4 are the unsprung mass of the four suspensions, respectively. K_t is the equivalent stiffness of the tire. k_f and c_f are the spring stiffness and the damping coefficient of the front suspensions, respectively. k_r and c_r are the spring stiffness and the damping coefficient of rear suspensions, respectively. Z_{10}, Z_{20}, Z_{30}, and Z_{40} are the vertical displacements of the connection between the vehicle body and the four suspensions. F_{10}, F_{20}, F_{30}, and F_{40} are the forces of the four suspensions, respectively. Z_1, Z_2, Z_3, and Z_4 are the vertical displacements of four unsprung masses, respectively. Z_{b3} and Z_{b4} are the vertical displacements of the mechatronic inerter of the left rear suspension and the right rear suspension. Q_1, Q_2, Q_3, and Q_4 are the displacement inputs of the four wheels. Z_a is the vertical displacement of the sprung mass. θ is the body roll angle, and I_x is the body roll moment of inertia. φ is the body pitch angle, and I_y is the body pitch moment of inertia. l_1 and l_2 are the distance from the front axle and rear axle to the body centroid, respectively. d is the wheelbase, b is the inertial coefficient of the mechatronic inerter, and v is the vehicle speed. k_t and k_e are the inductive torque constant and the inductive voltage constant of the rotating motor, respectively. Z is the impedance of the external circuit of the rotating motor. u_3 and u_4 are the forces of the series branch of the mechatronic inerter and the damper, respectively. P is the lead of the ball screw.

In this paper, a real vehicle was used for the test, the simulation parameters refer to real vehicle data, and the main dimensional parameters of the vehicle were obtained according to the vehicle user manual, as shown in Table 1.

Table 1. Main parameters of the vehicle model.

Parameters	Symbol	Unit	Value
Sprung mass	m_a	kg	1659
Unsprung mass of left and right front wheels	m_1, m_2	kg	47.5
Unsprung mass of left and right rear wheels	m_3, m_4	kg	42.5
Spring stiffness of front axle suspension	k_f	$kN{\cdot}m^{-1}$	25
Spring stiffness of rear axle suspension	k_r	$kN{\cdot}m^{-1}$	22
Damping coefficient of front axle suspension	c_f	$kN{\cdot}s{\cdot}m^{-1}$	1.8
Damping coefficient of rear axle suspension	c_r	$kN{\cdot}s{\cdot}m^{-1}$	1.5
Equivalent stiffness of tire	K_t	$kN{\cdot}m^{-1}$	192
Distance from front axle to body centroid	l_1	m	1.28
Distance from rear axle to body centroid	l_2	m	1.43
Wheelbase	d	m	1.62
Inertance of rear suspension	b	kg	308
Body roll moment of inertia	I_x	$kg{\cdot}m^2$	1088
Body pitch moment of inertia	I_y	$kg{\cdot}m^2$	3032

2.2. Bridge Network

Common bridge networks include a bridge rectifier circuit, half bridge circuit, full bridge circuit, wheatstone bridge, balanced bridge, and unbalanced bridge. The bridge network used in this paper was an unbalanced bridge circuit. The most common bridge network consists of five resistance elements, as shown in Figure 3a. And Figure 3b is the equivalent network.

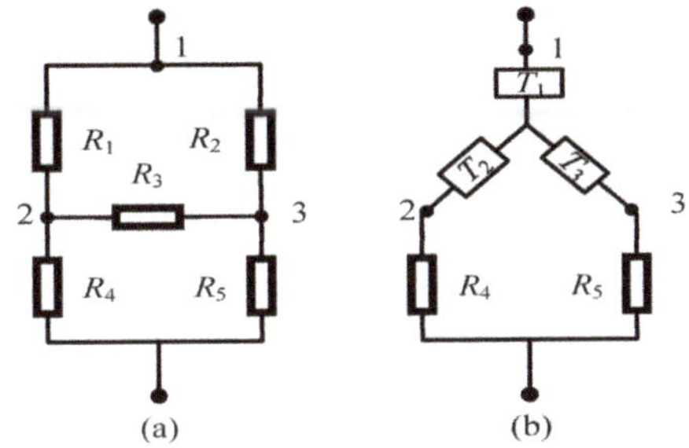

Figure 3. (**a**). bridge network consists of five resistance elements. (**b**). Equivalent network of the bridge network.

Where R_1, R_2, R_3, R_4, and R_5 are resistors, T_1, T_2, and T_3 are equivalent impedances.

In Figure 3, when the Δ structure of Figure 3a is transformed into the Y structure of Figure 3b, the equivalent transformation equation is:

$$\begin{cases} T_1 = \frac{R_1 R_2}{R_1+R_2+R_3} \\ T_2 = \frac{R_1 R_3}{R_1+R_2+R_3} \\ T_3 = \frac{R_2 R_3}{R_1+R_2+R_3} \end{cases} \tag{7}$$

Therefore, the impedance expression of Figure 3a is:

$$Z = T_1 + \frac{(T_2 + R_4)(T_3 + R_5)}{T_2 + R_4 + T_3 + R_5} \tag{8}$$

A bridge network composed of resistance, capacitance, and inductance has many kinds of structures. In this paper, the three most basic bridge networks are selected. Considering the special solution method of a bridge network, the selection principles of the bridge network studied in this paper were as follows. First of all, a resistor, a capacitor, and an inductor can constitute three kinds of Δ structures and three kinds of Y structures, respectively. Since the impedance of the resistance has no effect on the total impedance order [41], the remaining two elements of the bridge network are selected as resistors to form a five-element bridge network with three resistors, one capacitor, and one inductor, as shown in Figure 4.

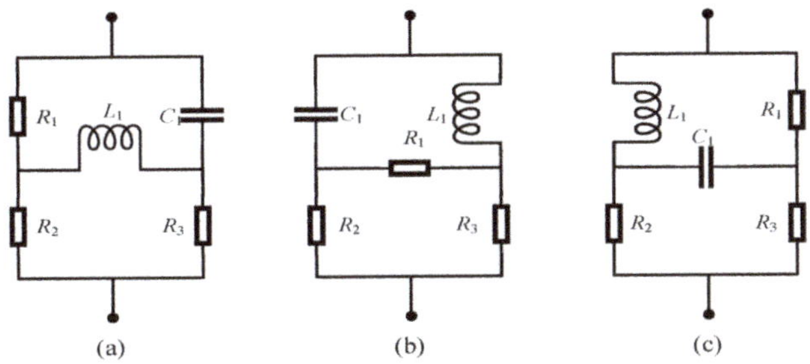

Figure 4. (**a**–**c**) are five-element bridge networks of three different structures.

According to the impedance transformation method shown in the Equation (8), the impedance transfer function of Figure 4a can be expressed as:

$$\frac{1}{Z_a(s)} = \frac{A_1 s^4 + B_1 s^3 + C_1 s^2 + D_1 s + E_1}{F_1 s^4 + G_1 s^3 + H_1 s^2 + I_1 s + J_1} \tag{9}$$

$$\begin{aligned}
A_1 &= (R_1 + R_2 + R_3)L_1^2 C_1^2, \\
B_1 &= (R_1R_2C_1 + R_1R_3C_1 + L_1)L_1C_1 + (R_1 + R_2 + R_3)R_1L_1C_1^2, \\
C_1 &= (R_1R_2C_1 + R_1R_3C_1 + L_1)R_1C_1 + (R_1 + 2R_2 + 2R_3)L_1C_1, \\
D_1 &= 2R_1C_1(R_2 + R_3) + L_1, \\
E_1 &= R_2 + R_3, \\
F_1 &= (R_1 + R_2)R_3{L_1}^2{C_1}^2, \\
G_1 &= (R_1 + R_2)(R_1R_3C_1 + L_1)L_1C_1 + R_1R_2R_3L_1C_1^2, \\
H_1 &= (R_1 + R_2 + R_3)R_1L_1C_1 + (R_1 + R_2)R_3L_1C_1 + (R_1R_3C_1 + L_1)R_1R_2C_1 + R_2R_3L_1C_1, \\
I_1 &= (R_1R_2C_1 + R_1R_3C_1 + L_1)R_1 + R_1R_2R_3C_1 + (R_1R_3C_1 + L_1)R_2, \\
J_1 &= (R_2 + R_3)R_1 + R_2R_3.
\end{aligned} \tag{10}$$

Similarly, it was calculated that the orders of the impedance transfer functions of the bridge network (b) and the bridge network (c) were also double fourth order, which will not be repeated here.

2.3. Series-Parallel Network

Correspondingly, three five-element series-parallel networks are shown in Figure 5.

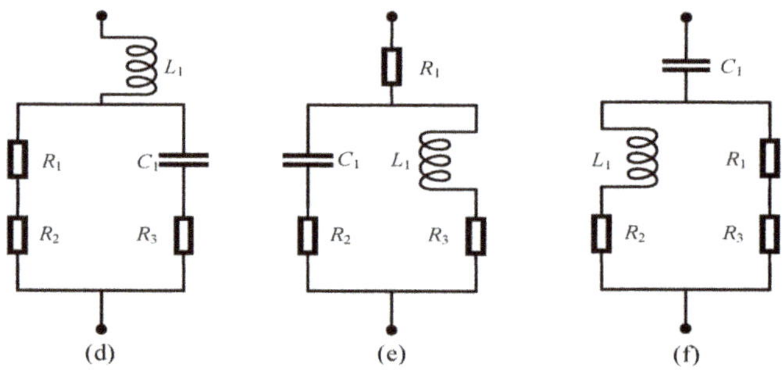

Figure 5. (**d**–**f**) are five-element series-parallel networks of three different structures.

From electrical knowledge, the impedances of the three series-parallel networks in Figure 5d–f were all of biquadratic order. With the same number of components, the bridge network can achieve a higher order impedance than the series-parallel network.

3. Optimization of the Inertial Suspension Parameters

In order to achieve the best performance of the mechatronic inertial suspension, the parameters of the suspension needed to be optimized. In this paper, the particle swarm optimization (PSO) algorithm, which is suitable for multi-objective environments, was selected to optimize the structural parameters of the external electric circuit of the mechatronic inerter. The PSO algorithm is an intelligent optimization algorithm, whose basic idea is to initialize a group of random particles (random solutions) and find the optimal solution through iteration. The specific optimization process is shown in the Figure 6. To begin, the particle is initialized, the fit value of the particle is then compared with the best location it passes through, and the speed and position of the particle is updated. The process ends when the termination condition is met in each iteration, the particles update their position attributes by tracking the individual extremum and global extremum, and finally find the optimal particle [42]. The PSO algorithm has the advantages of fast search speed and memory, due to the need to adjust few parameters, so the structure is simple, and it is suitable for solving practical engineering problems. Particle velocities and positions are updated in accordance with the following two formulas:

$$V^{k+1} = \lambda V^k + d_1 r_1 ({P^k}_{id} - X^k) + d_2 r_2 ({P^k}_{gd} - X^k) \tag{11}$$

$$X^{k+1} = X^k + V^{k+1} \tag{12}$$

where λ is the inertia factor, its value affects the global and individual optimization ability. V is the velocity of the particle. X is the current position of the particle, and k is the current number of iterations. d_1 and d_2 are non-negative constants, called acceleration factors, and the general range is between 0 and 4. r_1 and r_2 are random numbers between (0, 1). P_{id} and P_{gd} are the individual extremum and global extremum, respectively.

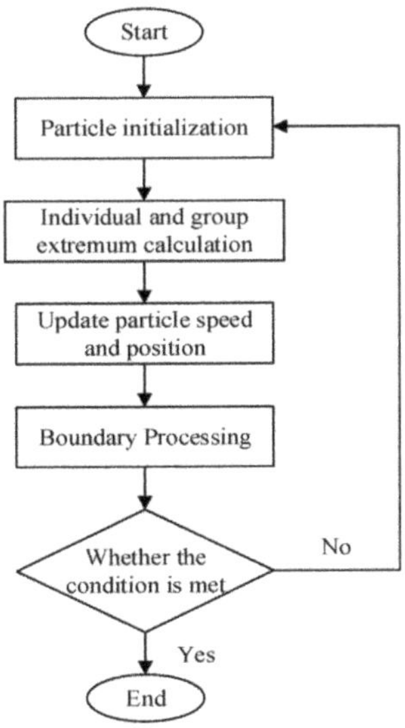

Figure 6. Parameter optimization flow chart.

This paper mainly studies the influence of the bridge network and the series-parallel network on the mechatronic inertial suspension, so three resistors (R_1, R_2, R_3), one capacitor (C_1), and one inductor (L_1) of the external electrical circuit are taken as individuals to be solved. Moreover, the performance indicators have different units and orders of magnitude, so it is necessary to establish a unified objective function. The performance indexes of the mechatronic inertial suspension are divided by the corresponding indexes of the passive suspension, and the sum of their quotients is taken as the objective function. In this paper, f is the objective function of optimization, which is obtained by weighting the following parameters. The influence of different units of evaluation indexes is ignored; meanwhile, the improvement of suspension performance is studied by quantifying the objective function Therefore, the optimization of evaluation indexes of the ride comfort and the road friendliness is transformed into the minimum value problem of the unified objective function. The smaller the value of the optimization objective function, the better the optimization effect, and the performance improvement is obvious. In the optimization process, the road condition and running speed are set as C grade and 20 km/h, respectively, and the number of iterations is 100. Due to the mechatronic inertial suspension being used to replace the rear suspension system, the relevant evaluation indexes of the rear suspension are mainly used as the optimization objectives. The expression of the unified objective function and constraint conditions are as follows:

$$f = \frac{BA(P)}{BA_{\text{pas}}} + \frac{LRSWS(P)}{LRSWS_{\text{pas}}} + \frac{LRDTL(P)}{LRDTL_{\text{pas}}} + \frac{RRSWS(P)}{RRSWS_{\text{pas}}} + \frac{RRDTL(P)}{RRDTL_{\text{pas}}} \tag{13}$$

$$P = \begin{bmatrix} C_1 & L_1 & R_1 & R_2 & R_3 \end{bmatrix} \tag{14}$$

$$s.t.\begin{cases} BA(P) < BA_{pass} \\ LRSWS(P) < LRSWS_{\text{pas}} \\ LRDTL(P) < LRDTL_{pass} \\ RRSWS(P) < RRSWS_{\text{pas}} \\ RRDTL(P) < RRDTL_{pass} \\ LM < P < UM \end{cases} \tag{15}$$

$$\begin{aligned} LM &= [0,\ 0,\ 0,\ 0,\ 0] \\ UM &= [10,\ 100,\ 5000,\ 5000,\ 5000] \end{aligned} \tag{16}$$

where *BA(P)*, *LRSWS(P)*, *LRDTL(P)*, *RRSWS(P)*, and *RRDTL(P)* indicate the root mean square (RMS) of the centroid acceleration, working space of the left rear wheel suspension, dynamic tire load of the left rear wheel, working space of the right rear wheel suspension, and dynamic tire load of the right rear wheel of the optimized mechatronic inertial suspension, respectively. The above parameters were used to evaluate the ride comfort and road friendliness of the vehicle. BA_{pas}, $LRSWS_{\text{pas}}$, $LRDTL_{\text{pas}}$, $RRSWS_{\text{pas}}$, and $RRDTL_{\text{pas}}$ are the RMS of the corresponding performance indexes of the traditional passive suspension. *P* is the set of parameters to be optimized. *LM* and *UM* are the upper and lower bounds of these parameters, and these parameters will affect the handling stability of the vehicle.

After many iterations, the optimized parameters of the proposed bridge network and series-parallel network were revealed, as shown in Table 2.

Table 2. Optimized parameters of the electrical network.

Parameters	Bridge Network			Series-Parallel Network		
	(a)	(b)	(c)	(d)	(e)	(f)
Capacitance C_1 (mF)	8	3	8.4	6.3	2.5	7.5
Inductance L_1 (mH)	18.8	17.5	3.5	16	13.7	15
Resistance R_1 (Ω)	2908	2553	857	2976	2856	2598
Resistance R_2 (Ω)	2984	2824	3000	2768	2708	2714
Resistance R_3 (Ω)	2992	2996	1350	2748	2158	2944

4. Discussion

In this section, numerical simulations are carried out to verify the effectiveness of the mechatronic inertial suspension based on the bridge network (a).

4.1. Road Input

Assuming the vehicle is driving at a speed of *u* on a grade C road, the random road input is expressed as:

$$\dot{z}_r(t) = -0.111[uz_r(t) + 40\sqrt{G_q(n_0)uw(t)}] \tag{17}$$

where $z_r(t)$ is the vertical displacement of the random road input, $w(t)$ is the white noise with mean value of 0, and $G_q(n_0)$ is the road roughness ($2.56/10^4$ m^3).

In the simulation model, four road inputs were required. The random road input of the left front wheel is the same as that of the left rear wheel, but the random road inputs of left front wheel and right front wheel are slightly different. In addition, the time when the front wheel and the rear wheel receive the road input is different in the simulation process. Therefore, the wheelbase divided by the speed was the value of the delay in the simulation, and the left rear wheel and the right rear wheel have a delay compared with the left front wheel and the right front wheel, respectively.

In this paper, the road inputs of the left front wheel and the right front wheel are almost the same; thus, those of the left rear wheel and the right rear wheel are also virtually identical. To make this paper concise and clear, the centroid acceleration, vehicle body pitch angular acceleration, vehicle body roll angular acceleration, working spaces of the left front wheel suspension and the left rear wheel suspension, and the dynamic tire loads of the left front wheel and the left rear wheel were selected as the performance evaluation indexes.

4.2. Performance Analysis of Mechatronic Inertial Suspension

The parameters in Tables 1 and 2 were input into the proposed mechatronic inertial suspension based on a bridge electrical network and a series-parallel electrical network for simulation. The vehicle speed was 20 m/s. The RMS values of performance indexes were obtained, as shown in Tables 3 and 4.

Table 3. RMS comparison of mechatronic inertial suspension based on the bridge network.

Suspension Performance Index	Passive Suspension	Bridge Network		
		(a)	(b)	(c)
RMS of centroid acceleration (m·s^{-2})	1.8792	1.7902	1.8023	1.8010
RMS of body roll angular acceleration (rad·s^{-2})	0.1059	0.1044	0.1037	0.1037
RMS of body pitch angular acceleration (rad·s^{-2})	1.3827	1.3440	1.3532	1.3539
RMS of working space of left front suspension (m)	0.0266	0.0256	0.0257	0.0257
RMS of dynamic tire load of left front wheel (kN)	1.9287	1.8665	1.8781	1.8781
RMS of working space of left rear suspension (m)	0.0271	0.0201	0.0202	0.0203
RMS of dynamic tire load of left rear wheel (kN)	1.8907	1.7389	1.7497	1.7495

Table 4. RMS comparison of mechatronic inertial suspension based on the series-parallel network.

Suspension Performance Index	Passive Suspension	Series-Parallel Network		
		(d)	(e)	(f)
RMS of centroid acceleration (m·s^{-2})	1.8792	1.8354	1.8378	1.8446
RMS of body roll angular acceleration (rad·s^{-2})	0.1059	0.1131	0.1125	0.1125
RMS of body pitch angular acceleration (rad·s^{-2})	1.3827	1.4571	1.4591	1.4642
RMS of working space of left front suspension (m)	0.0266	0.0253	0.0252	0.0254
RMS of dynamic tire load of left front wheel (kN)	1.9287	1.8444	1.8441	1.8528
RMS of working space of left rear suspension (m)	0.0271	0.0229	0.0233	0.0230
RMS of dynamic tire load of left rear wheel (kN)	1.8907	1.7855	1.7897	1.7935

From Tables 3 and 4, it can be noted that among three mechatronic inertial suspensions based on the bridge network, the bridge network (a) had the best improvement in vehicle performance. Similarly, the series-parallel network (d) was the best in its group. Moreover, compared with the series-parallel network (d), the mechatronic inertial suspension based on the bridge network (a) had a greater effect on improving the performance of the vehicle. According to Section 2, it can be concluded that with the same number of electrical components, the bridge network could achieve a higher order impedance compared with the series-parallel network. Therefore, the bridge network (a) was selected as the external electrical network for the mechatronic inerter for the following part.

5. Experimental Research

To further validate the vibration isolation performance of the vehicle mechatronic inertial suspension system employing a bridge electrical network (a), a test of a real vehicle was carried out on the road.

5.1. Structure Selection and Real Vehicle Installation

The key components of mechatronic inerter include the ball–screw pair, flywheel, and rotating motor. The specific parameters were determined, as shown in Table 5.

Table 5. Parameters of the key components.

Parameter	Value	Parameter	Value
Nominal shaft diameter d_0 (mm)	16	Rated power P (W)	2000
Lead P (mm)	5	Rated speed n_e (r·min^{-1})	3000
Center distance of balls on both sides d_p (mm)	16.75	Maximum speed n_m (r·min^{-1})	6000
Groove diameter d_c (mm)	13.5	Rated torque T_e (N·m)	5.88
Number of columns × Number of turns	1 × 2.65	Rated voltage U_e (V)	310
Effective stroke l_0 (mm)	120	Rated current I_e (A)	6
Lead screw stiffness k_l (N·µm^{-1})	130	Inductive torque constant k_t (N·m/A)	0.98
Dynamic rated load c_a (kN)	5.4	Inductive voltage constant k_e (V·s/rad)	0.98
Static rated load c_{oa} (kN)	13.3	Allowable stress σ_p (N·mm^{-2})	150
Dynamic load coefficient k_s	2	Radius of flywheel r (mm)	30
Static load coefficient k_d	3	Thickness of flywheel h (mm)	20

After completing the parameter design of the ball–screw pair, flywheel, and rotating motor, a mechatronic inerter based on a bridge network (a) was developed and connected in series with the damper. Then they were installed in the rear suspension of the test vehicle, to prepare for the subsequent road test, to comprehensively analyze the performance advantages of a mechatronic inertial suspension based on a bridge network (a). In addition, three axis acceleration sensors, a PCB acceleration sensor, a SICK laser displacement sensor, and Siemens LMS Test Lab data acquisition instrument were used to collect the centroid acceleration, roll angular acceleration, pitch angular acceleration, suspension working space, and dynamic tire load signals of the test vehicle.

The test instruments and real vehicle test are shown in Figure 7.

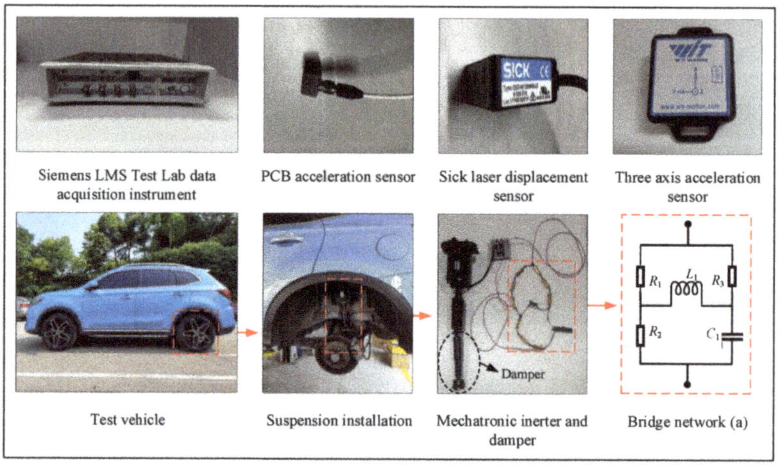

Figure 7. Test instruments and real vehicle test.

5.2. Random Road Input

It was assumed that the test vehicle ran at a speed of 20 km/h on a C grade road. To ensure the consistency of the random road input, the test vehicle was tested on the same road section and ran along the white solid line of the road. The sampling interval was 0.005 s, and the sampling time was 10 s. The result of time domain is shown in Figure 8 and Table 6.

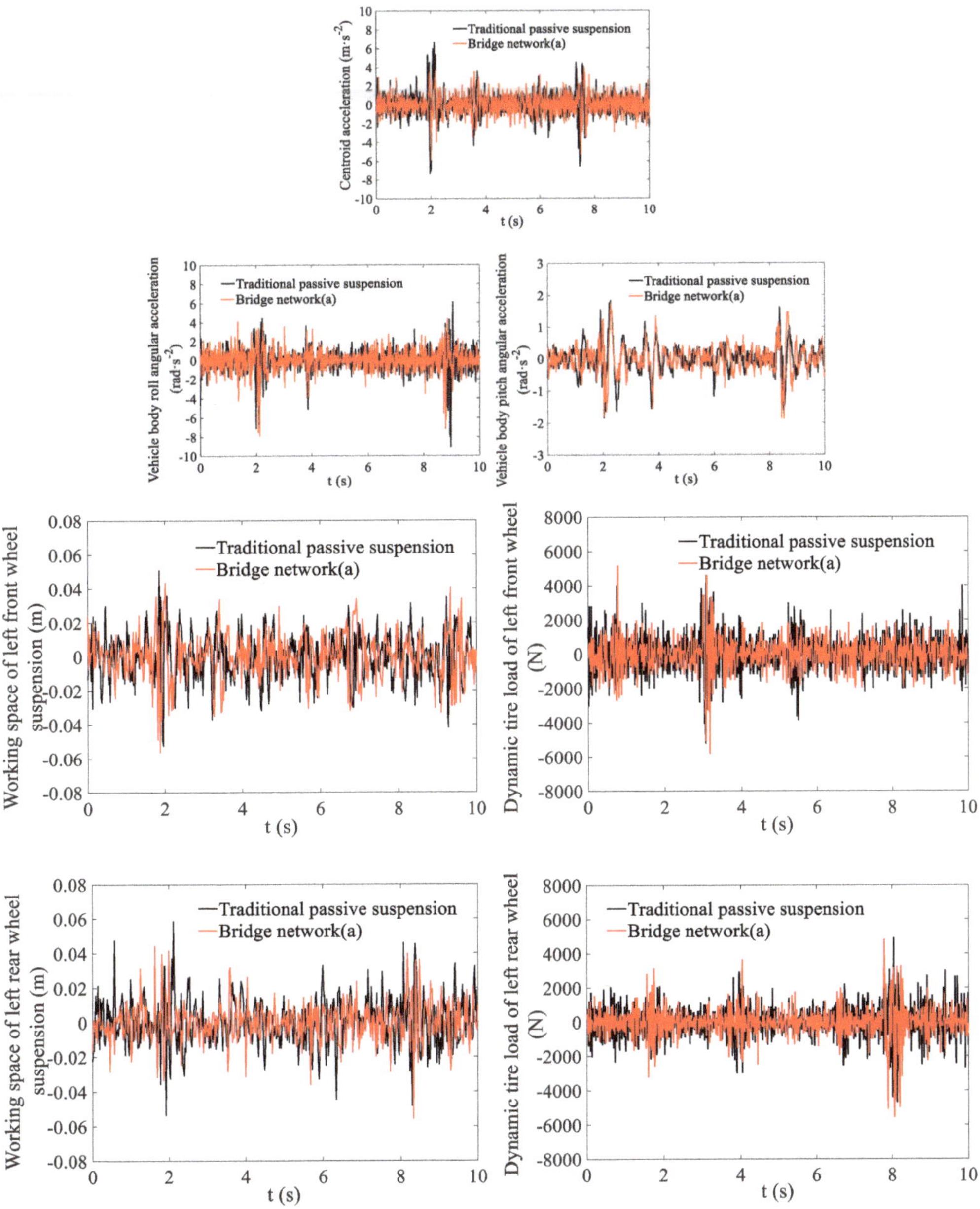

Figure 8. Performance comparison under random road conditions with a vehicle speed of 20 km/h.

Table 6. Performance index with a random input.

Performance Index	RMS of Passive Suspension	Bridge Network (a)	
		RMS	Improvement (%)
Centroid acceleration (m·s^{-2})	1.2656	1.2428	1.8
Vehicle body roll angular acceleration (rad·s^{-2})	1.1139	1.0526	5.5
Vehicle body pitch angular acceleration (rad·s^{-2})	0.4426	0.4643	−4.9
Working space of left front suspension (m)	0.0142	0.0138	2.5
Dynamic tire load of left front wheel (N)	1022	1000	2.2
Working space of left rear suspension (m)	0.0145	0.0114	21.1
Dynamic tire load of left rear wheel (N)	981	919	6.3

It can be seen from Table 6 and Figure 8 that, compared with the passive suspension, the RMS values of centroid acceleration and vehicle body roll angular acceleration of the mechatronic inertial suspension based on a bridge network (a) were reduced by 1.8% and 5.5%, respectively. While, the RMS values of the suspension working space and dynamic tire load of the left rear suspension, comparing the passive suspension with the mechatronic inertial suspension based on a bridge network (a), were decreased by 21.1% and 6.3%. The above improvements improved the ride comfort of the vehicle. However, the RMS value of vehicle body pitch angular acceleration was increased by 4.9%, and this increase of vehicle body pitch angular acceleration made vehicle handling stability slightly worse. Meanwhile, considering that only the rear suspension system adopted the proposed mechatronic inertial suspension based on a bridge network, it was found that the RMS values of the suspension working space and dynamic tire load of the left front suspension were improved by 2.5% and 2.2%, respectively, which is less than the improvement effect of the rear suspension system. It can be concluded from the above results that the bridge network (a) could improve the ride comfort of the vehicle, but the effect was not sufficiently significant, and the handling stability was not improved.

5.3. Pulse Road Input

A comparison of the peak value of performance indexes with a pulse road input and when the vehicle speed was 20 km/h is shown in Table 7 and Figure 9.

Table 7. Performance index with a pulse road input.

Performance Index	Peak to Peak of Passive Suspension	Bridge Network (a)	
		Peak to Peak	Improvement (%)
Centroid acceleration (m·s^{-2})	7.2110	7.1722	0.5
Vehicle body roll angular acceleration (rad·s^{-2})	4.9686	4.5532	8.4
Vehicle body pitch angular acceleration (rad·s^{-2})	9.6225	10.1903	-5.9
Working space of left front suspension (m)	0.1128	0.1102	2.3
Dynamic tire load of left front wheel (N)	7200	7038	2.2
Working space of left rear suspension (m)	0.1234	0.1008	18.3
Dynamic tire load of left rear wheel (N)	7047	6568	6.8

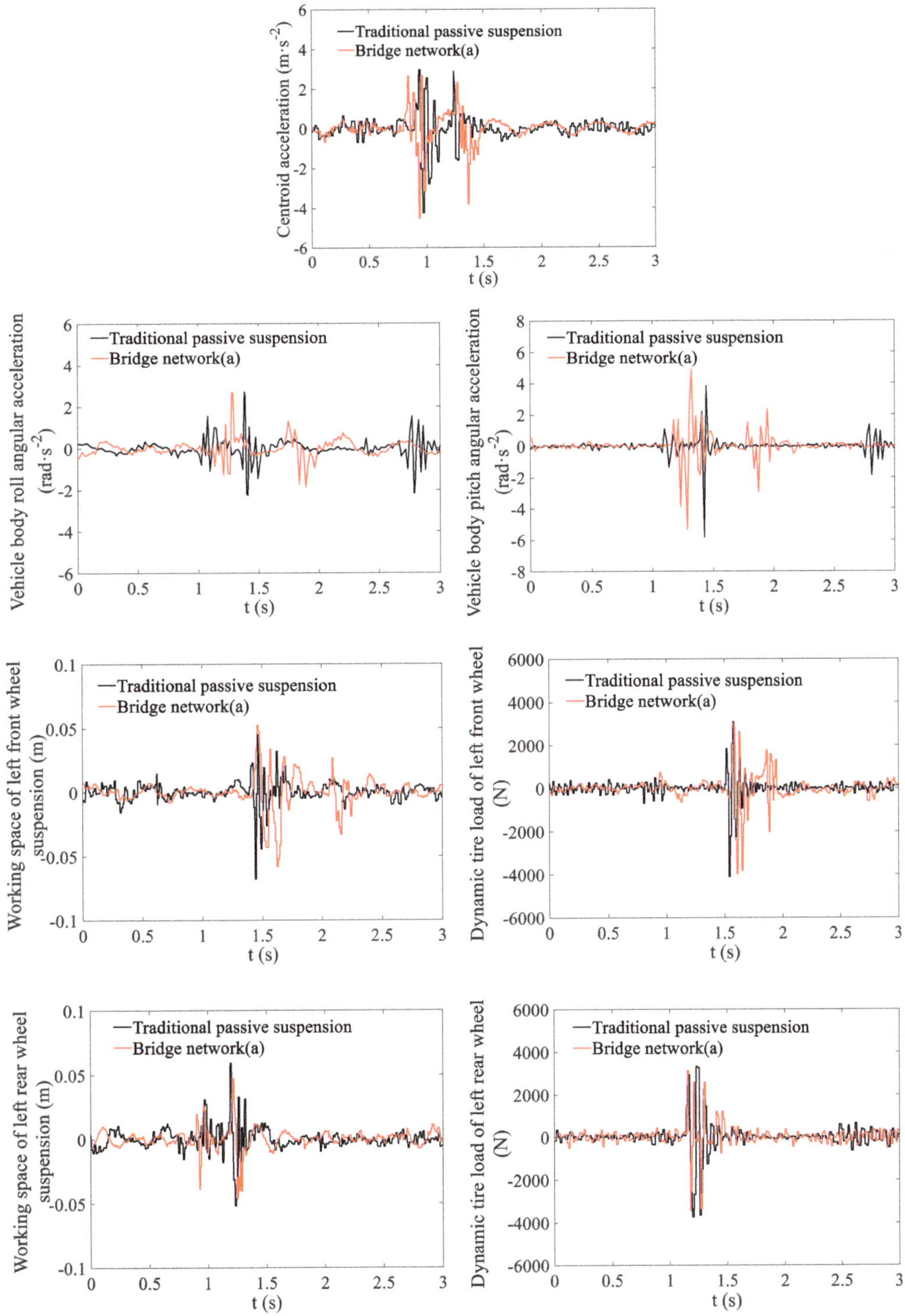

Figure 9. Comparison diagram with a pulse road input with a vehicle speed of 20 km/h.

From Table 7 and Figure 9, we can see that, for the centroid acceleration, the improvement of the peak to peak of the mechatronic inertial suspension based on the bridge network (a) was much small than that of the passive suspension, from 7.2110 ($m \cdot s^{-2}$) to 7.1722 ($m \cdot s^{-2}$), which is only 0.5% and helped to improve the vehicle ride comfort. However, for the vehicle body roll angular acceleration, the improvement was apparent, from 4.9686 ($rad \cdot s^{-2}$) to 4.5532 ($rad \cdot s^{-2}$), and the degree of reduction was 8.4%. For the vehicle body pitch angular acceleration, the peak to peak of the mechatronic inertial suspension was relatively higher than that of the passive suspension, which increased from 9.6225 ($rad \cdot s^{-2}$) to 10.1903 ($rad \cdot s^{-2}$) (5.9%), and there was no improvement in the vehicle handling stability. The peak to peak values of the suspension working space and dynamic tire load of the left front suspension, comparing the passive suspension with the mechatronic inertial suspension based on a bridge network (a), were reduced by 2.3% and 2.2%, respectively. The performance improvement of the left rear suspension was obvious compared to the left front suspension. The suspension working space decreased from 0.1234 (m) to 0.1008 (m) (18.3%), and the dynamic tire load decreased from 7047 (N) to 6568 (N) (6.8%). These improvements helped to improve the vehicle ride comfort and road friendliness.

In summary, the proposed mechatronic inertial suspension based on a bridge network (a) can better realize a high-order suspension impedance and has better working performance, to effectively improve the ride comfort and vibration isolation performance of vehicles.

6. Conclusions

In this paper, the optimal design of mechatronic inertial suspension with an external electrical network was studied. First, an optimization design method of the external circuit of vehicle mechatronic inertial suspension was proposed using a bridge network. Then, a whole-vehicle dynamic model, considering the vertical motion, roll motion, and pitch motion of vehicle body, was established, and the basic structure and impedance transfer function of the bridge network and series-parallel network were analyzed. The particle swarm optimization algorithm was used to optimize component parameters of three bridge networks and three series-parallel networks. Then, a performance comparison and analysis of the mechatronic inertial suspension using a bridge network and series-parallel network was studied, and the bridge network (a) and series parallel-network (d) were selected. On this basis, the bridge network (a) and series-parallel network (d) were compared in the time domain and frequency domain, and the bridge network (a) with better vibration suppression ability was selected as the electrical network for the mechatronic inerter. Finally, a road test with a real vehicle was carried out with a random road input and pulse road input, respectively.

The results show that, under the random input condition, compared with the passive suspension, the mechatronic inertial suspension based on a bridge network (a) could reduce the RMS value of centroid acceleration by 1.8%, the RMS value of the body roll angular acceleration by 5.5%, and the RMS values of the suspension working space and dynamic tire load of the left rear wheel suspension by 21.1% and 6.3%, respectively, which effectively improved the ride comfort of the vehicle.

This article provides a research direction for the passive optimization design of mechatronic inertial suspension and verified its effectiveness and feasibility, laying the foundation for further improvements of the vibration isolation of vehicle mechatronic inertial suspension.

Author Contributions: Conceptualization, X.Y.; methodology, Y.S.; software, T.Z.; validation, X.L.; formal analysis, T.H.; investigation, T.H.; writing—original draft preparation, T.Z.; writing—review and editing, Y.S.; supervision, X.Y. All authors have read and agreed to the published version of the manuscript.

Funding: The authors disclose receipt of the following financial support for the research, authorship, and/or publication of this article: This research was funded by the National Natural Science

Foundation of China under Grant 52072157 and 52002156, the Natural Science Foundation of Jiangsu Province under Grant BK20200911 and Natural Science Foundation of Beijing Municipality under Grant 3214045.

Data Availability Statement: Not applicable.

Acknowledgments: The authors would like to thank the editor and the anonymous reviewers for their careful reading and helpful comments.

Conflicts of Interest: No potential conflict of interest are reported by the authors.

References

1. Lee, D.; Jin, S.W.; Rhee, E.-J.; Lee, C. Practical Damper Velocity Estimation for Semi-Active Suspension Control. *Int. J. Automot. Technol.* **2021**, *22*, 499–506. [CrossRef]
2. Lee, G.-W.; Hyun, M.; Kang, D.-O.; Heo, S.-J. High-efficiency Active Suspension based on Continuous Damping Control. *Int. J. Automot. Technol.* **2022**, *23*, 31–40. [CrossRef]
3. Smith, M. Synthesis of mechanical networks: The inerter. *IEEE Trans. Autom. Control* **2002**, *47*, 1648–1662. [CrossRef]
4. Papageorgiou, C.; Houghton, N.E.; Smith, M.C. Experimental testing and analysis of inerter devices. *J. Dyn. Syst. Meas. Control.-Trans. Asme* **2009**, *131*, 101–116. [CrossRef]
5. Faraj, R.; Jankowski, L.; Graczykowski, C.; Holnicki-Szulc, J. Can the inerter be a successful shock-absorber? The case of a ball-screw inerter with a variable thread lead. *J. Frankl. Inst.* **2019**, *356*, 7855–7872. [CrossRef]
6. Liu, X.; Jiang, J.Z.; Titurus, B.; Harrison, A. Model identification methodology for fluid-based inerters. *Mech. Syst. Signal Process.* **2018**, *106*, 479–494. [CrossRef]
7. Swift, S.J.; Smith, M.C.; Glover, A.R.; Papageorgiosu, C.; Gartner, B.; Houghton, N.E. Design and modelling of a fluid inerter. *Int. J. Control.* **2013**, *86*, 2035–2051. [CrossRef]
8. Shen, Y.; Chen, L.; Liu, Y.; Zhang, X. Modeling and Optimization of Vehicle Suspension Employing a Nonlinear Fluid Inerter. *Shock Vib.* **2016**, *2016*, 1–9. [CrossRef]
9. Zhang, H.; Shen, Y.; Yang, H. Impact of coil factors on a hydraulic electric inerter based vehicle suspension. *J. Theor. Appl. Mech.* **2020**, *58*, 711–722. [CrossRef]
10. Yang, X.; Song, H.; Shen, Y.; Liu, Y.; He, T. Control of the Vehicle Inertial Suspension Based on the Mixed Skyhook and Power-Driven-Damper Strategy. *IEEE Access* **2020**, *8*, 217473–217482. [CrossRef]
11. Li, X.; Li, F.; Shang, D. Dynamic Characteristics Analysis of ISD Suspension System under Different WoSrking Conditions. *Mathematics* **2021**, *9*, 1345. [CrossRef]
12. Zhao, Z.; Zhang, R.; Wierschem, N.E.; Jiang, Y.; Pan, C. Displacement mitigation-oriented design and mechanism for inerter-based isolation system. *J. Vib. Control.* **2021**, *27*, 1991–2003. [CrossRef]
13. Shi, A.; Shen, Y.; Wang, J. Parameter optimization of a grounded dynamic vibration absorber with lever and inerter. *J. Low Freq. Noise, Vib. Act. Control* **2022**, *41*, 784–798. [CrossRef]
14. Zhang, R.F.; Zhao, Z.P.; Liu, X.C.; Zhang, L.X. Optimal design of inerter systems for the force-transmission suppression of oscillating structures. *Earthq. Eng. Eng. Vib.* **2022**, *21*, 441–454.
15. Dai, J.; Wang, Y.; Wei, M.; Zhang, W.; Zhu, J.; Jin, H.; Jiang, C. Dynamic characteristic analysis of the inerter-based piecewise vibration isolator under base excitation. *Acta Mech.* **2022**, *233*, 513–533. [CrossRef]
16. Lewis, T.D.; Jiang, J.Z.; Neild, S.A.; Gong, C.N.; Iwnicki, S.D. Using an inerter-based suspension to improve both passenger comfort and track wear in railway vehicles. *Veh. Syst. Dyn.* **2020**, *58*, 472–493. [CrossRef]
17. Xia, Z.H.; Zhou, J.S.; Lian, J.Y.; Ding, S.; Gong, D.; Sun, W.; Sun, Y. Online detection and control of car body low-frequency swaying in railway vehicles. *Veh. Syst. Dyn.* **2021**, *59*, 70–100. [CrossRef]
18. Wang, Y.; Li, H.X.; Meng, H.D.; Wang, Y. Dynamic characteristics of an underframe inerter-based suspended equipment for high speed trains. *J. Vib. Shock.* **2022**, *41*, 246–254.
19. Zhang, H.; Ye, Z.; Chen, X.; Yao, W. Seismic response mitigation of girder displacement of cable-stayed bridge using inerter systems. *Structures* **2022**, *39*, 928–944. [CrossRef]
20. Xu, K.; Bi, K.; Han, Q.; Li, X.; Du, X. Using tuned mass damper inerter to mitigate vortex-induced vibration of long-span bridges: Analytical study. *Eng. Struct.* **2019**, *182*, 101–111. [CrossRef]
21. Zhao, Z.; Chen, Q.; Zhang, R.; Jiang, Y.; Pan, C. A negative stiffness inerter system (NSIS) for earthquake protection purposes. *Smart Struct. Syst.* **2020**, *26*, 481–493.
22. Li, Y.; Li, S.; Chen, Z. Optimal design and effectiveness evaluation for inerter-based devices on mitigating seismic responses of base isolated structures. *Earthq. Eng. Eng. Vib.* **2021**, *20*, 1021–1032. [CrossRef]
23. Xu, S.; He, B. A compliance modeling method of flexible rotary joint for collaborative robot using passive network synthesis theory. *Proc. Inst. Mech. Eng. Part C J. Mech. Eng. Sci.* **2021**, *236*, 4038–4048. [CrossRef]
24. Yang, X.; He, T.; Shen, Y.; Liu, Y.; Yan, L. Research on predictive coordinated control of ride comfort and road friendliness for heavy vehicle ISD suspension based on the hybrid-hook damping strategy. *Proc. Inst. Mech. Eng. Part D J. Automob. Eng.* **2022**. [CrossRef]

25. Smith, M.C.; Wang, F.-C. Performance Benefits in Passive Vehicle Suspensions Employing Inerters. *Veh. Syst. Dyn.* **2004**, *42*, 235–257. [CrossRef]
26. Papageorgiou, C.; Smith, M. Positive real synthesis using matrix inequalities for mechanical networks: Application to vehicle suspension. *IEEE Trans. Control Syst. Technol.* **2006**, *14*, 423–435. [CrossRef]
27. Wang, F.-C.; Chan, H.-A. Vehicle suspensions with a mechatronic network strut. *Veh. Syst. Dyn.* **2011**, *49*, 811–830. [CrossRef]
28. López-Martínez, J.; Martínez, J.; García-Vallejo, D.; Alcayde, A.; Montoya, F.G. A new electromechanical analogy approach based on electrostatic coupling for vertical dynamic analysis of planar vehicle models. *IEEE Access* **2021**, *9*, 119492–119502. [CrossRef]
29. Chen, M.Z.; Papageorgiou, C.; Scheibe, F.; Wang, F.-C.; Smith, M.C. The missing mechanical circuit element. *IEEE Circuits Syst. Mag.* **2009**, *9*, 10–26. [CrossRef]
30. Li, Y.A.; Cheng, Z.; Hu, N.Q.; Yang, Y.; Zhuo, X. Modeling, design and experiments of a ball-screw inerter with mechanical diodes. *J. Sound Vib.* **2021**, *504*, 116121.
31. Yang, X.; Yan, L.; Shen, Y.; Liu, Y.; Liu, C. Optimal Design and Dynamic Control of an ISD Vehicle Suspension Based on an ADD Positive Real Network. *IEEE Access* **2020**, *8*, 94294–94306. [CrossRef]
32. Shen, Y.; Hua, J.; Fan, W.; Liu, Y.; Yang, X.; Chen, L. Optimal design and dynamic performance analysis of a fractional-order electrical network-based vehicle mechatronic ISD suspension. *Mech. Syst. Signal Process.* **2023**, *184*, 2592–2601. [CrossRef]
33. Ning, D.; Du, H.; Zhang, N.; Sun, S.; Li, W. Controllable Electrically Interconnected Suspension System for Improving Vehicle Vibration Performance. *IEEE/ASME Trans. Mechatron.* **2020**, *25*, 859–871. [CrossRef]
34. Hu, Y.L.; Chen, M.Z.Q. Low-complexity passive vehicle suspension design based on element-number-restricted networks and low-order admittance networks. *J. Dyn. Syst. Meas. Control.-Trans. Asme* **2018**, *140*, 101014. [CrossRef]
35. Li, Y.; Yang, X.; Shen, Y.; Liu, Y.; Wang, W. Optimal design and dynamic control of the HMDV inertial suspension based on the ground-hook positive real network. *Adv. Eng. Softw.* **2022**, *171*, 103171. [CrossRef]
36. Shen, Y.; Hua, J.; Wu, B.; Chen, Z.; Xiong, X.; Chen, L. Optimal design of the vehicle mechatronic ISD suspension system using the structure-immittance approach. *Proc. Inst. Mech. Eng. Part D J. Automob. Eng.* **2022**, *236*, 512–521. [CrossRef]
37. Zhong, R.; Bi, C.; Chen, Y.; Chen, Z.; Zhou, A.; Yang, Z.; Zhai, J. A Simplified Method for Extracting Parasitic Inductances of MOSFET-Based Half-Bridge Circuit. *IEEE Access* **2021**, *9*, 14122–14129. [CrossRef]
38. Advani, J.; Gupta, O. Networks for a Subclass of Minimum Biquartic Impedance Functions. *IEEE Trans. Circuit Theory* **1965**, *12*, 621–622. [CrossRef]
39. Lee, S.; Frisch, I. A Class of RLC Networks with Fewer Nonreactive Elements than the Brune Realization. *IEEE Trans. Circuit Theory* **1964**, *11*, 418–421. [CrossRef]
40. Yang, X.F.; Song, H.; Shen, Y.J.; Liu, Y.L. Study on adverse effect suppression of hub motor driven vehicles with inertial suspensions. *Proc. Inst. Mech. Eng. Part D J. Automob. Eng.* **2022**, *236*, 767–779. [CrossRef]
41. Wang, F.C.; Chan, H.A. Mechatronic Suspension Design and Its Applications to Vehicle Suspension Control. *IEEE Conf. Decis. Control.* **2008**, *12*, 3769–3774.
42. Clerc, M.; Kennedy, J. The particle swarm-explosion, stability, and convergence in a multidimensional complex space. *IEEE Trans. Evol. Comput.* **2002**, *6*, 58–73. [CrossRef]

World Electric Vehicle Journal

MDPI

Article

Model-Based Fault Diagnosis of Actuators in Electronically Controlled Air Suspension System

Xinwei Jiang, Xing Xu * and Haiqiang Shan

Automotive Engineering Research Institute, Jiangsu University, 301 Xuefu Road, Zhenjiang 212013, China
* Correspondence: xuxing@ujs.edu.cn

Abstract: The air suspension adjusts the height of the vehicle body through charging and bleeding air to meet the high performance of the vehicle, which needs a reliable electronic control system. Through fault tree analysis of the electronically controlled air suspension (ECAS) system and considering the correlation between the duty cycle and flow rate of the air spring solenoid valve, the fault model of the solenoid valve is constructed, and the fault diagnosis design method of the ECAS system solenoid valve based on multiple extended Kalman filter banks (EKFs) is proposed. An adaptive threshold is used to realize fault diagnosis, and active fault-tolerant control is carried out based on an analytical model. The real controller based on d2p rapid prototyping technology and the vehicle model based on AMESim are further verified on the hardware-in-the-loop (HiL) simulation test platform and compared with the pure simulation results. The test results show that the fault diagnosis and fault-tolerant control algorithm can work normally in the actual controller, and can effectively realize the fault diagnosis and fault-tolerant control of the actuator in the vehicle ECAS system.

Keywords: electronically controlled air suspension; solenoid valve; extended Kalman filter bank; fault diagnosis; fault-tolerant control

Citation: Jiang, X.; Xu, X.; Shan, H. Model-Based Fault Diagnosis of Actuators in Electronically Controlled Air Suspension System. *World Electr. Veh. J.* **2022**, *13*, 219. https://doi.org/10.3390/wevj13110219

Academic Editor: Joeri Van Mierlo

Received: 13 October 2022
Accepted: 16 November 2022
Published: 21 November 2022

1. Introduction

Air suspension can improve vehicle ride comfort and road friendliness, and its natural frequency is low and has variable stiffness characteristics [1–5]. However, the general air suspension cannot adjust the suspension stiffness and damping according to the load change. The natural frequency and controllability of the electronically controlled air suspension (ECAS) are low, which can further improve the vehicle ride comfort and control stability [6–8]. In recent years, research on ECAS has mainly focused on improving comfort and stability. In 2019, Rui modeled the ECAS system according to its nonlinear characteristics and designed an adaptive sliding mode control strategy. The method effectively improves the stability of the system by considering the parameter uncertainty [9]. In 2021, Ma et al. designed an integrated control strategy to solve the problems of small stiffness adjustment range and poor roll stability of traditional ECAS systems. The handling stability and anti-roll performance of the vehicle are improved [10]. In 2021, Hu et al. conducted research on the hybrid control of body height and attitude of the ECAS system. They built a vehicle model based on mixed logic dynamics and designed the switching strategy of the solenoid valve. The coordinated control between the ECAS system body height and attitude is well solved, and good vibration isolation performance and stability are achieved [11].

However, the ECAS system is highly dependent on the reliability of each component. Sensors and actuators are very important components [12,13]. The structure of the ECAS system is shown in Figure 1. The actuator is the air spring solenoid valve. If any of the four solenoid valves fail, the ride comfort and handling stability of the whole vehicle will be severely affected [14]. Therefore, it is necessary to consider the reliability of the actuator of the electronically controlled air suspension system.

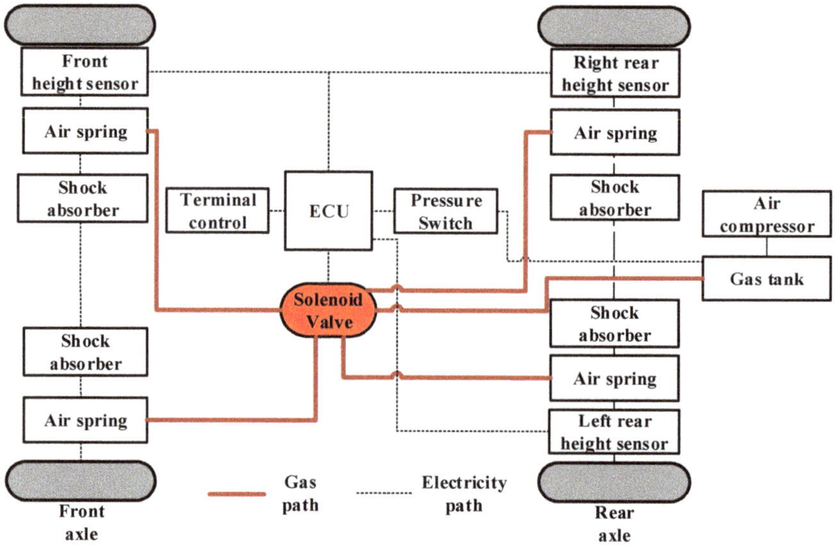

Figure 1. ECAS system.

The fault diagnosis technology in the reliability research of the control system is widely used in the control system of various industries. The technology can effectively improve the reliability and maintainability of the system. The technology was first proposed in the 1990s by Frank, who divided the fault diagnosis methods into three categories based on an analytical model, expert knowledge, and signal processing [15]. In 2009, Zhou further developed the method from qualitative and quantitative perspectives [16]. In 2020, Feng proposed an SVM model and effectively distinguished different fault conditions of trains by using the support vector machine method [17]. The fault diagnosis method based on qualitative analysis is mainly divided into the graph theory method, the expert system method, and the qualitative simulation method. Among them, the graph theory method includes the symbol-directed graph method [18] and the fault tree method [19]. The main principle is to judge the fault according to the logical causal relationship. This method can be understood easily and is widely adopted.

The fault location and type can be determined by fault diagnosis. In this way, fault-tolerant control (FTC) can be implemented. Alwi et al. classified fault-tolerant control methods in detail [20]. Fault-tolerant control is usually divided into passive FTC (PFTC) and active FTC (AFTC). The common methods of passive fault-tolerant control include H methods based on ∞ control theory [21] and sliding mode control theory [22]. The characteristic of passive fault-tolerant control is that there is no need for fault diagnosis, and the controller parameters are not changed, so it is easy to implement, but the fault-tolerant control is limited. The active fault-tolerant control adjusts the controller parameters online or configures the controller structure units online based on the fault diagnosis information to realize the system stability, which is different from the passive fault-tolerant control. Active fault-tolerant control methods can be divided into two types: planning type and online adjustment type [23]. In the planning of FTC, the controller is designed in advance for all possible faults of the system, and the corresponding controller is activated when the corresponding fault occurs. On the other hand, the online adjustment of the controller is mainly through adaptive control or control signal redistribution to achieve fault-tolerant control [24,25]. In 2020, Pang et al. designed a fault-tolerant controller based on the nonlinear suspension system, which combines the state feedback observer and the H-infinity observer. The purpose of fault-tolerant control is achieved by comparing the system state under the fault-free state to compensate [26]. In 2021, Xue et al. considered the failure of a 1/4 active suspension actuator under the change of sprung mass. They used the finite-element neural network method to approximate the observer fault, and turned it into a linear matrix inequality problem. This method can obtain the results faster and more accurately, and can well adapt to the changing spring mass [27]. The above methods are very effective, but the actuator fault diagnosis and fault-tolerant control for the ECAS system are rarely used. In order to fill the shortage in this field, this research will use the extended Kalman filter bank with an adaptive threshold as the observer, and use the online adjustment method to continuously improve the body height. Compared with the existing research, this method has stronger adaptability and accuracy.

At present, research on the fault diagnosis and fault-tolerant control of the ECAS system is scarce. In addition, research on ECAS vehicle height control rarely focuses on how to ensure effective control in the case of actuator failure. However, the ECAS system with actuator failure has many problems to be solved, such as risk analysis, fault diagnosis architecture, and fault-tolerant control strategy design.

The contribution of this paper is to use the extended Kalman filter bank based on an adaptive threshold to study the fault diagnosis and fault-tolerant control of the air spring solenoid valve. This method can change the threshold according to the system input, and can effectively reduce the probability of missed diagnosis and misdiagnosis. After fault diagnosis and isolation, corresponding active fault-tolerant methods are adopted for different types of faults. Among them, the method of online adjustment of controller parameters is used to improve the height and attitude control effect of the constant gain fault. In addition, a hardware-in-the-loop simulation platform is built to finally test the effectiveness and accuracy of the above methods. The HiL platform can realize the connection between the real controller and the simulation model of the controlled object, so as to form a complete loop to test the actual operation and feasibility of fault diagnosis and fault-tolerant control algorithm model in the real controller. This paper provides a research idea for the operation stability and fault-tolerant control method of electronically controlled air suspension.

The article is organized as follows. In the next section, the model of the height adjustment system of the electronically controlled air suspension is established. After analyzing the fault mode by fault tree method, the controlled object and actuator model are established. In Section 3. The fault diagnosis mechanism based on the EKF group is designed and verified by simulation. In Section 4, active fault-tolerant control is designed based on the previous section. In Section 5, a hardware-in-the-loop simulation test bed is built to verify the fault diagnosis and active fault-tolerant control system. Finally, conclusions are drawn in Section 6.

2. Modeling and Failure Analysis of ECAS System

2.1. Modeling of Vehicle Height Adjustment System in ECAS System

To facilitate the design of the fault diagnosis filter, the vehicle ECAS system is simplified, and the corresponding vehicle height adjustment model of the vehicle ECAS system is established. This provides a basic platform for the implementation of actuator fault diagnosis. The vehicle ECAS dynamics model is established based on the following assumptions: (1) The sprung mass is a rigid body, and only vertical, pitch, and roll motions are considered; (2) Tyre damping is ignored; (3) The speed characteristic of the shock absorber is linear; (4) The ECAS system does not collide with the buffer block during the working process; (5) The rigidity of the frame and the body is sufficiently large, regardless of the vibration modes caused by the elasticity of the frame. Therefore, the ECAS dynamics model of the whole vehicle as shown in Figure 2 is established.

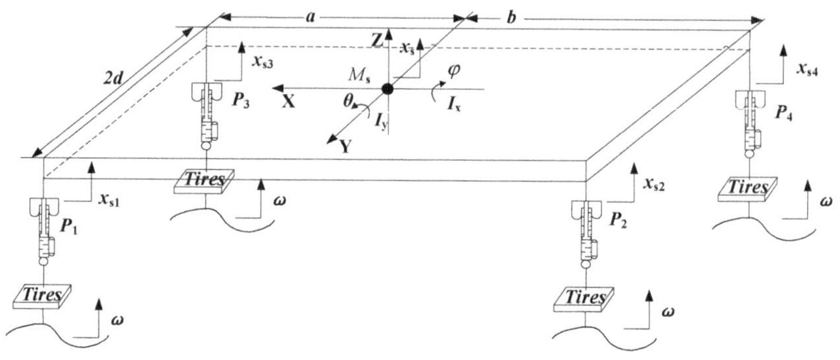

Figure 2. Dynamic model of the whole vehicle based on ECAS.

The ECAS vehicle dynamics equation is built according to the vehicle dynamics characteristics and related geometric relationships. Among them, the vehicle body vertical, pitch, and roll motion equations are as follows:

$$\begin{cases} M_s\ddot{x}_s = F_1 + F_2 + F_3 + F_4 \\ I_y\ddot{\theta} = b(F_2 + F_4) - a(F_1 + F_3) \\ I_x\ddot{\varphi} = (F_1 + F_2 - F_3 - F_4)d \end{cases} \tag{1}$$

where M_s is the sprung mass of the whole vehicle, kg; $\ddot{x}_s$ is the vertical acceleration of the body center of mass; F_1, F_2, F_3, and F_4 are the forces on the front left, rear left, front right, and rear right body, respectively, N; I_y is the moment of inertia of the car body around the Y-axis, km·m^2; $\ddot{\theta}$ is the pitch angular acceleration, rad/s^2; a and b are the distance from the center of mass of the car body to the front and rear axles, m; I_x is the moment of inertia of the car body around the X-axis, km·m^2; $\ddot{\varphi}$ is the roll angular acceleration, rad/s^2; and d is the 1/2 tread, m.

After establishing the vehicle dynamics model, a single-group air spring model is built. Assuming that the heat exchange during the gas flow is negligible, the charging and discharging process of the air spring can be regarded as a variable-volume adiabatic process. According to the first law of thermodynamics, when the solenoid valve is opened, the air spring opening inflates and discharges the variable mass model as follows:

$$\kappa RT\frac{dm}{dt} = \kappa P\frac{dv}{dt} + V\frac{dP}{dt} \tag{2}$$

When the solenoid valve is closed, $\frac{dm}{dt} = 0$, then Equation (4) can be rewritten as:

$$\frac{dP}{dt} = -\frac{\kappa PdV}{Vdt} \tag{3}$$

Research has shown that when the height of the diaphragm air spring changes near the working position, its effective area and volume change rate can be regarded as fixed values. Therefore, the volume change of the air spring can be approximated as the spring vertical displacement change under the volume change rate, namely:

$$V = V_0 + \Delta V(x_{s1} - x_{u1}) \tag{4}$$

where V_0 is the initial volume of the air spring, m^3; and ΔV is the rate of change of air spring volume, m^3/m.

Combining Equations (2)–(4) can derive a complete air spring charging and discharging process model as follow:

$$V\dot{P} = -\kappa P\Delta V(\dot{x}_{s1} - \dot{x}_{u1}) + \kappa RTq_m \tag{5}$$

where V is the air spring volume, m^3; k is the adiabatic coefficient of air; R is the gas constant, N·m/(kg·K); T is the internal temperature of the air spring, °C; and q_m is the gas mass flow rate when charging and discharging the air spring, kg/s.

According to the established vehicle ECAS dynamics model and the air spring charging and discharging process model, the mathematical model of the vehicle height adjustment of the vehicle ECAS system can be derived. Since the focus of the ECAS system model is the charging and discharging process of the air spring, the unsprung mass vibration and road input are combined as system noise. Then, the mathematical expression of the vehicle height adjustment of the complete vehicle ECAS system can be finally simplified as:

$$\begin{cases} M_s\ddot{x}_s = F_1 + F_2 + F_3 + F_4 \\ I_x\ddot{\varphi} = (F_1 + F_2 - F_3 - F_4)d \\ I_y\ddot{\theta} = b(F_2 + F_4) - a(F_1 + F_3) \\ V_1\dot{P}_1 = -\kappa\Delta VP_1\dot{x}_{s1} + \kappa RTq_{m1} \\ V_2\dot{P}_2 = -\kappa\Delta VP_2\dot{x}_{s2} + \kappa RTq_{m2} \\ V_3\dot{P}_3 = -\kappa\Delta VP_3\dot{x}_{s3} + \kappa RTq_{m3} \\ V_4\dot{P}_4 = -\kappa\Delta VP_4\dot{x}_{s4} + \kappa RTq_{m4} \end{cases} \tag{6}$$

where

$$\begin{cases} F_1 = (P_1 - P_a)A_e - m_{s1}g - C_1\dot{x}_{s1} - k_1\omega_1 \\ F_2 = (P_2 - P_a)A_e - m_{s2}g - C_2\dot{x}_{s2} - k_2\omega_2 \\ F_3 = (P_3 - P_a)A_e - m_{s3}g - C_3\dot{x}_{s3} - k_3\omega_3 \\ F_4 = (P_4 - P_a)A_e - m_{s4}g - C_4\dot{x}_{s4} - k_4\omega_4 \end{cases}, \begin{cases} V_1x_{s1} \\ V_2 = V_{20} + \Delta Vx_{s2} \\ V_3 = V_{30} + \Delta Vx_{s3} \\ V_4 = V_{40} + \Delta Vx_{s4} \end{cases}, \begin{cases} x_{s1} = x_s - a\theta + d\varphi \\ x_{s2} = x_s + b\theta + d\varphi \\ x_{s3} = x_s - a\theta - d\varphi \\ x_{s4} = x_s + b\theta - d\varphi \end{cases}$$

2.2. Actuator Failure Analysis and Modeling

Through the analysis of the potential failure modes of the components of the system, the reasons for the failures and their effects are drawn. The corresponding detection method is designed to greatly improve the reliability of the system.

To analyze the failure mode of the system, a fault tree analysis method is introduced here. The fault tree analysis method belongs to the method of graphical deduction. Through top-down or bottom-up logical deductive reasoning, the reasons leading to system failure are analyzed and expressed in a tree diagram.

In the fault tree analysis, the causal relationship between the faults is represented by event symbols, logic gate symbols, etc. Table 1 shows the basic symbols and their meanings used in the fault tree analysis method.

Figure 3 shows the air spring solenoid valve fault tree. It can be seen from the fault tree that the failure of the air spring solenoid valve of the ECAS system is mainly caused by the failure of the front left, front right, rear left, and rear right air spring solenoid valves. The main reasons for the failure of the above four solenoid valves include short circuits, open circuits, plugging of the valve core, spring fatigue, and leakage of the valve core.

Table 1. Basic symbols and meanings of fault tree.

Event	Symbol	Description
Basic event		The lowest level event that does not need to be ascertained
No extended Events		Events that have little impact on the top event or whose cause cannot be known temporarily
Result Event		Contains top and middle events that are always at the outputs
Transfer symbol		This event indicates information transfer and avoids drawing repetition
Logic symbol: and gate		Output events only occur when all input events occur
Logical symbols: or doors		As long as one of the input events occurs, the output event occurs

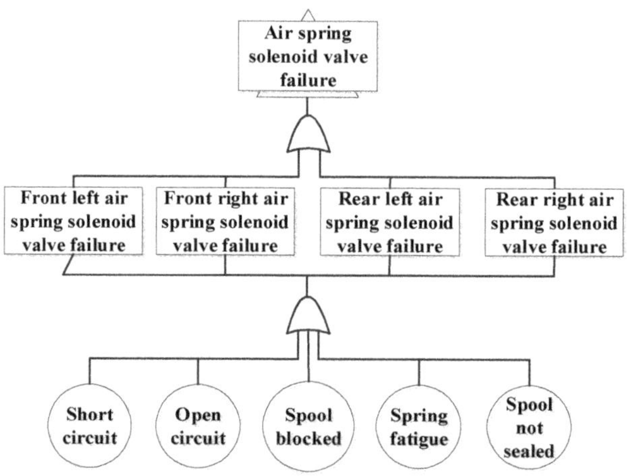

Figure 3. Fault tree of solenoid valves for air spring system.

After the failure mode analysis of the vehicle ECAS system, mathematical modeling is mainly carried out for the different failures of the actuator, namely the air spring solenoid valve.

The solenoid valve of the air spring is a high-speed switching solenoid valve, using the pulse-width modulation (PWM) control method. The relationship between the duty cycle and the flow rate is:

$$q_m = C_d \cdot A \cdot D \cdot \sqrt{\frac{2\Delta P}{\rho}} \tag{7}$$

where C_d is the flow coefficient; A is the flow area of the valve port, m^2; D is the duty cycle; ΔP is the difference between input pressure and output pressure, P_a; and ρ is the gas density, kg/m^3.

There are two common failure modes of solenoid valves. One is that the valve core is stuck due to the open circuit of the solenoid valve and cannot be opened normally. The flow through the solenoid valve becomes zero. Another failure is caused by the increase of internal friction of the solenoid valve. The core cannot reach the maximum displacement; that is, the valve port is not fully opened, resulting in a gain loss in the flow through the solenoid valve. Therefore, from the flow point of view, the fault behavior of the ECAS system actuator is defined as stuck and constant gain. In the case of failure, the flow area of the valve port is expressed as follows:

$$A_f = n \cdot A + \beta \tag{8}$$

where n is the fault gain factor, and β is the dead value for failure.

Then, the stuck fault of the ECAS system actuator can be defined as $n = \beta = 0$, and the constant gain fault is $n \neq 0$ and $\beta \neq 0$.

According to Equations (7) and (8), when the actuator fails the control input of the ECAS system u_f is as follow:

$$u_f = q_m = n \cdot u_i + \delta = C_d \cdot n \cdot A \cdot D \cdot \sqrt{\frac{2\Delta P}{\rho}} + C_d \cdot \beta \cdot D \cdot \sqrt{\frac{2\Delta P}{\rho}} \tag{9}$$

where δ is the input fault stuck value for the control, subscript $i = 1 - 4$.

The fault vector can be defined as:

$$F = \begin{bmatrix} f_1 \\ f_2 \\ f_3 \\ f_4 \end{bmatrix} = \begin{bmatrix} (n_1 - 1)u_1 + \delta_1 \\ (n_2 - 1)u_2 + \delta_2 \\ (n_3 - 1)u_3 + \delta_3 \\ (n_4 - 1)u_4 + \delta_4 \end{bmatrix} \tag{10}$$

Fault control input U_f can be expressed as:

$$U_f = \begin{bmatrix} U_{f1} \\ U_{f2} \\ U_{f3} \\ U_{f4} \end{bmatrix} = \begin{bmatrix} u_1 \\ u_2 \\ u_3 \\ u_4 \end{bmatrix} + \begin{bmatrix} (n_1 - 1)u_1 + \delta_1 \\ (n_2 - 1)u_2 + \delta_2 \\ (n_3 - 1)u_3 + \delta_3 \\ (n_4 - 1)u_4 + \delta_4 \end{bmatrix} = U + F \tag{11}$$

3. Fault Diagnosis of ECAS System Based on Adaptive Threshold

3.1. Fault Diagnosis System Architecture Based on KEFs

When the actuator of the ECAS system fails, the controller may adjust the charging and bleeding process of the air spring abnormally. Therefore, the actuator fault detection and isolation strategy shown in Figure 4 is designed for the vehicle ECAS system.

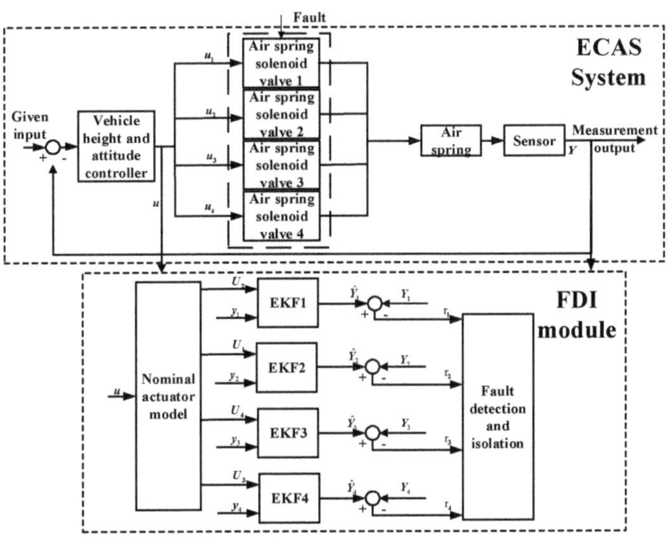

Figure 4. Actuator fault detection & isolation strategy of the vehicle ECAS system.

The extended Kalman filter (EKF) algorithm is widely used for state estimation in nonlinear systems. The main idea is to use Taylor's formula to transform the nonlinear model into a linear model and then perform Kalman filtering. Therefore, in view of the nonlinear characteristics of the ECAS system, the extended Kalman filter algorithm is used to design the fault diagnosis filter.

To design EKF1, selecting the state variable $X_1 = \left[\theta\dot{\theta}\varphi\ \dot{\varphi}x_s\dot{x}_s x_2\dot{x}_2 P_2\right]^T$, defining measurement output $Y_1 = [x_2\ddot{x}_2 P_2]^T$, control input $U_2 = [qm_2]^T$; For EKF2, selecting state variable $X_2 = \left[\theta\dot{\theta}\varphi\ \dot{\varphi}x_s\dot{x}_s x_1\dot{x}_1 P_1\right]^T$, defining measurement output $Y_2 = [x_1\ddot{x}_1 P_1]^T$, control input $U_1 = [qm_1]^T$; For EKF3, selecting state variable $X_3 = \left[\theta\dot{\theta}\varphi\ \dot{\varphi}x_s\dot{x}_s x_4\dot{x}_4 P_4\right]^T$, defining measurement output $Y_3 = [x_4\ddot{x}_4 P_4]^T$, control input $U_4 = [qm_4]^T$; For EKF4, selecting state variable $X_4 = \left[\theta\dot{\theta}\varphi\ \dot{\varphi}x_s\dot{x}_s x_3\dot{x}_3 P_3\right]^T$, defining measurement output $Y_4 = [x_3\ddot{x}_3 P_3]^T$, and control input $U_3 = [qm_3]^T$.

According to Equation (5) and the selected state variables, measurement output, and control input, write the corresponding system and measurement equations, respectively. The general form of the system and measurement equation is as follows:

$$\begin{cases} \dot{X} = f(X) + g(X)U + q(X)\omega \\ Y = h(X) + v \end{cases} \tag{12}$$

$f(X)$, $g(X)$, $q(X)$ and $h(X)$ derived from EKF1, EKF2, EKF3, and EKF4, respectively.

According to the Equation (11) design filter equation, the general form of the equation is:

$$\begin{cases} \dot{\hat{X}} = f(\hat{X}) + g(\hat{X})U + L_k(Y - \hat{Y}) \\ \hat{Y} = h(\hat{X}) \end{cases} \tag{13}$$

where $\hat{X}$ is the estimate for state variables, $\hat{Y}$ is the estimate for measurement output, Y is the output for measurement, U is the input matrix for control, and L_k is the filter gain matrix. Process noise ω and measurement noise v are mutually uncorrelated Gaussian white noise. The probability distribution characteristics are as follows:

$$\begin{cases} E(\omega_k) = 0,\ Cov(\omega_k,\ \omega_j) = Q_k\delta_{kj} \\ E(v_k) = 0,\ Cov(v_k,\ v_j) = R_k\delta_{kj} \\ Cov(\omega_k,\ v_j) = 0 \end{cases} \tag{14}$$

where Q_k is the process noise covariance matrix, and R_k is the measurement noise covariance matrix. The EKF algorithm is shown in Figure 5, where $F_k = \frac{\partial f(x_k)}{\partial x_k}|x_k = \hat{x}_k$, $H_k = \frac{\partial h(x_k)}{\partial x_k}|x_k = \hat{x}_k$, $\Gamma_k = \frac{\partial f(x_k)}{\partial \omega_k}|x_k = \hat{x}_k$, $\hat{z}_k = h(\hat{x}_{k,\,k-1})$.

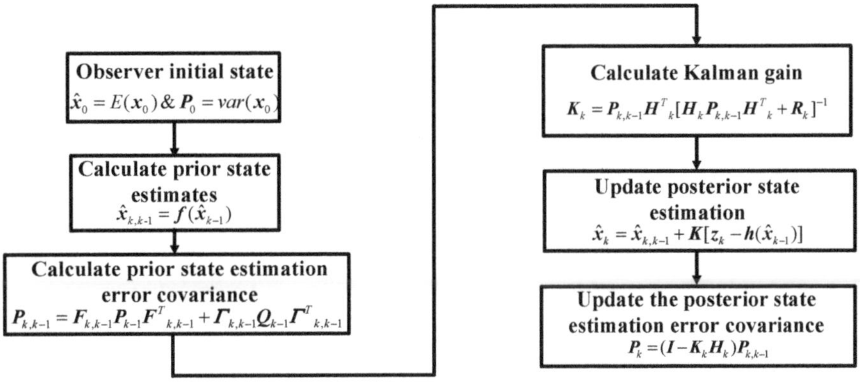

Figure 5. EKF algorithm flow.

3.2. Calculation of Adaptive Threshold

Threshold selection is an important step to achieve fault detection and isolation. The adaptive threshold changes with system input, which can effectively reduce the probability of missed diagnosis and misdiagnosis. In this section, an adaptive threshold is designed according to the system model error and system input, so as to minimize the probability of missed diagnosis and misdiagnosis. The system model error includes linearization error and parameter uncertainty error. Firstly, the system equation of EKF1 is taken as an example to illustrate the analysis of the linearization error and parameter uncertainty error

1. Linearization error

The ECAS system equation has nonlinear characteristics. In the design process of the extended Kalman filter, the system equations need to be linearized, resulting in linearization errors. Equation (12) can be written in the form of $\dot{X} = AX + BU\ X$. Thus, the state transition matrix A and the control input coefficient matrix B are obtained.

$$A = \begin{bmatrix} 0 & 1 & 0 & 0 & 0 & 0 & 0 & 0 & 0 \\ 0 & 0 & 0 & 0 & 0 & 0 & 0 & \frac{-bC_2}{I_y} & \frac{bA_e}{I_y} \\ 0 & 0 & 0 & 1 & 0 & 0 & 0 & 0 & 0 \\ 0 & 0 & 0 & 0 & 0 & 0 & 0 & \frac{-C_2d}{I_x} & \frac{dA_e}{I_x} \\ 0 & 0 & 0 & 0 & 0 & 1 & 0 & 0 & 0 \\ 0 & 0 & 0 & 0 & 0 & 0 & 0 & \frac{-C_2}{m_s} & \frac{A_e}{m_s} \\ 0 & 0 & 0 & 0 & 0 & 0 & 0 & 1 & 0 \\ 0 & 0 & 0 & 0 & 0 & 0 & 0 & \frac{-C_2}{m_{s2}} & \frac{A_e}{m_{s2}} \\ 0 & 0 & 0 & 0 & 0 & 0 & 0 & \frac{-\kappa\Delta VP_0}{V_{20}} & 0 \end{bmatrix} \tag{15}$$

$$B = \begin{bmatrix} 0 & 0 & 0 & 0 & 0 & 0 & 0 & 0 & \frac{\kappa RT}{V_{20}} \end{bmatrix}^T \tag{16}$$

Thus, the linearization error matrix can be obtained as:

$$\Delta A_1 = \begin{bmatrix} 0 & 0 & 0 & 0 & 0 & 0 & 0 & 0 & 0 \\ 0 & 0 & 0 & 0 & 0 & 0 & 0 & 0 & 0 \\ 0 & 0 & 0 & 0 & 0 & 0 & 0 & 0 & 0 \\ 0 & 0 & 0 & 0 & 0 & 0 & 0 & 0 & 0 \\ 0 & 0 & 0 & 0 & 0 & 0 & 0 & 0 & 0 \\ 0 & 0 & 0 & 0 & 0 & 0 & 0 & 0 & 0 \\ 0 & 0 & 0 & 0 & 0 & 0 & 0 & 0 & 0 \\ 0 & 0 & 0 & 0 & 0 & 0 & 0 & 0 & 0 \\ 0 & 0 & 0 & 0 & 0 & 0 & 0 & \Delta A_{11} & 0 \end{bmatrix} \tag{17}$$

$$\Delta B_1 = \begin{bmatrix} 0 & 0 & 0 & 0 & 0 & 0 & 0 & 0 & \Delta B_{11} \end{bmatrix}^T \tag{18}$$

where $\Delta A_{11} = \frac{-\kappa \Delta V P_2}{V_{20}+\Delta V x(7)} + \frac{\kappa \Delta V P_0}{V_{20}}$, and $\Delta B_{11} = \frac{\kappa RT}{V_{20}} - \frac{\kappa RT}{V_{20}+\Delta V x(7)}$.

2. Parameter uncertainty error

In the ECAS vehicle model, the uncertain parameters mainly include the sprung mass and the damping of the shock absorber. The sprung mass has the characteristics of uneven distribution and changes with vehicle masses. The damping value of the shock absorber is also different under different working temperatures. Therefore, the parameter uncertainty is introduced as follows:

$$\begin{cases} \Delta m_{s1} = m_{s1} - m_{s1}{}^{real} \\ \Delta m_{s2} = m_{s2} - m_{s2}{}^{real} \\ \Delta m_{s3} = m_{s3} - m_{s3}{}^{real} \\ \Delta m_{s4} = m_{s4} - m_{s4}{}^{real} \\ \Delta m_{s} = m_{s} - m_{s}{}^{real} \end{cases} \tag{19}$$

$$\begin{cases} \Delta C_1 = C_1 - C_1{}^{real} \\ \Delta C_2 = C_2 - C_2{}^{real} \\ \Delta C_3 = C_3 - C_3{}^{real} \\ \Delta C_4 = C_4 - C_4{}^{real} \end{cases} \tag{20}$$

The parameter value with real superscript represents the actual parameter value or the floating limit of the parameter. From this, the parameter uncertainty error matrix is derived as follows:

$$\Delta A_2 = \begin{bmatrix} 0 & 0 & 0 & 0 & 0 & 0 & 0 & 0 & 0 \\ 0 & 0 & 0 & 0 & 0 & 0 & 0 & \Delta A_{21} & 0 \\ 0 & 0 & 0 & 0 & 0 & 0 & 0 & 0 & 0 \\ 0 & 0 & 0 & 0 & 0 & 0 & 0 & \Delta A_{22} & 0 \\ 0 & 0 & 0 & 0 & 0 & 0 & 0 & 0 & 0 \\ 0 & 0 & 0 & 0 & 0 & 0 & 0 & \Delta A_{23} & \Delta A_{24} \\ 0 & 0 & 0 & 0 & 0 & 0 & 0 & 0 & 0 \\ 0 & 0 & 0 & 0 & 0 & 0 & 0 & \Delta A_{25} & \Delta A_{26} \\ 0 & 0 & 0 & 0 & 0 & 0 & 0 & 0 & 0 \end{bmatrix} \tag{21}$$

$$\Delta B_2 = \begin{bmatrix} 0 & 0 & 0 & 0 & 0 & 0 & 0 & 0 & 0 \end{bmatrix}^T \tag{22}$$

where $\Delta A_{21} = \frac{-bC_2}{I_y} + \frac{bC_2{}^{real}}{I_y}$, $\Delta A_{22} = \frac{-C_2 d}{I_x} + \frac{C_2{}^{real} d}{I_x}$, $\Delta A_{23} = \frac{-C_2}{m_s} + \frac{C_2{}^{real}}{m_s{}^{real}}$, $\Delta A_{24} = \frac{A_e}{m_s} - \frac{A_e}{m_s{}^{real}}$, $\Delta A_{25} = \frac{-C_2}{m_{s2}} + \frac{C_2{}^{real}}{m_{s2}{}^{real}}$, and $\Delta A_{26} = \frac{A_e}{m_{s2}} - \frac{A_e}{m_{s2}{}^{real}}$

The model error matrix is introduced, and the system input is considered to determine the adaptive threshold. Then, the system can be expressed to:

$$\dot{X} = (A + \Delta A)X + (B + \Delta B)U \tag{23}$$

where $\Delta A = \Delta A_1 + \Delta A_2$, and $\Delta B = \Delta B_1 + \Delta B_2$.

Model error can be defined as $\varepsilon = X - \hat{X}$. Equations (13) and (23) are combined, so that the model error can be expressed as follow:

$$\dot{\varepsilon} = (A + L_k C)\varepsilon + (\Delta A_1 + \Delta A_2)X + (\Delta B_1 + \Delta B_2)U \tag{24}$$

Integrating Equation (24), the error ε can be deduced as:

$$\begin{aligned} \varepsilon = \; & \varepsilon^{(A+L_k C)t}\varepsilon(0) + \int_0^t \varepsilon^{(A+L_k C)(t-\tau)}(\Delta A_1 + \Delta A_2)X(\tau)d\tau \\ & + \int_0^t \varepsilon^{(A+L_k C)(t-\tau)}(\Delta B_1 + \Delta B_2)U(\tau)d\tau \end{aligned} \tag{25}$$

The adaptive threshold for fault detection is given from the above equation as:

$$h = \varepsilon + c \tag{26}$$

where c calculates the acceptable deviation.

3.3. Fault Detection

The output residual is defined according to fault detection and isolation strategy as $r = Y - \hat{Y}$. For the vehicle ECAS system, the residual characteristic description shown in Table 2 can be obtained.

Table 2. Characterization of residuals.

Residual	$r_1^{(1)}$	$r_2^{(1)}$	$r_3^{(1)}$	$r_4^{(1)}$	$r_1^{(2)}$	$r_2^{(2)}$	$r_3^{(2)}$	$r_4^{(2)}$	$r_1^{(3)}$	$r_2^{(3)}$	$r_3^{(3)}$	$r_4^{(3)}$
No Failure	0	0	0	0	0	0	0	0	0	0	0	0
Actuator 1 failure	0	1	0	0	0	1	0	0	0	1	0	0
Actuator 2 failure	1	0	0	0	1	0	0	0	1	0	0	0
Actuator 3 failure	0	0	0	1	0	0	0	1	0	0	0	1
Actuator 4 failure	0	0	1	0	0	0	1	0	0	0	1	0

In Table 2, $r_i^{(j)}$ is the output estimated residual, i is the extended Kalman filter number ($i = 1 - 4$), and j is the measurement output number. $j = 1, 2, 3$ represent the height change of the air spring, the vertical acceleration at the four corners of the body, and the internal air pressure of the air spring, respectively. Actuators 1–4 represent the front left, rear left, front right, and rear right air spring solenoid valves, respectively. Taking the residual r as the fault detection indicator, there are three fault detection indicators, including the air spring height estimation residual $r_i^{(1)}$ and the vertical acceleration estimation residual error at the four corners of the car body $r_i^{(2)}$, and the air spring pressure estimation residual $r_i^{(3)}$. Each extended Kalman filter will produce the above three fault detection indicators (that is, $r_i^{(1)}$, $r_i^{(2)}$ and $r_i^{(3)}$). Comparing the fault detection index with the adaptive threshold h, it can be detected whether the actuator has a fault.

$$\begin{cases} index \geq h, \text{Failure} \\ index < h, \text{Nofailure} \end{cases} \tag{27}$$

According to Equation (27), when the fault detection index value is greater than or equal to the detection threshold, the actuator has failed. The corresponding detection index $r_i^{(j)} = 1$; when the detection index value is less than the detection threshold, the actuator has not failed. The corresponding detection index $r_i^{(j)}$ is equal to 0.

As long as one of the three fault detection indicators exceeds the threshold, it is considered that a fault has occurred. The advantage of setting three fault detection indicators is to further reduce the missed diagnosis rate and increase the reliability and effectiveness of fault detection. By looking up Table 2, you can not only know whether the actuator is malfunctioning, but also determine the location of the malfunctioning actuator; that is, fault detection and isolation are realized.

3.4. Simulation and Analysis

To verify the proposed fault diagnosis program, four fault behaviors shown in Table 3 are selected. Fault 1 is a stuck fault, and Faults 2–4 are constant gain faults. In the fault behavior, n represents the gain coefficient, and δ represents the fault stuck value. The flow area of the valve port can be calculated from Equation (8). Therefore, the fault behavior represents the flow area of the valve port, corresponding to four fault behaviors.

Actuators 1–4, respectively, represent the front left, rear left, front right, and rear right air spring solenoid valves. The fault diagnosis module was started at the same time as the vehicle height adjustment was started at 5 s, and the fault occurred at 8 s. The first 5 s is the process of the air spring model in AMEsim gradually returning to a steady state. The simulation results of Faults 1 and 2 detections are shown in Figures 6 and 7.

Table 3. Description of fault behavior.

Fault Number		1	2	3	4
Moment of failure/s			8		
Fault behavior	n	0	0.2	0.4	0.6
	δ	0	0	0	0

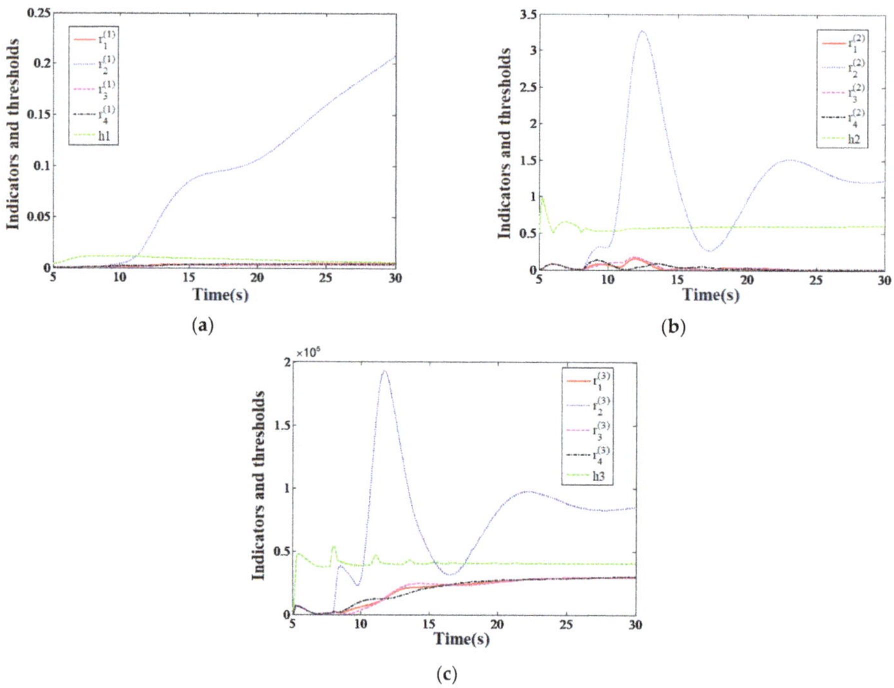

Figure 6. Changes of fault detection indicators under Fault 1: (**a**) Fault detection index $r_i^{(1)}$ and threshold h1; (**b**) Fault detection index $r_i^{(2)}$ and threshold h2; (**c**) Fault detection index $r_i^{(3)}$ and threshold h3.

As shown in Figure 6, before the occurrence of Fault 1, the estimated residuals of the three fault detection indicators, namely displacement, acceleration, and air pressure output, are smaller than the adaptive threshold. The fault occurs at the 8th second, and the residual error $r_2^{(1)}$, $r_2^{(2)}$ and $r_2^{(3)}$ output by EKF2 all exceeds the adaptive threshold. The residual outputs by other filters still fluctuate around zero or are less than the adaptive threshold. According to Figure 7 and the residual characteristics in Table 2, the front left air spring solenoid valve is malfunctioning. The fault detection time is 11.1 s, 10.4 s, and 10.1 s, respectively. If one of $r_2^{(1)}$, $r_2^{(2)}$ and $r_2^{(3)}$ exceeds the threshold, it is considered that a fault has occurred. Therefore, the actuator failure was detected at 10.1 s.

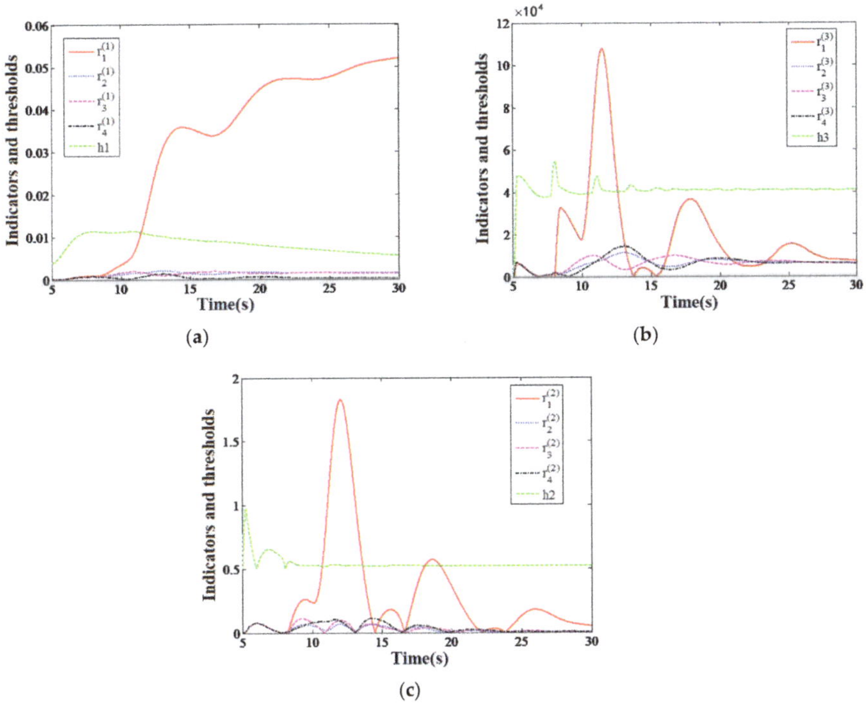

Figure 7. Change of fault detection index under Fault 2: (**a**) Fault detection index $r_i^{(1)}$ and threshold h1; (**b**) Fault detection index $r_i^{(2)}$ and threshold h2; (**c**) Fault detection index $r_i^{(3)}$ and threshold h3.

As shown in Figure 7, before Fault 2 occurs, the fault detection indicators are all less than the adaptive threshold. After the fault occurs, the output residuals of EKF1 $r_1^{(1)}$, $r_1^{(2)}$, $r_1^{(3)}$ all exceed the adaptive threshold. The residual outputs by other filters are still smaller than the adaptive threshold, and the fault detection time is 11.4 s, 10.6 s, and 10.4 s, respectively. Therefore, the actuator failure is detected at 10.4 s. According to Figure 7 and the residual characteristic Table 2, it can be judged that the rear left air spring solenoid valve is malfunctioning.

Similarly, simulation verification was performed for Failures 3 and 4. According to the simulation results and the residual characteristics in Table 2, it can be accurately known that the fault occurred in the front right and rear right air spring solenoid valves. In summary, the ECAS system fault diagnosis system based on the adaptive threshold is accurate and effective. It can correctly judge whether there is a fault and the location of the corresponding faulty solenoid valve.

4. Design and Simulation of Active Fault Tolerant Control

4.1. Design of Active Fault Tolerant Control

An active fault-tolerant control strategy is designed for the actuator failure of the vehicle ECAS system, shown in Figure 8. The active fault-tolerant control decision-making module (the content of this module is shown in Table 4) judges whether a fault has occurred according to the information sent by the fault detection and isolation module. If there is no fault, use the original controller. If a fault is detected, the fault estimation value sent by the fault estimation module is judged as whether it is stuck or a constant gain fault. If it is a stuck fault, the vehicle height adjustment is stopped immediately. If it is a constant gain fault, the estimated value $\hat{A}_1$ of the valve port area calculated from the estimated value of the fault is sent to the controller. The original valve port area parameter A_0 of the controller is replaced so that the duty ratio of the solenoid valve is increased, and the height change rate of the air spring corresponding to the faulty solenoid valve is increased. This can improve the inconsistency of air spring changes at the four corners and, ultimately, increase the vehicle height adjustment speed, as well as improve the body posture.

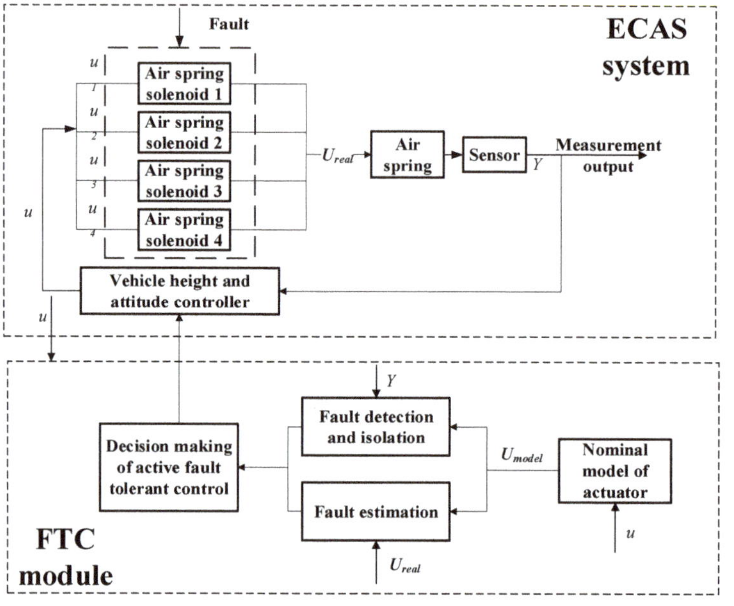

Figure 8. Active fault tolerant control strategy of vehicle ECA system.

Table 4. Contents of active fault tolerant control decision module.

Fault Type	Fault Tolerance Measures
No fault	The original controller is adopted
Constant gain fault	Online adjustment of controller parameters
Stuck fault	Close all solenoid valves and stop height adjustment

4.2. Simulation of Active Fault Tolerant Control

Taking Fault 1 (stuck) and Fault 2 (constant gain) shown in Table 3 as examples, the simulation results are as follows.

From Figure 9b,c, it can be seen that under normal circumstances, the pitch and roll angles of the vehicle body during the vehicle height adjustment process (5 s to 22.5 s) are well controlled. In the case of Fault 1, the height of the front left air spring stops increasing because the front left air spring solenoid valve is stuck. Its changes cannot be synchronized with other air springs. This causes the pitch angle and roll angle during the height adjustment process to rapidly increase to about 0.0055 rad and −0.0066 rad, respectively, and the body attitude deteriorates. Under fault-tolerant control, the vehicle height adjustment stops because all air spring solenoid valves are closed. Therefore, the pitch angle and roll angle of the body are stabilized at about 0.0005 rad and −0.0012 rad, to avoid further deterioration of the body attitude.

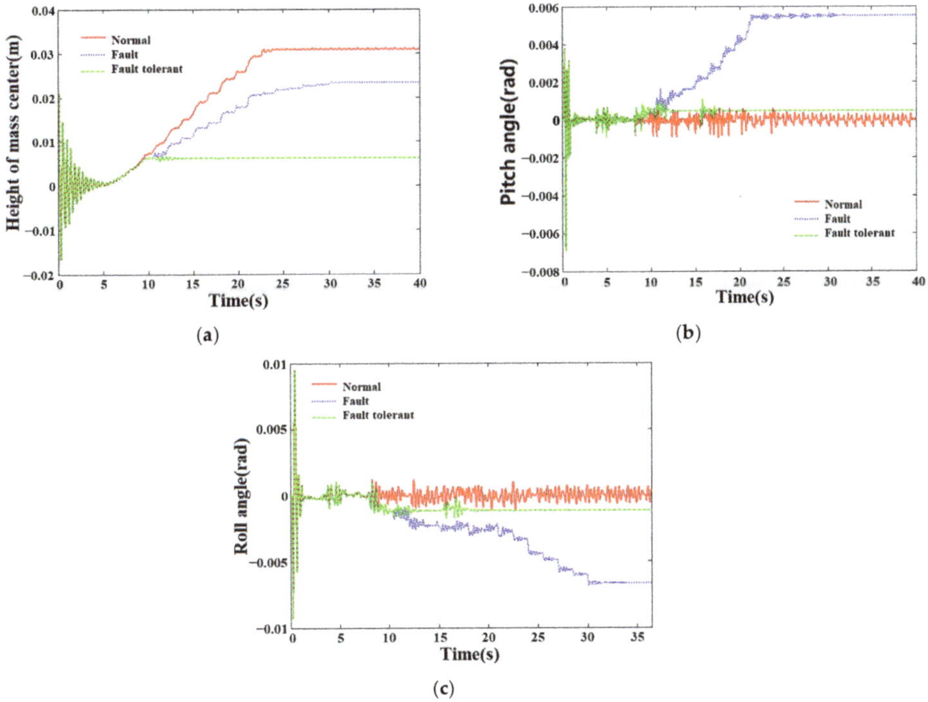

Figure 9. Performance comparison before and after fault tolerant control under Fault 1 (stuck): (**a**) Relative height of mass center; (**b**) Pitch angle; (**c**) Roll angle.

It can be seen from Figure 10 that when Fault 2 occurs, the vehicle height adjustment speed without fault-tolerant control decreases, and the pitch and roll angle peaks increase. This is due to the fact that the solenoid valve port of the rear left air spring cannot be fully opened, and a constant gain failure has occurred, resulting in a decrease in flow. After performing fault-tolerant control, the duty ratio of the rear left air spring solenoid valve is increased by adjusting the controller parameters online. After entering, the flow rate of the left air spring increases, and the vehicle height adjustment time decreases. Compared with fault-tolerant control, the vehicle height adjustment time is improved by about 15.3%. The peak pitch angle and roll angle are reduced, and the improvement in the peak pitch angle of the body is about 43.8%. The peak roll angle improvement rate is about 37.5%.

Faults 3 and 4 are also constant gain faults. Since the air spring solenoid valve cannot be fully opened, the flow into the solenoid valve is reduced. The simulation results show that the vehicle height adjustment speed decreases, and the pitch angle and roll angle peaks increase when there is no fault-tolerant control. After performing fault-tolerant control, the vehicle height adjustment time range, pitch angle, and roll angle peak are all improved.

In summary, when the actuator fails, the designed active fault-tolerant control method can effectively improve the vehicle height adjustment and attitude control performance under the fault compared to the case of no fault-tolerant control.

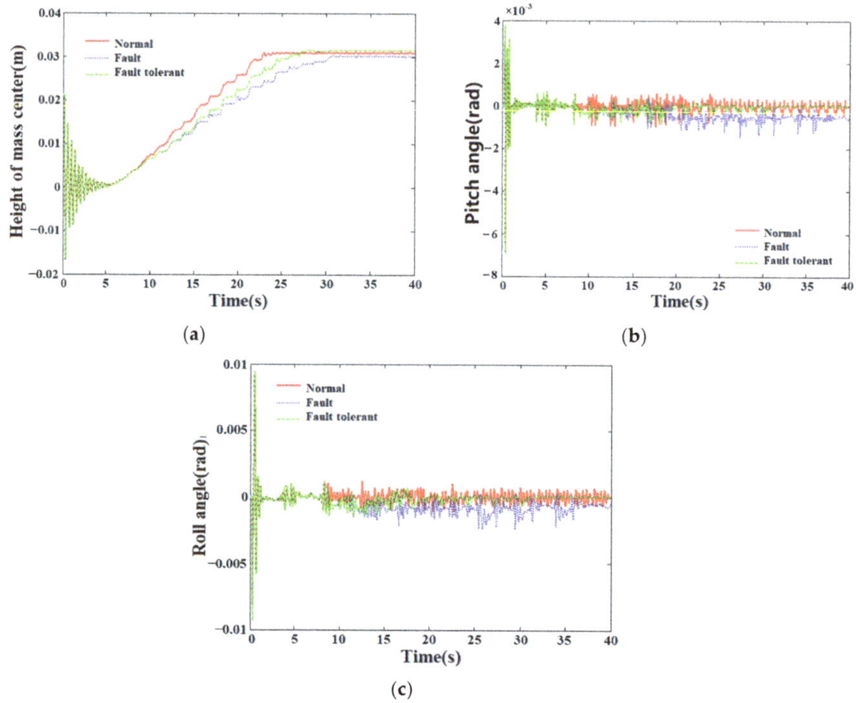

Figure 10. Performance comparison before and after fault tolerant control under Fault 2 (constant gain): (**a**) Relative height of mass center; (**b**) Pitch angle; (**c**) Roll angle.

5. Hardware In-Loop Simulation System

5.1. Hardware Platform

The fault-tolerant control HiL test platform of the vehicle ECAS system simulates the input and output signals through various boards. The connection between the real controller and the controlled object simulation model can be realized to form a complete loop. To verify the actual operation and feasibility of the fault-tolerant control algorithm model in the real controller, the hardware test is shown in Figure 11.

Under the existing hardware platform, building an HiL test system that is mainly divided into three steps includes establishing the controlled object model, developing the control model, and creating the system engineering file and user interface.

The controlled object model of the vehicle ECAS system is obtained by specifying relevant settings based on the AMESim model, including (1) online parameter setting, (2) observation variable setting, and (3) external interface setting. After applying the corresponding settings, you can compile and generate *.dll file, and load it into the system project file. The control model is built based on D2P rapid control prototyping technology, and NI VeriStand software is used to create project files and user interfaces.

NI VeriStand software is a software environment for configuring real-time test applications. Its functions include configuring the operating system, board, vehicle model, input/output interface with the actual controller, and user interface. NI VeriStand software carries out hardware-in-the-loop simulation by interacting with the signal of the actual controller. Researchers can monitor the simulation process in real time through the upper computer. The test principle of the HiL platform is shown in Figure 12.

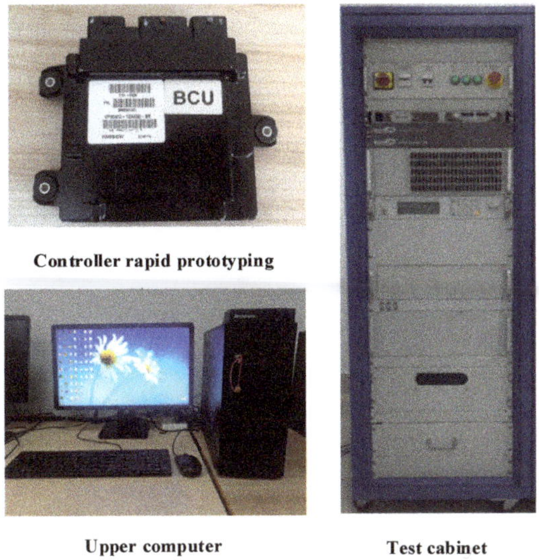

Figure 11. Hardware composition of fault-tolerant control HiL test platform for ECAS system.

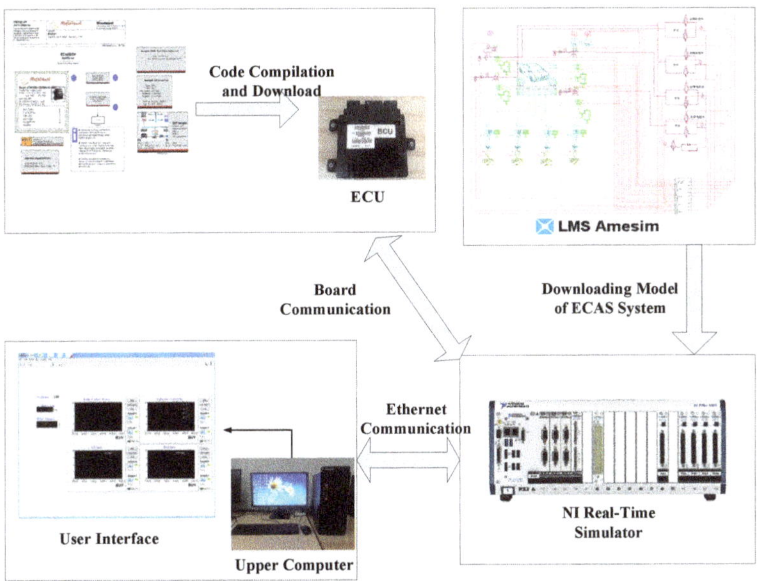

Figure 12. The working principle of HiL test platform.

5.2. Results and Analysis

Fault Behaviors 1 and 2 in Table 3 are selected for the fault-tolerant control HiL test, and the results are shown in Figures 13 and 14.

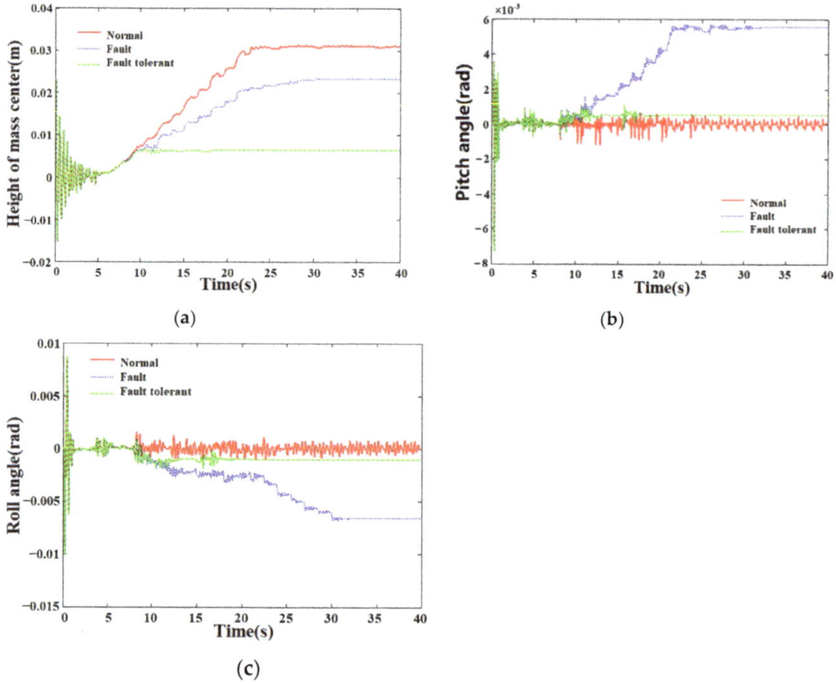

Figure 13. Performance comparison before and after fault tolerant control under Fault 1: (**a**) Relative height of mass center; (**b**) Pitch angle; (**c**) Roll angle.

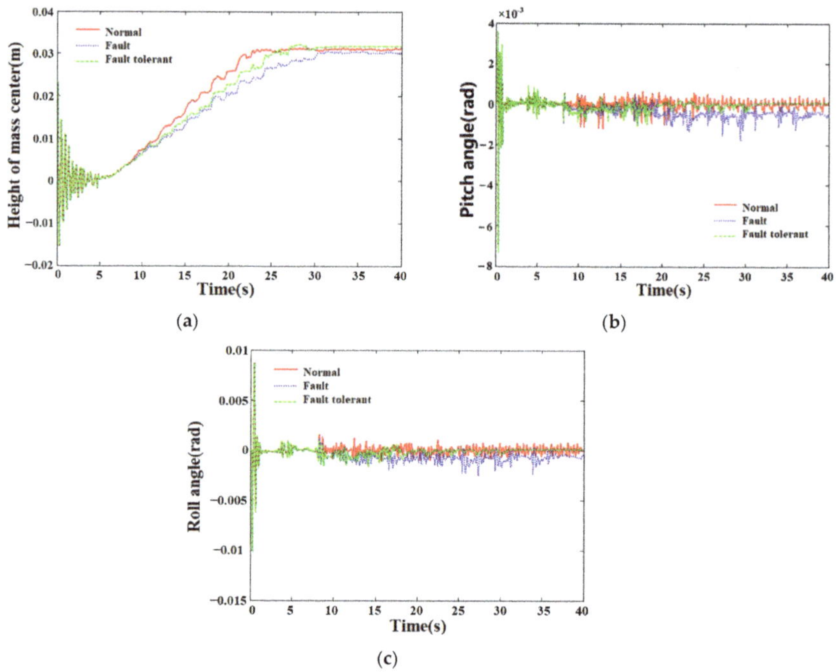

Figure 14. Performance comparison before and after fault tolerant control under Fault 2: (**a**) Relative height of mass center; (**b**) Pitch angle; (**c**) Roll angle.

As shown in Figure 13, after the vehicle height starts to increase (that is, after 5 s), it will reach the target height after about 17.5 s (that is, the 22.5 s) normally. After the current left air spring solenoid valve has a stuck fault, if there is no fault-tolerant control, the relative height of the body's center of mass will continue to increase to about 0.0232 m, and the pitch angle and roll angle will rapidly increase to about 0.0056 rad and −0.0066 rad, respectively. Under fault-tolerant control, about 2.1 s after the fault occurs, the air spring solenoid valve is closed, and the vehicle height adjustment stops. At this time, the pitch angle and roll angle of the body are stabilized at about 0.0005 rad and −0.0011 rad. Compared with the model-in-loop (MiL) simulation of the Fault 1 situation in Section 3.4, it can be found that the data results are similar. For Fault 2, the HiL test curve is shown in Figure 14.

The data results of the MiL and HiL simulation tests are given in Tables 5 and 6, respectively. By comparing Tables 5 and 6, it can be found that the HiL test data results of faults are basically consistent with the MiL simulation data results.

Table 5. Analysis of simulation results of Faults 2–4.

Fault Number	Performance Index	Failure (No Fault Tolerance)	Fault Tolerant Control	Improvement Range
2	Height adjustment time (s)	25.5	21.6	15.3%
	Peak pitch angle (RAD)		0.0009	43.8%
	Peak roll angle (RAD)	0.0024	0.0015	37.5%
3	Height adjustment time (s)	24.2	20.4	15.7%
	Peak pitch angle (RAD)	0.0094	0.001	89.4%
	Peak roll angle (RAD)	0.0063	0.0013	79.4%
4	Height adjustment time (s)	20.8	19.2	7.7%
	Peak pitch angle (RAD)	0.0027	0.0011	59.3%
	Peak roll angle (RAD)	0.0017	0.0012	29.4%

Table 6. HiL test result analysis of Faults 2–4.

Fault Number	Performance Index	Failure (No Fault Tolerance)	Fault Tolerant Control	Improvement Range
2	Height adjustment time (s)	25.5	21.6	15.3%
	Peak pitch angle (RAD)	0.0018	0.001	44.4%
	Peak roll angle (RAD)	0.0025	0.0016	36%
3	Height adjustment time (s)	24.2	20.4	15.7%
	Peak pitch angle (RAD)	0.0091	0.00098	89.2%
	Peak roll angle (RAD)	0.0061	0.0016	73.8%
4	Height adjustment time (s)	20.8	19.2	7.7%
	Peak pitch angle (RAD)	0.003	0.0012	60%
	Peak roll angle (RAD)	0.0017	0.0013	23.5%

In summary, the HiL test results of fault diagnosis and fault-tolerant control strategy are basically consistent with the MiL simulation results in Section 3.4. It shows that the fault diagnosis and fault-tolerant control model designed in the actual controller can operate normally. It can effectively realize the fault detection isolation of the controlled object under the actuator failure and the improvement of the vehicle height adjustment and attitude control performance.

6. Conclusions

In this paper, the fault diagnosis and active fault-tolerant control of the ECAS system under actuator fault are studied. Based on the fault model, an extended Kalman filter bank with an adaptive threshold is designed for fault diagnosis. In addition, online adjustment is adopted for fault-tolerant control.

Firstly, the ECAS vehicle model is simplified, and the mathematical model of vehicle height regulation is established as the basis of fault-tolerant control. The faults are classified into constant gain faults and stuck faults by the fault tree method. According to the fault type, an accurate fault mathematical model is established. Then, an adaptive threshold extended Kalman filter bank is designed as the observer. Each residual is compared with the adaptive threshold. Therefore, the fault location and type can be accurately located. This method improves the accuracy and speed of diagnosis. Then, based on the method of model analysis, the fault-tolerant control of the ECAS system under fault is successfully carried out by closing the air spring solenoid valve or adjusting it online.

The designed observer, controller, and vehicle model are run on the simulation platform. The validity of the above methods is verified by comparing the relationship between detection modes and adaptive thresholds under the four proposed fault behaviors. Finally, in order to verify the control effect on the actual vehicle, an HiL semi-physical test platform was built. Such a test platform can combine the actual controller with the simulation model. The test results prove that the fault diagnosis and fault-tolerant control methods proposed in this study can be used in actual controllers. At the same time, it can accurately diagnose the location and type of faults, so as to carry out effective active fault-tolerant control.

Author Contributions: Conceptualization, X.J. and X.X.; methodology, H.S.; software, X.J. and H.S.; validation, H.S., X.J. and X.X.; investigation, X.J. and X.X.; data curation, H.S.; writing—original draft preparation, X.J.; writing—review and editing, H.S. and X.J.; supervision, X.X.; project administration, X.X.; funding acquisition, X.X.; All authors have read and agreed to the published version of the manuscript.

Funding: This work is supported by the National Natural Science Foundation of China (Grant No. 51875256).

Data Availability Statement: Not applicable.

Conflicts of Interest: The authors declare no conflict of interest.

References

1. Yu, Z.; Deng, Y.; Zhang, Z.; Zhou, T. Control for car ride comfort based on electronic-controlled air suspension. *Tract. Farm Transp.* **2014**, *41*, 57–60.
2. Zhou, J.; Wang, K.; Liao, Y.; Tang, A. Study on road-friendliness of air suspension for truck. *Technol. Highw. Transp.* **2017**, *33*, 122–125.
3. Shao, J.; Wang, L.; Luo, N. Application of air suspension system in heavy truck. *Auto Manuf. Eng.* **2018**, *3*, 59–61.
4. Chen, L.; Xu, X.; Liang, C.; Jiang, X.; Wang, F. Semi-active control of a new quasi-zero stiffness air suspension for commercial vehicles based on H2H∞ state feedback. *J. Vib. Control* **2022**. [CrossRef]
5. Liu, H.; Xu, X.; Xie, J.; Jiang, X. Uncertainty ana1ysis and optimization of quasi zero stiffness air suspension based on po1ynomia1 chaos method. *Chin. J. Mech. Eng.* **2022**, *35*, 1–19.
6. Yang, L. *Research on Fault Tolerant Control Strategy of Vehicle Active Suspension*; Hefei University of Technology: Hefei, China, 2013.
7. Ricciardi, V.; Ivanov, V.; Dhaens, M.; Vandersmissen, B.; Geraerts, M.; Savitski, D.; Augsburg, K. Ride blending control for electric vehicles. *World Electr. Veh. J.* **2019**, *10*, 36. [CrossRef]
8. Adeleke, O.P.; Li, Y.; Chen, Q.; Zhou, W.; Xu, X.; Cui, X. Torque distribution based on dynamic programming algorithm for four in-wheel motor drive electric vehicle considering energy efficiency optimization. *World Electr. Veh. J.* **2022**, *13*, 181. [CrossRef]
9. Rui, B. Nonlinear adaptive sliding-mode control of the electronically controlled air suspension system. *Int. J. Adv. Robot. Syst.* **2019**, *16*, 443–475. [CrossRef]
10. Ma, Y.; Yan, T.; Zhao, Y. Research on Integrated Control Strategy of a New-Type Electronically Controlled Air Suspension System. *Automot. Eng.* **2021**, *43*, 1394–1401.
11. Hu, Q.; Lu, W.; Jiang, J. Design of a vehicle height and body posture adjustment hybrid automaton of electronically controlled air suspension. *Int. J. Adapt. Control. Signal Process.* **2021**, *35*, 1879–1897. [CrossRef]
12. Xu, X. *Research on Nonlinear System Control of ECAS Ride Height*; Jiangsu University: Zhen Jiang, China, 2010.
13. Gao, D.; Wang, Y.; Zheng, X.; Yang, Q. A fault warning method for electric vehicle charging process based on adaptive deep belief network. *World Electr. Veh. J.* **2021**, *12*, 265. [CrossRef]
14. Zhao, Y.; Yan, T.; Zhang, Q. Study on fault detection and isolation of height sensor for electronic air suspension system. *J. Qingdao Univ. (Eng. Technol. Ed.)* **2021**, *36*, 40–46.
15. Frank, P.M. Fault diagnosis in dynamic systems using analytical and knowledge-based redundancy: A survey and some new results. *Automatic* **1990**, *26*, 459–474. [CrossRef]
16. Zhou, D.; Hu, Y. Active fault diagnosis for dynamic systems. *Acta Autom. Sin.* **2009**, *35*, 748–758. [CrossRef]
17. Feng, Z.; Wu, P. Research on bogie fault diagnosis technology based on support vector machine. *Machinery* **2020**, *8*, 37–43.
18. Liu, Y.; Meng, Q.; Zeng, M.; Zeng, M.; Ma, S. Fault diagnosis method based on probability extended SDG and fault index. In Proceedings of the 2016 12th World Congress on Intelligent Control and Automation (WCICA), Guilin, China, 12–15 June 2016; pp. 2868–2873.

19. Li, S. *Research of Vibration Fault Diagnosis for Hydropower Unit Based on Fault-Tree Analysis*; Nanchang Institute of Technology: Nanchang, China, 2015.
20. Alwi, H.; Edwards, C.; Tan, C.P. *Fault Detection and Fault-Tolerant Control Using Sliding Modes*; Springer: London, UK, 2011.
21. Ding, R.; Xiao, L. Robust fault tolerant control for aircraft longitudinal attitude sensor faults. *J. Shanghai Inst. Technol. Nat. Sci.* **2015**, *15*, 167–172.
22. Liu, S.; Zhou, H.; Luo, X.; Deng, W.; Ying, Y. Nonlinear sliding fault control for active suspension. *J. East China Univ. Technol. (Nat. Sci.)* **2016**, *39*, 293–297.
23. Yu, J. Research on fault tolerant control strategy of automobile active suspension. *Automob. Appl. Technol.* **2021**, *46*, 86–89.
24. Liang, J.; Wang, Q.; Dong, C. An adaptive fuzzy estimator-based satellite fault-tolerant control system. *J. Astronaut.* **2010**, *31*, 1970–1975.
25. Wang, B.; Zhang, Y. An adaptive fault-tolerant sliding mode control allocation scheme for multirotor helicopter subject to simultaneous actuator faults. *IEEE Trans. Ind. Electron.* **2017**, *65*, 4227–4236. [CrossRef]
26. Pang, H.; Liu, X.; Shang, Y.; Yao, R. A hybrid fault-tolerant control for nonlinear active suspension systems subjected to actuator faults and road disturbances. *Complexity* **2020**, *2020*, 1874212. [CrossRef]
27. Xue, W.; Jin, P.; Li, K. Parameter-dependent actuator fault estimation for vehicle active suspension systems based on RBFNN. *Proc. Inst. Mech.Eng. Part D J. Automob. Eng.* **2021**, *235*, 2540–2550. [CrossRef]

World Electric Vehicle Journal

MDPI

Article

Vehicle Safety Planning Control Method Based on Variable Gauss Safety Field

Zixuan Zhu [1], Chenglong Teng [2], Yingfeng Cai [3,*], Long Chen [1], Yubo Lian [2] and Hai Wang [1]

1 Automotive Engineering Research Institute, Jiangsu University, Zhenjiang 212013, China
2 BYD Auto Industry Co., Ltd., Shenzhen 518118, China
3 School of Automotive and Traffic Engineering, Jiangsu University, Zhenjiang 212013, China
* Correspondence: caicaixiao0304@126.com

Abstract: The existing intelligent vehicle trajectory-planning methods have limitations in terms of efficiency and safety. To overcome these limitations, this paper proposes an automatic driving trajectory-planning method based on a variable Gaussian safety field. Firstly, the time series bird's-eye view is used as the input state quantity of the network, which improves the effectiveness of the trajectory planning policy network in extracting the features of the surrounding traffic environment. Then, the policy gradient algorithm is used to generate the planned trajectory of the autonomous vehicle, which improves the planning efficiency. The variable Gaussian safety field is used as the reward function of the trajectory planning part and the evaluation index of the control part, which improves the safety of the reinforcement learning vehicle tracking algorithm. The proposed algorithm is verified using the simulator. The obtained results show that the proposed algorithm has excellent trajectory planning ability in the highway scene and can achieve high safety and high precision tracking control.

Keywords: autonomous driving; planning algorithm; variable Gaussian safety field; reinforcement learning; policy gradient

Citation: Zhu, Z.; Teng, C.; Cai, Y.; Chen, L.; Lian, Y.; Wang, H. Vehicle Safety Planning Control Method Based on Variable Gauss Safety Field. *World Electr. Veh. J.* **2022**, *13*, 203. https://doi.org/10.3390/wevj13110203

Academic Editors: Yong Li, Xing Xu, Lin Zhang, Yechen Qin and Yang Lu

Received: 12 October 2022
Accepted: 26 October 2022
Published: 31 October 2022

1. Introduction

In recent years, autonomous driving technology has developed rapidly due to its significant economic potential and advantages in improving traffic efficiency and driving safety. Various methods have been proposed to solve the decision-making problem of autonomous vehicles in highway driving tasks. Most studies have considered decision making as a control problem. As an unavoidable part of the autonomous driving system, trajectory planning is of great significance to the study of the autonomous vehicle. Avoiding the surrounding obstacles accurately and driving safely and efficiently based on the upper perception and prediction results are the basic requirements for automobile driving. Therefore, most autonomous driving researchers are now focusing on more intelligent, safe and efficient trajectory-planning methods.

The existing trajectory-planning methods are generally divided into four categories: potential field methods [1], sample-based methods [2], search-based methods [3], and optimization-based methods [4]. A potential field method simulates the movement of a controlled object in space into a forced movement of a particle in a virtual force field and plans the future trajectory of a vehicle by calculating the combined force field to which the vehicle is subjected. However, this method relies on accurate modeling of the environment, which will put the training into the dilemma of the local optimal solution and increase the computational cost. The sampling-based methods are mainly divided into fast random search tree (RRT) and probability path map (PRM) methods. The probability map path method is based on the graph structure, converts the continuous space into a discrete space, and uses the search algorithms such as A* to find paths on the route map to improve search efficiency. However, this method needs to solve the boundary value problem and does not focus on generating paths in the process of building the graph. The search-based planning algorithms mainly refer to map search methods, including A*, D*, and the corresponding variants. This kind of algorithm is widely used in the field of robot motion planning, but its planned path does not consider the geometric constraints of the road and has poor smoothness. Qi Xuanxuan et al. [5] introduced simulated annealing to optimize the expansion of nodes and heuristic functions, and guided the algorithm to search for the target point, which improved the inefficiency of the traditional A* algorithm but still fell into the dilemma of a suboptimal solution. To improve sampling efficiency and avoid suboptimal dilemmas for agents, Claussmann et al. [6] classified the spatial configuration for route planning into three main categories: sampling [7], connection unit [8], and raster representation (Lattice) [9]. The raster representation can be used to predict and plan based on the moving obstacles around the vehicle while considering the kinematic constraints. However, the raster method is difficult to sample completely and can only sample better driving tracks. It is also difficult for the complete search method to consider the dynamic constraints of the automobile. The trajectory planning based on the optimization method has higher computational power requirements for the vehicle computer, and the optimization delay between each frame is large. In summary, most of the existing traditional trajectory-planning methods have relatively stable security performance and excellent computational efficiency. However, they focus only on the generation of the optimal path and can fall into the suboptimal dilemma.

In recent years, deep reinforcement learning (DRL) has shown satisfactory performance in both trajectory planning and trajectory tracking control. Feher et al. [10] trained deep deterministic policy gradient (DDPG) agents to generate waypoints for vehicle tracking and achieved good results. However, the algorithm only focused on the lateral trajectory and provided a suboptimal solution. Several studies have used original sensor measurements to generate turn angles and throttle values [11–16] in an end-to-end manner. The deep deterministic actor-critic (DDAC) algorithm [11,12] can keep the vehicle as far as possible on the center line of the lane and has achieved satisfactory results. However, this algorithm only considers the lateral control, not the longitudinal vehicle following. Lingli Yu et al. [15,16] proposed to use the DDPG algorithm to reduce the dependence on sample data. Their method had more continuous corner control and less lateral error when a vehicle was traveling. Although better results have been shown in the simulation environment, the agent is still affected by turn and throttle fluctuations and does not consider safety issues when interacting with other vehicles in highway conditions resulting in poor stability and safety.

To solve the above-mentioned problems, a vehicle safety planning and control method based on the variable Gauss safety field is designed in this paper. A planning model is constructed using a time series bird's-eye view as a state quantity and policy gradient algorithm. The timeliness and security of the planning model are verified by experiments. The reinforcement learning method of multi-task partitioning is used to partition and train the whole automatic driving trajectory tracking control task. Compared with the general end-to-end reinforcement learning auto-driving method, the multi-task partitioned training method reduces the training duration by dividing the entire auto-driving tracking control task into several sub-tasks and improves the noise input method in the longitudinal control module to further improve the training efficiency and provide a smoother driving experience. Meanwhile, protecting traffic participants is the most important topic in driving theory. Wang et al. [17,18] proposed the driving safety field theory modeling method and developed a collision warning algorithm, field experiments were conducted to verify the proposed algorithm. However, the whole framework contains several factors of driver, vehicle, and road, which bring great difficulties to practical application. To improve the practicability of safety field theory, a variable Gaussian safety field model is proposed to reveal the dynamic field characteristics of vertices. We use the variable Gaussian safety field model as the reward function of the planning module and combined with the constraint and evaluation index of the control module. The model combines a Gaussian field in both directions to form an envelope and varies with the vehicle speed angle. While ensuring reasonable trajectory generation, the interaction of the ego vehicle with the surrounding vehicles is utilized to actively avoid the surrounding vehicles when they enter the Gaussian field, which improves the safety performance of the vehicle in high-speed scenarios such as highways. The simulation results in CARLA show that the vehicle safety planning control method based on the variable Gauss safety field has good planning efficiency and better safety compared with the traditional algorithms.

The main contributions of this paper are as follows:

(1) An automatic driving trajectory-planning method based on time series bird's-eye view and policy gradient algorithm is designed. The policy gradient algorithm is used to improve the ability of automatic driving vehicle trajectory planning and the efficiency of Lattice sampling method for trajectory planning. The time series bird's-eye view combined with the policy gradient algorithm can enhance the ability of feature extraction of the policy network, make the network convergence easier, and improve the feasibility of the method.

(2) The variable Gauss security field is added as the evaluation index of the reward function and control part to improve the security of trajectory and control effect.

2. Route Planning Algorithm

The goal of trajectory planning for autonomous driving is to find the optimal trajectory in advance for a vehicle. On the one hand, it is necessary to ensure the safety of the vehicle; On the other hand, getting to the destination through obstacles as soon as possible, reducing traffic pressure and improving driving efficiency are also important criteria to measure the effectiveness of the planned trajectory. Figure 1 shows that the trajectory planning module plays a key role in the overall auto-driving system.

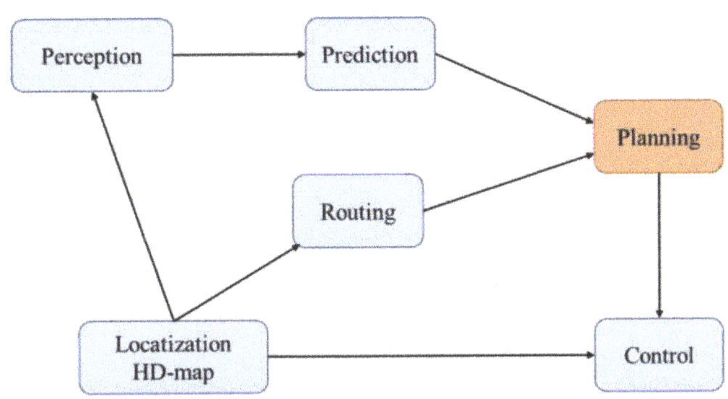

Figure 1. Autopilot system flow chart.

2.1. Time Series Bird's-Eye View and Strategic Network

The agents of reinforcement learning obtain the state input through interaction with the surrounding complex traffic environment to conduct effective learning training. One of the difficulties of the existing reinforcement learning algorithm is obtaining effective state features from complex environments. Overly redundant states will increase the learning difficulty of the agent. It is particularly important to make it easier for an agent to extract valid features. Therefore, this paper designs a policy network and corresponding time series bird's-eye view as the state quantity of the reinforcement learning, enabling the network to extract better environmental features.

2.1.1. Policy Network State Quantity

For an effective policy network for reinforcement learning, it is essential to obtain the perceptual information including lane lines, pedestrians, vehicles, and obstacles from the surrounding environment as well as the predictive tracks for the next few moments including dynamic obstacles.

The sequential bird's-eye view significantly improves the learning efficiency of the policy network. Figure 2 shows the time series bird's-eye view matrix diagram.

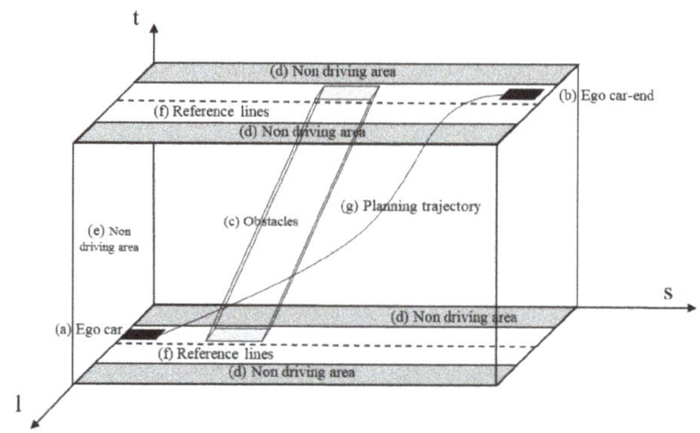

Figure 2. A time series bird's-eye view matrix diagram.

The bird's-eye view is a three-dimensional matrix composed of lateral displacement, vertical displacement and time. The specific elements in the matrix diagram shown in Figure 2 include (a) the current position status of the ego vehicle, (b) the ego vehicle, (c) obstacles, (d) the non-driving area and (e) the exercisable area, (f) the reference line, (g) the planned trajectory.

The generation of the time series bird's-eye view includes the following two steps: (1) According to the perception module of the autonomous vehicle, obtain the surrounding environmental information, including dynamic and static obstacles and lane lines. The prediction module is used to obtain the position information of dynamic obstacles in the future time of $0 \sim t_{end}$. (2) The information obtained from the perception module and the information is used to generate a bird's-eye view of features in three dimensions: horizontal, vertical and time.

The size of the three-dimensional the time series bird's-eye view matrix is (40, 400, 80). The first dimension 40 represents the horizontal range of 10 m on the left and right of the reference line, with the horizontal displacement interval of 0.5 m; The second dimension 400 represents the longitudinal 200 m forward range with the ego vehicle as the origin, the longitudinal displacement interval is 0.5 m, and the third dimension 80 represents the time range within the next 8 s, the time interval is 1 s. The (c) obstacles and (d) the non-driving area are represented by -1 in the time series bird's-eye view matrix; (e) the exercisable area is represented by 0 in the time series bird's-eye view matrix; (f) the reference line is represented by 1 in the time series bird's-eye view. In the matrix, the reference line represents higher priority than (c) obstacles, (d) the non-driving area and (e) the exercisable area. At the same time, (a) the current position status of the ego vehicle, (b) the ego vehicle, and (g) the planned trajectory are not specifically represented in the time series bird's-eye view matrix.

Figure 3 shows the vertical view of a time series bird's-eye view with a green rectangle representing the vehicles on the highway and a dashed grey line representing the driveway sidelines.

Figure 3. Time series bird's-eye view top view.

The generation of a time series bird's-eye view includes the following two steps: (1) Obtain the surrounding environment information, including dynamic and static obstacles, and lane lines, according to the perception module of the automobile. Obtain dynamic obstacles using prediction module in the future $0 \sim t_{end}$ location information within the end. (2) Generate cross-sectional, vertical, and temporal feature bird's-eye views using the information obtained from the perception and prediction modules. Then, train using the bird's-eye view as the state input.

2.1.2. Strategic Network Structure

Figure 4 shows the structure of the policy network $\pi_\theta(z,a)$. The network includes a convolution feature extraction network consisting of one convolution layer and a fully connected network consisting of three fully connected layers. Where z is the input state quantity of the policy network, including the time series bird's-eye view matrix and the history track of the vehicle, θ denote the weights and offset parameters for the network and a is the output of the policy network, that is, the final state of the planning trajectory $a = \left\{s,\dot{s},\ddot{s},l,\dot{l},\ddot{l},t\right\}$, where s, $\dot{s}$ and $\ddot{s}$ are the final longitudinal position, the end-of-longitudinal speed, and the acceleration of the longitudinal end state of the vehicle, respectively, while l, $\dot{l}$ and $\ddot{l}$ are the lateral end state position, the lateral end-state speed and the acceleration of the lateral end state of the vehicle, respectively. The input of the convolution feature extraction network is the time series aerial view matrix and the output is the final extracted environmental feature information. The input of the fully connected network is the convolution feature. The environmental feature information and the historical track information of the vehicle are extracted from the network output.

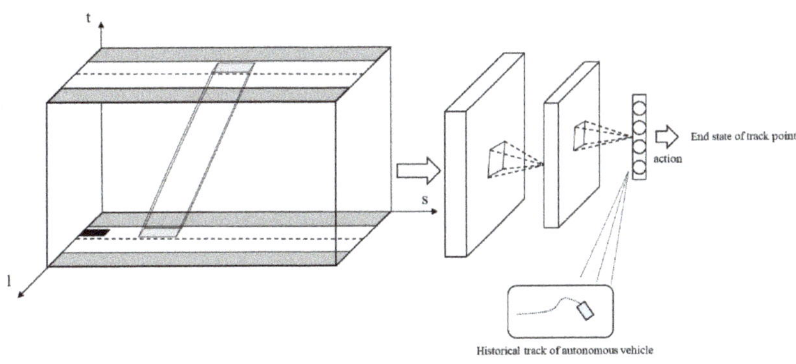

Figure 4. Strategic network structure diagram.

2.2. Variable Gauss Safety Field Theory

Since reinforcement learning explores policies and rewards by making agents constantly try and error, the security of reinforcement learning is lower than the other methods. Improving the security of reinforcement learning remains the focus of research. The variable Gauss security field model based on risk center transfer further improves the security of trajectory planning and control methods and serves as the reward function of the trajectory planning part and the constraint boundary of the control part.

Figure 5 shows that a static vehicle is abstracted as a rectangle with a length of l_v, a width of w_v, and the risk center $O(x_0, y_0)$ is its geometric center. The static security field of the vehicle is described by a two-dimensional Gaussian function as:

$$S_{sta} = C_a \cdot \exp\left(-\frac{(x-x_0)^2}{a_x^2} - \frac{(y-y_0)^2}{b_y^2}\right) \tag{1}$$

where C_a is the field strength factor, a_x and b_y represent the function of vehicle shape. The main control parameter for the shape of a static safety field is anisotropy:

$$\varepsilon = \frac{a_x^2 - b_y^2}{a_x^2 + b_y^2} = \frac{\phi^2 - 1}{\phi^2 + 1} \tag{2}$$

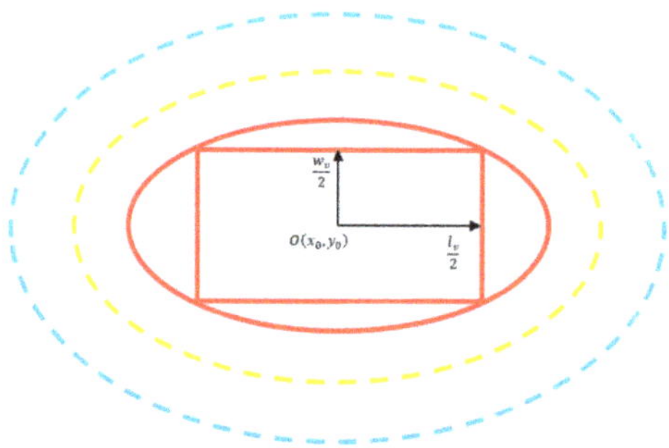

Figure 5. Static safety field overhead projection.

Parameter ε equivalently expressed in aspect ratio $\varnothing = a_x/b_y = l_v/w_v$.

The direction of the safety field is a vector from the risk center whose isoelectric line is projected upward into a series of ellipses. In Figure 5, the red rectangle represents the vehicle, the area in the solid red rectangle is called the core domain, the area between the red and the yellow ellipses is called the restriction domain, the area between the yellow and the blue ellipses is called the expansion domain, and each area represents a different risk state. The sizes of these different domains are related to the shape and motion of the vehicle and can be determined based on the parameters a_x, b_y of the Gaussian function (1). The Gauss security field is variable. The aspect ratio of the virtual vehicle will change with the change of the vehicle motion state and will significantly change the core, restriction and extension domains of the Gauss security field.

Figure 6 shows the overhead projection of the dynamic safety field. It can be seen that when the vehicle is in motion, the risk center will transfer following the vector $k_v\vec{v}$, the new risk center becomes $O'(x_0', y_0')$ and there are:

$$\begin{cases} x_0' = x_0 + k_v\left|\vec{v}\right|\cos\beta \\ y_0' = y_0 + k_v\left|\vec{v}\right|\sin\beta \end{cases} \tag{3}$$

where $\vec{v}$ is the velocity vector of the vehicle motion, k_v is the regulator and $0 < k_v < 1$ or $-1 < k_v < 0$, the sign corresponds to the front and back directions of the movement. β is the transferred angle between the vector and the x-axis.

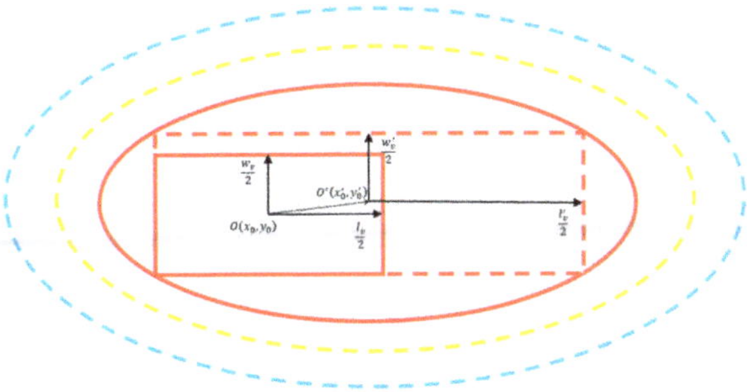

Figure 6. Overhead projection of dynamic safety field.

A virtual vehicle is formed with a length of l'_v and width of w'_v under the transfer of the risk center, whose geometric center is (x'_0, y'_0), which establishes its dynamic security field as:

$$S_{dyn} = C_a \cdot \exp(-\frac{(x - x'_0)^2}{(a'_x)^2} - \frac{(y - y'_0)^2}{(b'_y)^2}) \tag{4}$$

where a'_x and b'_y are parameters related to vehicle shape and motion state. The new aspect ratio is expressed as $\varnothing' = a_x'/b_y' = l_v'/w_v'$.

2.3. Improved Lattice Programming Algorithm Based on Strategic Gradient Algorithm

The traditional Lattice programming algorithm achieves trajectory planning by sampling the target vertically and horizontally. This method will lead to the dilemma of a suboptimal solution for the sample-fitting trajectory, and it would be difficult to obtain the optimal trajectory. However, too many sampling points will lead to complex and inefficient calculations.

The Lattice algorithm is improved by using the policy gradient algorithm to directly obtain the optimal final state sample points as shown in Figure 7. This improved method abandons sampling with high time complexity and cost function evaluation for each alternate trajectory, which considerably improves the timeliness of the algorithm. Although the training process of reinforcement learning has better universality than the general rule-based planning algorithm, the design of the reward function based on the final control effect will make it more suitable for complex traffic scenes and complex vehicle dynamic features.

2.3.1. Track Planning Agent Design

The trajectory output by general dynamic programming, Monte Carlo sampling and time series difference methods will have a complete state action sequence $< s_0, a_0, s_1, a_1 \cdots s_{end-1}, a_{end-1}, s_{end} >$ and a trajectory consists of several state–action pairs as shown in Figure 8. Different actions a in each step will inevitably lead to changes in the overall trajectory. This will necessarily result in an exponential increase in the complexity of the solution as the length of the trajectory will increase. The simplified trajectory τ is composed of the start state s_0, action a and end state s_{end}. In the start state s_0, executing action a produces a unique trajectory τ, reaching the end state s_{end}.

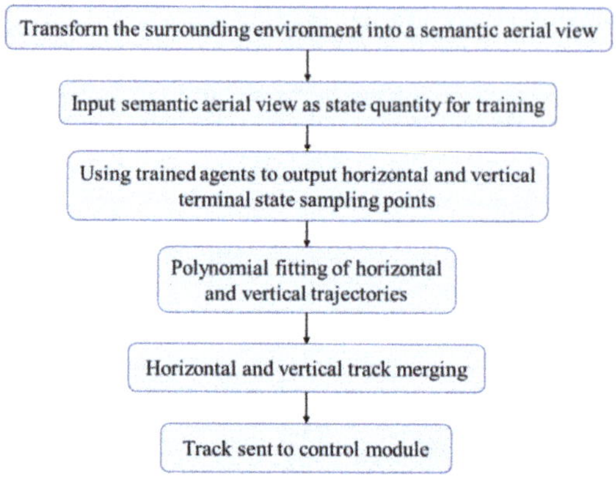

Figure 7. Lattice sampling process improvement.

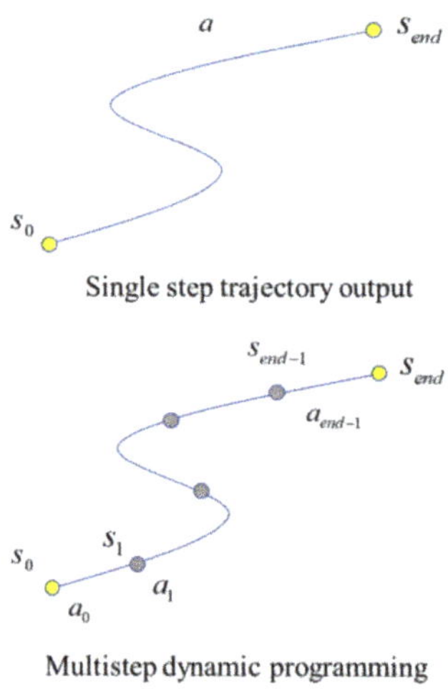

Figure 8. Diagrams of single-step and multi-step dynamic planning trajectory outputs.

In practice, the policy gradient algorithm is used instead of the last state sampling process in the Lattice algorithm. The end state of the track is used as the action space A:

$$A = \left\{ s_{end}, \dot{s}_{end}, \ddot{s}_{end}, l_{end}, \dot{l}_{end}, \ddot{l}_{end} \right\} \tag{5}$$

Policy network $\pi_\theta(z, a)$ maximizes the expected return of the output trajectory as an optimization objective:

$$J(\pi) = \sum_\tau p(\tau, \theta) \cdot r(\tau) \tag{6}$$

where z denotes the state features of the surrounding traffic environment, a is the network output action, θ is a network parameter, $p = (\tau, \theta)$ is the probability of executing action a and outputting track τ under parameter θ and state z, and $r(\tau)$ is the reward function of trajectory τ.

The gradient rise method is used to optimize $\pi_\theta(z, a)$ from Equation (6):

$$\theta = \theta + \alpha \cdot \nabla_\theta J(\pi) \tag{7}$$

To calculate the derivative of the optimization objective with respect to network parameter θ, the strategy gradient is derived as:

$$\begin{aligned}\nabla_\theta J(\theta) &= \nabla_\theta \textstyle\sum_\tau p(\tau,\theta)\cdot r(\tau)\\ &= \textstyle\sum_\tau \nabla_\theta p(\tau,\theta)\cdot r(\tau)\\ &= \textstyle\sum_\tau \frac{p(\tau,\theta)}{p(\tau,\theta)}\nabla_\theta p(\tau,\theta)\cdot r(\tau)\\ &= \textstyle\sum_\tau p(\tau,\theta)\nabla_\theta \log p(\tau,\theta)\cdot r(\tau)\end{aligned} \quad (8)$$

To improve the efficiency of training, during the training process, the agent continuously stores the experience data $< z, a, \tau, r >$ from the interaction with the environment in real-time into the experience pool (Memory). The Monte Carlo method is also used to randomly extract the mini-batch-sized empirical data from the experience pool for training:

$$\nabla_\theta J(\theta) \approx \frac{1}{n}\sum_{i=1}^{n} \nabla_\theta \log p(\tau,\theta)\cdot r(\tau) \quad (9)$$

From Formula (9), the update direction of the final policy parameters θ is:

$$\theta = \theta + \alpha\cdot\frac{1}{n}\sum_{i=1}^{n} \nabla_\theta \log p(\tau,\theta)\cdot r(\tau) \quad (10)$$

To enhance the agent's exploring ability in unfamiliar state space and avoid the agent falling into local optimal space during training, the output of the policy network $\pi_\theta(z,a)$ will conform to normal distribution. It consists of two parts: mean $\mu(z,a)$ and variance $\sigma(z,a)$:

$$\pi_\theta(z,a) = \frac{1}{\sqrt{2\pi}\cdot\sigma(z,\theta)}\exp\left(-\frac{(z-\mu(z,\theta))^2}{2\sigma^2(z,\theta)}\right) \quad (11)$$

During the learning process of the policy network $\pi_\theta(z,a)$, the mean $\mu(z,a)$ and the variance $\sigma(z,a)$ of the output keep approaching $arg_{max}Q(z,a)$ and 0, respectively, and the probability of the agent taking random behavior exploration keeps decreasing. During training, the agent selects action $a = \left\{s,\dot{s},\ddot{s},l,\dot{l},\ddot{l},t\right\}$ from this normal distribution as the training output and executes it.

2.3.2. Reward Function Design

Reinforcement learning obtains the amount of state by interacting with the environment and evaluates the training agent by a reward function. The agents obtain higher returns by continuously optimizing their network of policies. Therefore, the design of the reward function is critical to the convergence of the agent, which affects the final decision-making results of the overall model. Moreover, a reasonable reward function design can also make the agent obtain more incentives from the environment and accelerate the convergence speed of the agent.

The reward function design for the trajectory planning section includes the following sections:

$$reward = k_1\cdot r_{speed} + k_2\cdot r_{acc} + k_3\cdot r_{lateral} + k_4\cdot r_{comfort} + k_5\cdot r_{additional} + k_6\cdot r_{safe} \quad (12)$$

In the formula, $r_{speed} = -\sum_{t<t_{total}} t\cdot\left(v_{target} - v_t\right)^2$ is the speed reward, its goal is to keep the speed at the target speed; $r_{acc} = -\sum_{t<t_{total}} \ddot{s}_t^2$ and $r_{comfort} = -\sum_{t<t_{total}} \ddot{l}_t^2$ are the longitudinal and lateral comfort rewards, respectively, their goals are to maintain low longitudinal acceleration and low lateral acceleration, respectively; $r_{lateral} = -\sum_{t<t_{total}} l_t^2$ is the lateral deviation reward, its goal is to maintain a small lateral deviation from the reference line; $r_{additoanal} = -\sum_{t<total}\left(s_t - s_t^{actual}\right)^2 + \left(l_t - l_t^{actual}\right)^2$ is the additional coupling reward, the objective is to maintain the coupling force between the planned trajectory and the controller and vehicle dynamics, and to maintain a better horizontal and vertical track-

ing accuracy of the vehicle during actual tracking; and r_{safe} is the safety reward. $k_1 \sim k_6$ is the proportion weight of each reward function. Where, $k_1 = 1.0$, $k_2 = 0.2$, $k_3 = 1.0$, $k_4 = 0.2$, $k_5 = 0.5$, and $k_6 = 1.0$. The value of $k_1 \sim k_6$ is obtained through debugging, and the specific value comparison is shown in Figure 9 below.

k1	k2	k3	k4	k5	k6	Mean-reward
1.0	0.2	1.0	0.2	0.5	1.0	0.829
1.0	0.2	0.5	0.2	0.5	1.0	0.647
1.0	0.1	1.0	0.1	0.5	1.0	0.532
1.0	0.5	1.0	0.5	0.5	1.0	0.247
1.0	0.2	1.0	0.2	0.8	1.0	0.073
1.0	0.1	1.0	0.1	0.5	1.0	0.532
1.0	0.2	1.0	0.2	0.2	1.0	−0.314
1.0	0.2	1.0	0.2	0.5	0.5	−2.012
1.0	0.2	0.5	0.2	0.5	0.5	−4.829

Figure 9. Comparison chart of proportional weights.

The design of r_{safe} is constrained by the variable Gaussian safety field, as shown below: When the vehicle is stationary:

$$\begin{cases} r_{safe,lat} = \begin{cases} -100 & if\ d_{lat} < l_v \\ 0 & if\ d_{lat} > l_v \end{cases} \\ r_{safe,lon} = \begin{cases} -100 & if\ d_{lon} < \frac{3l_v}{\sqrt{9-w_v^2}} \\ 0 & if\ d_{lon} > \frac{3l_v}{\sqrt{9-w_v^2}} \end{cases} \end{cases} \tag{13}$$

When the vehicle is moving:

$$\begin{cases} r_{safe,lat} = \begin{cases} -100 & if\ d_{lat} < l'_v \\ 0 & if\ d_{lat} > l'_v \end{cases} \\ r_{safe,lon} = \begin{cases} -100 & if\ d_{lon} < \frac{4l'_v}{\sqrt{16-(w'_v)^2}} \\ 0 & if\ d_{lon} > \frac{4l'_v}{\sqrt{16-(w'_v)^2}} \end{cases} \end{cases} \tag{14}$$

where $\begin{cases} w'_v = w_v + 2 \cdot k_v \cdot \left|\vec{v}\right| \cdot \sin\beta \\ l'_v = l_v + 2 \cdot k_v \cdot \left|\vec{v}\right| \cdot \cos\beta \end{cases}$, l_v and w_v are the length and the width of the agent, respectively, $\vec{v}$ is the speed vector of vehicle motion, k_v is the adjustment factor, and β is the angle between the transfer vector and the x-axis. After the actual vehicle test, $k_v = 0.35$.

3. Controller Design

The traditional trajectory planning module and the control module are simple upper and lower-level relationships. The trajectory planning module outputs the optimal trajectory and the controller tracks the control. Although this mode is simple and easy to operate, it cannot meet the real-time requirements in complex traffic environments. Figure 10 shows the relationship diagram of the proposed feedback design model. It can be seen from the figure that the trajectory planning agent based on the policy gradient algorithm, the trajectory tracking controller and the environment form a planning control environment closed loop. The proposed loop feedback design model will enable the agents to continuously learn to adapt to the environment and adapt to the trajectory tracking controller. This method effectively links the traffic environment, the planner and the controller, so that the output trajectory of the planner can effectively adapt to the dynamic features of the vehicle and the controller. To enable the agent to stably, efficiently and safely track the optimal trajectory output by the planner, and improve the efficiency, the training of the control part is divided into horizontal control and vertical control.

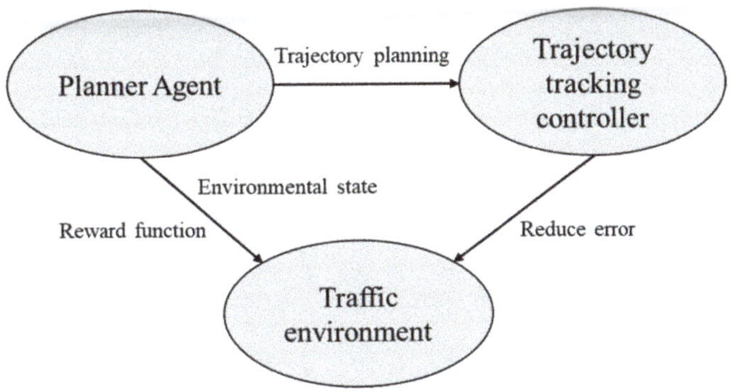

Figure 10. Planner/Controller/Environment relationship diagram.

3.1. Horizontal Trajectory Tracking Control Model Training

The goal of the traditional horizontal trajectory tracking task [19,20] is to enable vehicles to drive stably on the lane line without deviating, regardless of the state relationship with other vehicles. However, when the vehicle tracks and controls the track, the first consideration is the safety of the track, that is, it will not collide with other vehicles. Therefore, the variable Gaussian safety field is introduced as the evaluation index, and the state quantity and reward function are adjusted. The variables including the distance d_i from other vehicles, the lateral relative coordinate x_{i-v}, the coordinate (x_i, y_i) of the navigation point in the current vehicle coordinate system, the heading deviation φ and the speed v and acceleration $\dot{v}$ of the control vehicle are added as the state variables:

$$s^{lane-keep} = < d_0, d_1, \ldots, x_0, x_1, \ldots, x_{1-v}, x_{2-v}, \ldots, \varphi, v, \dot{v} > \quad (15)$$

The output action is only the steering wheel angle $a_{steer} \in [-1, 1]$. For the design of the reward function for lane keeping, the lateral error x_0 between the current vehicle coordinate and the lane centerline, the deviation φ of the heading angle and the relative distance d_i from other vehicles are considered as the evaluation index reward functions:

$$\begin{cases} r_{safe,lat} = -\log(\left|\frac{1}{\sqrt{2}}w'_v - d\right| + 1, 1.2) \\ r_{safe,lon} = -\log(\left|\frac{1}{\sqrt{2}}l'_v - d\right| + 1, 1.2) \\ r^{lane-keep} = -k_1 abs(x_0) - k_2 \sin\varphi \end{cases} \quad (16)$$

where $\begin{cases} w'_v = w_v + 2 \cdot k_v \cdot \left|\vec{v}\right| \cdot \sin\beta \\ l'_v = l_v + 2 \cdot k_v \cdot \left|\vec{v}\right| \cdot \cos\beta \end{cases}$, l_v and w_v are the length and the width of the agent, respectively, $\vec{v}$ is the speed vector of vehicle motion, k_v is the adjustment factor, and β is the angle between the transfer vector and the x-axis. After the actual vehicle test, $k_v = 0.35$.

If the lateral deviation of the current position of the autonomous vehicle is greater than the set maximum lateral deviation threshold value x_{0max} during the training, the current round of iterative training will be ended for the next round of training. Through the cumulative reward mechanism, agents that enhance learning continuously obtain higher reward reports. Hence, they can take more potential threats into account. However, the dynamic features of the vehicle will be hidden in the state quantity of the past few moments. Thus, it would be difficult to fully understand the current state of the intelligent vehicle only through the current state quantity. To enable the agent to better understand the dynamic features of the intelligent vehicle at the current time and output more reasonable trajectory tracking actions, the state quantities at the current time and at the past four times are stacked together as network inputs.

3.2. Training of Longitudinal Trajectory Tracking Control Model

To maintain an ideal distance between the ego vehicle and the vehicle in front without any collision with the vehicle in front, the ego vehicle is expected to cruise at a constant speed when there is no vehicle in front. When there are other vehicles in front of the ego vehicle, the road information is not considered, instead only the information of the current vehicle and the vehicle ahead is considered as the state quantity. Figure 11 describes the cruise mission status. The longitudinal trajectory tracking control task considers the speed v and acceleration $\dot{v}$ of the current vehicle, speed v_l and acceleration $\dot{v}_l$ of the vehicle in front, the distance d from the vehicle in front and the expected speed v_{des} of the current vehicle as the state variables:

$$s^{acc} = < y_0, y_1, \ldots, \varphi, v, \dot{v} > \tag{17}$$

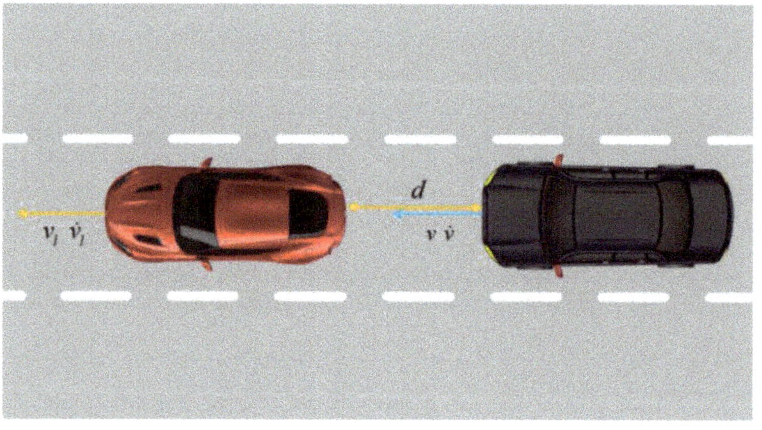

Figure 11. Description of cruise mission status.

Output action $a_{acc} \in [-1, 1]$ of the agent, including accelerator action $a_{throttle}$ and brake action a_{brake}:

$$\begin{cases} a_{throttle} = a_{acc}, a_{brake} = 0 & if\ a_{acc} \geq 0 \\ a_{throttle} = 0, a_{brake} = a_{acc} & if\ a_{acc} < 0 \end{cases} \tag{18}$$

For vertical control tasks, the reward function is designed as:

$$r^{acc} = \begin{cases} -k_5 abs(v - v_{des}) - k_6 abs(d - d_{des}) & if\ d > d_{safe} \\ -100 & if\ d \leq d_{safe} \end{cases} \tag{19}$$

where d_{des} and d_{safe} are the expected and safe distances from the vehicle in front, respectively. When the distance between the intelligent vehicle and the vehicle in front is less than the safe distance, the reward is -100 and the current interaction is stopped to start the next round of interaction. During longitudinal training, the speed v_l of the vehicle in front and the expected speed v_{des} of the current vehicle are randomly given each round, so that the training model can be generalized to more complex situations.

The traditional training mostly uses Gaussian noise or Ornstein Uhlenbeck (OU) noise to promote agents to actively explore the environment at the beginning of training. However, unnecessary exploration will prolong the training time of agents. Therefore, in this paper, a Multi-Head Actor network structure is designed for the tasks with convex solution space in longitudinal control tasks. The main function of the proposed structure is to make the output action noisy. Action noise reflects the uncertainty measure of the optimal solution of the current policy. The Multi-head Actor network structure is used to construct this uncertainty measurement method.

The output of the Online Actor network is connected to multiple Head networks. To reflect the difference of each Head network, the initialization and training sampling experience pool of each Head network are independent and the way to converge to the optimal solution space is also different. Therefore, the variance of the Head network output action is used to estimate the uncertainty measure of the output action of the Actor network as:

$$\begin{cases} N_t = \{k \cdot \mathrm{var}(\mu(s_t|\theta_{\mu online})) & if\ k \cdot \mathrm{var}(\mu_{\theta_{\mu_{online}}}(s_t|\theta_{\mu online})) < N_{threshold} \\ \quad N_{threshold} & else \end{cases} \tag{20}$$

where N_t and $N_{threshold}$ are the real-time action noise and the threshold noise, respectively, θ is the adopted policy, $\mu\left(s_t|\theta_{\mu online}\right)$ is the deterministic action of the network output, and k is the weight parameter.

Similar to the horizontal control part, the vertical control part also selects the current state quantity of the agent and the state quantity of the past four times as the network input, making the network easier to converge and having high training efficiency.

4. Experiment and Analysis

The simulation experiment is based on the open-source autopilot simulator CARLA, which supports the development, training and validation of autopilot systems. In addition to open-source code and API protocol, CARLA also provides open mathematical assets (urban layout, buildings and vehicles) that can be freely invoked. CARLA works through the client mode. It has a specific python API interface that can realize simulation environment configuration, environment interaction and vehicle control through interface code. CARLA is suitable as a training platform for automatic driving reinforcement learning. The simulation training was completed under the environment of TOWN06 and TOWN04 in CARLA 0.9.9. Figure 12 shows the specific CARLA simulation scenario.

Figure 12. CARLA simulation diagram.

4.1. Trajectory Planning Experiment Based on PG Algorithm

When training the trajectory planning module, other obstacle vehicles were randomly generated for each round of training to enable the trained agents to target complex traffic conditions. In a random environment, the average reward of each round was used to evaluate the training effect of the agent. When the agent reached the specified number of steps or encounters a collision, it directly started the next round of training. To avoid randomness, the final training results were obtained by averaging the five training results. The training results are shown in Figure 13. The red curve is the average reward, and the red-shaded part is the sliding average of the five training rewards. Due to the strong randomness of the training environment, the rewards show a strong jitter with the change of the round. The rewards show an overall upward trend with the change of rounds, indicating that the agents are increasingly adapting to the changing traffic environment to obtain higher rewards during the training process. After 100 rounds, the variance of rewards tends to decrease, and the training results of agents become more stable.

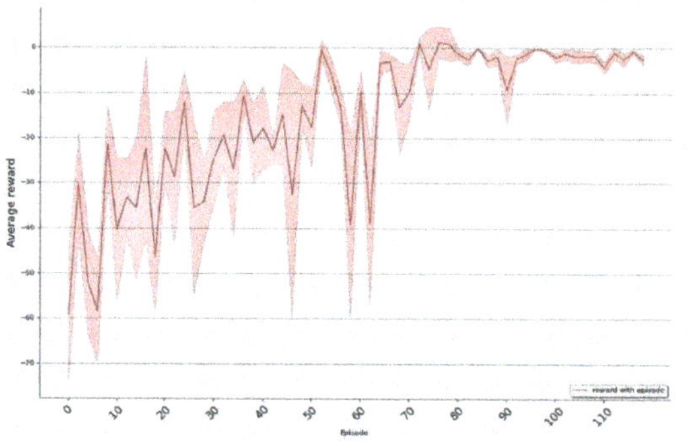

Figure 13. The change in average reward of the track planning module with the number of training wheels.

As shown in Figure 14, the red curve represents the reward curve of the planning method based on the time series bird's-eye view and the policy gradient algorithm proposed in this paper, and the blue curve represents the reward curve of the planning method using the DDPG algorithm. Because of the strong randomness of the training environment, the reward fluctuates greatly with the change of the round. In the comparison of average rewards, both curves are almost the same. However, it is obvious that the DDPG algorithm represented by the blue curve has convergence effect only after 100 rounds, while the planning method proposed in this paper starts to converge gradually after 70 rounds. Therefore, the proposed planning method has higher convergence efficiency and stability.

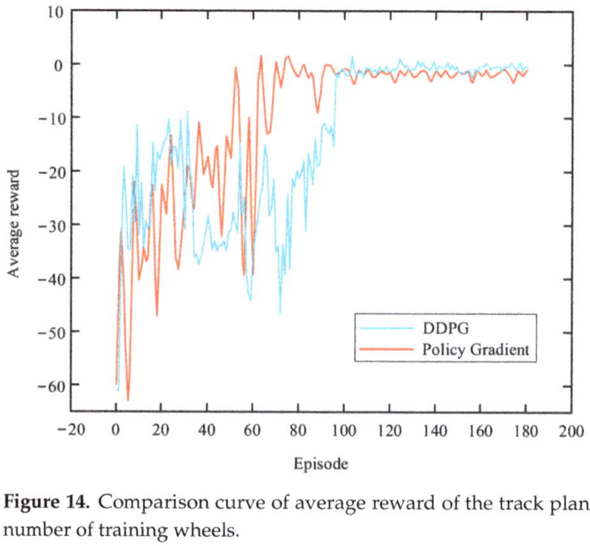

Figure 14. Comparison curve of average reward of the track planning module changing with the number of training wheels.

4.2. Safety Control Module Experiment

In the control module, due to the randomness of the steps that the autonomous vehicle can take during the training process, it is not suitable to use a single reward or a cumulative reward as the evaluation standard of the training effect of the agent at the current moment. Therefore, it is reasonable to take the average reward of each step of the current round as the evaluation standard of the training effect of the current round. The abscissa is the number of training rounds, and the ordinate is the average reward obtained in each round. Figure 15 shows the change in the training curve of the horizontal trajectory tracking task.

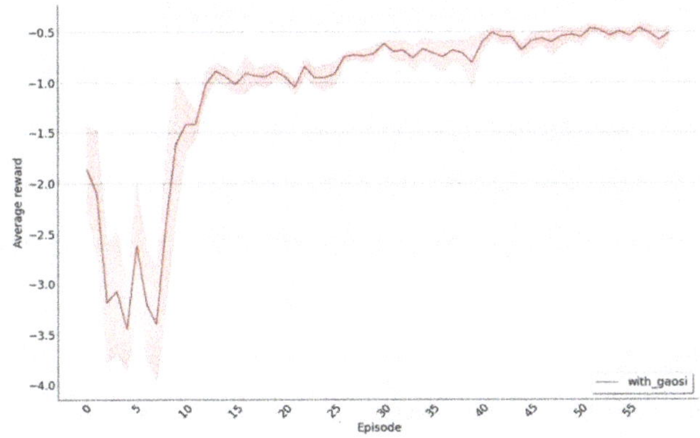

Figure 15. The change in average reward of the horizontal trajectory tracking task with the number of training wheels.

It can be seen from Figure 15 that in the first 15 rounds of the lateral trajectory tracking control task, the agent is still in the free exploration stage, and the reward curve fluctuates and does not converge. With the progress of training, the agent continuously optimizes its strategic network, makes more reasonable behavior, obtains higher rewards and optimizes its network again according to the rewards obtained from feedback, forming a virtuous circle. After 50 rounds, the reward curve begins to converge and achieves good training results.

In this paper, the variable Gaussian safety field is used as the constraint and evaluation index of the control part. Figure 16 shows the reward curve of the variable Gaussian safety field. The red curve represents the reward curve of the lateral tracking control considering the relationship with other vehicle state quantities under the variable Gaussian safety field. The blue curve represents the reward curve of the traditional lateral tracking control under the variable Gaussian safety field. In both cases, the average value of the five experiments is taken. Figure 16 clearly shows that the reward curve of the safety lateral tracking control method proposed in this paper is superior to the traditional lateral tracking control, with higher safety performance and greater response space to emergency conditions. At the beginning of several training rounds, since the agent did not interact with other vehicles in the opening exploration phase, the average reward was 0, as shown in Figure 16. From the sixth round, the agent interacts with other vehicles in the environment, the variable Gaussian safety field acts, and the reward curve changes.

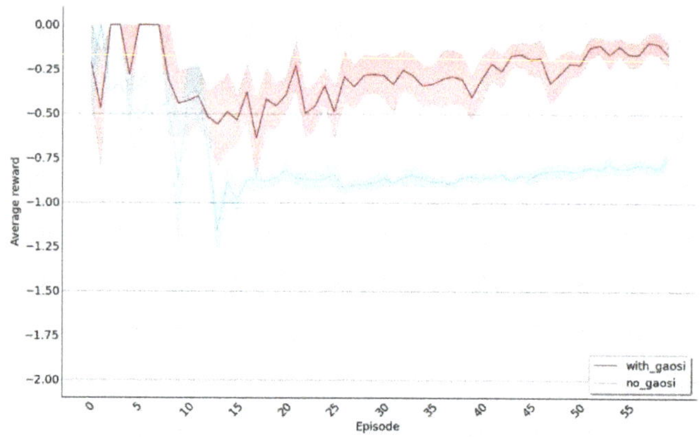

Figure 16. Reward curve of variable Gauss safety field.

Figure 17 shows the average reward of the longitudinal trajectory tracking control task over time. It can be seen that the average reward changes with the training times. The blue and red curves represent the average reward change curves of the agents with Gaussian noise and adaptive noise exploration, respectively, and the shaded part is the standard deviation of five experiments. Figure 17 shows that both types of agents have achieved good training results in the longitudinal trajectory tracking control task. Due to the randomness of the ego vehicle's speed and the state of the vehicle ahead in each training round, the average reward of the lateral trajectory tracking control task fluctuates to some extent. However, similar to the lateral trajectory tracking control task, the training effect of the adaptive noise detection method is better than that of the common noise attenuation method.

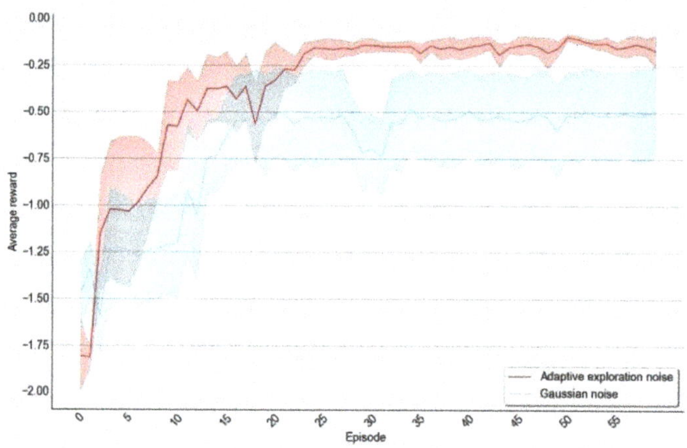

Figure 17. The change in the average reward of the longitudinal trajectory tracking task with the number of training wheels.

5. Conclusions

In this paper, a vehicle safety planning control method based on the variable Gaussian safety field is designed. The policy gradient algorithm is used to improve the driving safety of autonomous vehicles and make the driving trajectory of autonomous vehicles more intelligent. The spatiotemporal bird's-eye view proposed in combination with the policy gradient algorithm as a state variable can enhance the ability of feature extraction of the policy network and make the network convergence easier. The variable Gaussian safety field is added as the reward function of the trajectory planning module and the evaluation index of the control module to improve the safety and rationality of the output trajectory and tracking control, respectively. In the longitudinal control module, Gaussian noise input is improved to avoid repeated invalid exploration of agents and enhance training efficiency. Compared with the traditional planning control algorithm, the proposed method has the following advantages: (1) the spatiotemporal bird's-eye view is used as the input state of the policy network enabling the trajectory planning policy network to effectively extract the features of the surrounding traffic environment. The planning trajectory of autonomous vehicles is generated through reinforcement learning, which improves the trajectory planning ability of autonomous vehicles in complex scenes. The efficiency of the lattice sampling method for trajectory planning algorithm avoids invalid sampling in complex traffic scenes; (2) the variable Gaussian safety field is added as a reward function to improve the safety of trajectory and control effect; (3) the traditional noise input is improved and the multi-head actor network structure is designed to add noise in the output action and improve the training efficiency. The experimental results demonstrate and validate that the proposed framework is superior to the traditional methods.

At the same time, this paper does not consider the scenarios other than an expressway, and how to change lanes in an emergency. In the future, we will test and improve the algorithm in more complex environments, such as ramps and urban roads. From another point of view, the single vehicle will be extended to the fleet, and the driving efficiency and safety of the fleet on the expressway will be considered.

Author Contributions: Conceptualization, Z.Z. and Y.C.; methodology, Z.Z.; software, C.T.; validation, Z.Z., C.T. and Y.C.; formal analysis, Y.L.; investigation, H.W.; resources, L.C.; data curation, Z.Z.; writing—original draft preparation, Z.Z.; writing—review and editing, Z.Z.; visualization, Y.C.; supervision, Y.L.; project administration, H.W.; funding acquisition, Y.C. All authors have read and agreed to the published version of the manuscript.

Funding: This research was funded by [National Natural Science Foundation of China] grant number [U20A20333, 51875255, U20A20331, 52072160] and [Six talent peaks project in Jiangsu Province] grant number [2018-TD-GDZB-022] and [Key R&D projects in Jiangsu Province] grant number [BE2020083-3, BE2019010-2].

Data Availability Statement: Not applicable.

Conflicts of Interest: The authors declare no conflict of interest. And Yubo Lian is employee of BYD Company Limited. The paper reflects the views of the scientists, and not the company.

References

1. Hu, X.; Li, Z. An improved potential field method for robot path planning. *Mech. Sci. Technol. Aerosp. Eng.* **2017**, *36*, 1522–1529.
2. Fu, H.; Nie, W.; Wang, K.; Vagale, A.; Robin, T.B. Simulation and verification of path planning for autonomous vehicles based on sampling. *Auto Electr. Parts* **2021**, *9*, 13–15.
3. Chen, D.; Liu, X.; Liu, S. Improved A* algorithm based on two-way search for path planning of automated guided vehicle. *J. Comput. Appl.* **2021**, *41* (Suppl. S2), 309–313.
4. Peng, Y.; Liang, J. Q-learning path planning based on exploration /exploitation tradeoff optimization. *Comput. Technol. Dev.* **2022**, *32*, 1–7.
5. Qi, X.; Huang, J.; Cao, J. Path planning for unmanned vehicle based on improved A* algorithm. *J. Comput. Appl.* **2020**, *40*, 2021–2027.
6. Claussmann, L.; Revilloud, M.; Gruyer, D.; Glaser, S. A review of motion planning for highway autonomous driving. *IEEE Intell. Transp. Syst.* **2019**, *21*, 1826–1848. [CrossRef]
7. Li, X.; Sun, Z.; Zhu, Q.; Liu, D. A unified approach to local trajectory planning and control for autonomous driving along a reference path. In Proceedings of the 2014 IEEE International Conference on Mechatronics and Automation, Hongkong, China, 3–6 August 2014.

8. Yu, C.; Cherfaoui, V.; Bonnifait, P. Semantic evidential lane grids with prior maps for autonomous navigation. In Proceedings of the 2016 IEEE 19th International Conference on Intelligent Transportation Systems (ITSC), Rio de Janeiro, Brazil, 1–4 November 2016.
9. Werling, M.; Ziegler, J.; Kammel, S.; Thrun, S. Optimal trajectory generation for dynamic street scenarios in a frenet frame. In Proceedings of the 2010 IEEE International Conference on Robotics and Automation, Anchorage, Alaska, 3–8 May 2010.
10. Fehér, Á.; Aradi, S.; Hegedüs, F.; Bécsi, T.; Gáspár, P. Hybrid DDPG approach for vehicle motion planning. In Proceedings of the 16th International Conference on Informatics in Control, Automation and Robotics (ICINCO), Sapporo, Japan, 29–31 July 2019.
11. Kiran, B.R.; Sobh, I.; Talpaert, V.; Mannion, P.; Al Sallab, A.A.; Yogamani, S.; Perez, P. Deep reinforcement learning framework for autonomous driving: A survey. *IEEE Trans. Intell. Transp. Syst.* **2021**, *23*, 4909–4926. [CrossRef]
12. Roy, A.; Hossain, M.; Muromachi, Y. A deep reinforcement learning-based intelligent intervention framework for real-time proactive road safety management. *Accid. Anal. Prev.* **2022**, *165*, 106512. [CrossRef] [PubMed]
13. Aradi, S.; Becsi, T.; Gaspar, P. Policy gradient-based reinforcement learning approach for autonomous highway driving. In Proceedings of the 2018 IEEE Conference on Control Technology and Applications (CCTA), Copenhagen, Denmark, 21–24 August 2018.
14. Nageshrao, S.; Tseng, E.; Filev, D. Autonomous highway driving using deep reinforcement learning. In Proceedings of the 2019 IEEE International Conference on Systems, Man and Cybernetics (SMC), Bari, Italy, 6–9 October 2019.
15. Yu, L.; Shao, X.; Wei, Y.; Zhou, K. Intelligent land-vehicle model transfer trajectory planning method based on deep reinforcement learning. *Sensors* **2018**, *18*, 2905. [CrossRef] [PubMed]
16. Jaritz, M.; De Charette, R.; Toromanoff, M.; Perot, E.; Nashashibi, F. End-to-end race driving with deep reinforcement learning. In Proceedings of the 2018 IEEE International Conference on Robotics and Automation (ICRA), Brisbane, Australia, 21–25 May 2018.
17. Wang, J.; Wu, J.; Li, Y. The driving safety field based on driver-vehicle-road interactions. *IEEE Trans. Intell. Transp. Syst.* **2015**, *16*, 2203–2214. [CrossRef]
18. Wang, J.; Wu, J.; Zheng, X.; Ni, D.; Li, K. Driving safety field theory modeling and its application in pre-collision warning system. *Transp. Res. Part C* **2016**, *72*, 306–324. [CrossRef]
19. Zhe, X.; Hailiang, C.; Ziyu, L.; Enxin, S.; Qi, S.; Shengbo, L. Lateral trajectory following for automated vehicles at handling limits. *J. Mech. Eng.* **2020**, *56*, 138–145. [CrossRef]
20. Chen, H.; Chen, S.; Gong, J. A review on the research of lateral control for intelligent vehicles. *Acta Armamentarii* **2017**, *38*, 1203–1214.

World Electric Vehicle Journal

MDPI

Article

Trajectory Tracking Control of Intelligent X-by-Wire Vehicles

Zixu Wang [1], Yong Li [2], Chuyo Kaku [3] and Hongyu Zheng [1,*]

1 State Key Laboratory of Automotive Simulation and Control, Jilin University, Changchun 130022, China
2 Automotive Engineering Research Institute, Jiangsu University, Zhenjiang 212013, China
3 Jiangsu Chaoli Electric Co., Ltd., Danyang 212321, China
* Correspondence: zhenghy@jlu.edu.cn; Tel.: +86-13804467016

Abstract: Vehicle intelligence is an effective way to improve driving safety and comfort and reduce traffic accidents. The trajectory tracking control of unmanned vehicles is the core module of intelligent vehicles. As a redundant system, the X-by-wire electric vehicle has the advantage that the turning angles and driving torque of the four wheels can be precisely controlled and it has a higher degree of controllability and flexibility. In this paper, a trajectory tracking control algorithm based on a hierarchical control architecture is designed based on x-by-wire vehicles. The hierarchical control algorithm architecture includes the trajectory tracking layer, tire force distribution layer, and actuator control layer. The trajectory tracking layer uses the longitudinal force, lateral force, and yaw moment as the control variables; the model predictive control algorithm controls the vehicle to follow the desired trajectory. The tire force distribution layer is solved by transforming the tire force distribution problem into a quadratic programming problem with constraints. Based on the expected resultant force and resultant moment, the longitudinal force and lateral force of each tire in the vehicle coordinate system are obtained. The actuator control layer converts the coordinate system to obtain the longitudinal force and lateral force in the tire coordinate system, which uses the arctangent function tire model to solve the desired tire slip angle, and then obtains the vehicle steer angle and driving torque. To verify the effectiveness of the trajectory tracking control algorithm of the hierarchical control architecture, the proposed trajectory tracking control algorithm is simulated and verified through the variable speed double line change condition and the low road friction coefficient double line change condition. The simulation results show that the control algorithm proposed in this paper has the accuracy to follow the desired trajectory.Definition:

Keywords: x-by-wire vehicle; trajectory tracking control; model predictive control; hierarchical control

Citation: Wang, Z.; Li, Y.; Kaku, C.; Zheng, H. Trajectory Tracking Control of Intelligent X-by-Wire Vehicles. *World Electr. Veh. J.* **2022**, *13*, 205. https://doi.org/10.3390/wevj13110205

Academic Editor: Joeri Van Mierlo

Received: 21 September 2022
Accepted: 26 October 2022
Published: 1 November 2022

1. Introduction

The rapid development of control-by-wire technology provides a strong technical guarantee for the realization of electronic, intelligent, and electrified vehicles, and makes x-by-wire vehicles a current research hotspot [1,2]. The x-by-wire electric vehicle is a highly redundant system, whose steer angle and driving torque of the four wheels can be precisely controlled; thus, compared with traditional vehicles, it can theoretically achieve better control effects [3]. The x-by-wire electric vehicle has a variety of driving modes that can be switched, with better trajectory tracking performance and a larger turning curvature limit, which can easily realize in situ steering, differential steering, oblique driving, and other working conditions, and can meet the requirements of higher precision trajectory tracking and critical working conditions [4–6]. The high degree of controllability and flexibility and high execution ability of x-by-wire electric vehicles provide a platform for the research into driverless vehicles, which has great development potential.

The problem of trajectory tracking control is one of the three key technologies related to intelligent driving vehicles [7]. As far as the trajectory tracking control of traditional front-wheel steered vehicles is concerned, the relevant theories and research methods are relatively mature, the relevant simulation experiments have been quite abundant, and many control strategies have been verified by real vehicles in specific environments. The control strategies include the following: pure tracking control, traditional proportional-integral-derivative (PID) control, sliding mode control, dynamic feedback control, fuzzy control, and model predictive control (MPC) [8–10]. Al-Mayyahi et al. [11] proposed a PID-based fractional-order (FOPID) controller for the trajectory tracking problem, using two FOPID controllers to calculate the control input torque of the vehicle and using a particle swarm optimization algorithm to control the parameters in the FOPID controller. Kapania et al. [12] added the vehicle's side-slip characteristics to the feedforward controller to improve the dynamic characteristics of the tire within the friction limit while keeping the vehicle within the handling limit, minimizing lateral tracking deviation. Wu [13] proposed a back stepping sliding mode controller to reduce the chattering problem of the sliding mode control. Using the fuzzy control algorithm for the overdrive system, the simulation experiment proves that the back stepping sliding mode control method has higher control precision and a smoother control process than the traditional sliding mode control. Mallem et al. [14] proposed a fast terminal sliding mode dynamic inverse control method based on PID, which makes the position and direction of the movement around the desired trajectory asymptotically stable, and used a terminal sliding mode control method to ensure finite time convergence of the trajectory tracking error to zero. Rokonuzzaman et al. [15] proposed to use the large amount of data provided by the vehicle to design MPC with a neural-network-based vehicle learning dynamic model to improve the tracking performance; the results under various road conditions show that the proposed method outperforms the MPC of the traditional vehicle model. Funke et al. [16] proposed a new control structure based on model predictive control and feedback control that integrates path tracking, vehicle stabilization, and collision avoidance, and coordinated these conflicting goals through the priority of collision avoidance. Experimental data show that the controller drives safely within the operational limits of the vehicle.

The x-by-wire electric vehicle has the advantages of independent four-wheel drive/brake and steering control, so it has been widely studied by universities and enterprises. However, the research on trajectory tracking control mainly focuses on traditional vehicles and there are few studies on trajectory tracking control algorithms for x-by-wire electric vehicles.

Based on proportional-integral (PI) control and adaptive model predictive control, Ahn et al. [17] designed an integrated autonomous driving system independent of each wheel for vehicles equipped with four-wheel independent motors to improve vehicle stability and path tracking performance. Li et al. [7] proposed a sliding mode drive controller based on PID control and sliding mode control for 4WIS/4WID vehicles. Hiraoka et al. [18] proposed an automatic controller for four-wheel steering vehicle path tracking based on sliding mode control theory. Compared with active front-wheel steering, the four-wheel steering controller has a more stable and accurate trajectory tracking capability. Zheng et al. [19] designed a trajectory tracking strategy based on a hierarchical control method. The path tracking layer adopts a nonlinear state feedback controller, and a neural network proportional-integral-derivative controller is designed to track the desired path and obtain the desired yaw rate. Chen et al. [20] designed a new adaptive linear quadratic optimal regulator (LQR) as a coordinated controller for 4WIS/4WID electric vehicle stability control. According to different vehicle speeds and road conditions, the phase plane method is used to calculate the center of mass slip angle and stability margin.

On the other hand, the current research on the trajectory tracking control of x-by-wire electric vehicles rarely considers constraints such as tire adhesion. As each wheel of the x-by-wire vehicle can be independently controlled, a reasonable control algorithm and control strategy can achieve independent longitudinal and lateral control of each wheel, improving the accuracy of trajectory tracking and the driving stability of the vehicle.

This paper takes the x-by-wire electric vehicle as the research carrier and designs the trajectory tracking control strategy based on the hierarchical control architecture, which includes the trajectory tracking layer, the tire force distribution layer, and the actuator control layer. The main contribution of this paper is as follows:

1. Based on the hierarchical architecture, an x-by-wire electric vehicle trajectory tracking control strategy is established in which the controllers of each layer are modularly designed. The effectiveness of the algorithm is verified by simulation experiments.
2. In the tire force distribution layer, the regular octagon constraint is used to linearize the tire force constraint and the tire force distribution problem is transformed into a quadratic programming problem with constraints for solving, which improves the real-time performance of the algorithm.

The rest of this article is structured as follows. Section 2 introduces a nonlinear x-by-wire chassis vehicle model. Section 3 introduces the trajectory tracking controller based on a hierarchical architecture. Section 4 introduces the simulation under the DLC test. Section 5 presents the conclusion of the article.

2. Vehicle Dynamic Model and Tire Model

2.1. Vehicle Dynamic Model

The nonlinear dynamic model of the vehicle established in this paper is as follows:

$$\begin{cases} \dot{v}_y = -v_x r + \frac{1}{m}\sum F_y \\ \dot{v}_x = v_y r + \frac{1}{m}\sum F_x \\ \dot{r} = \frac{1}{I_z}\sum M_z \\ \dot{e}_\varphi = r - \dot{s}\kappa_r \\ \dot{e}_l = v_x \sin e_\varphi + v_y \cos e_\varphi \\ \dot{s} = \frac{v_x \cos e_\varphi - v_y \sin e_\varphi}{1 - l\kappa_r} \end{cases} \tag{1}$$

where m is the mass of the vehicle. I_z is the moment of inertia around the center of mass. s is the distance of the desired path. κ_r is the curvature of the desired path. r, v_x, and v_y are the yaw rate, longitudinal speed, and lateral speed of the vehicle, respectively. e_l and e_φ are the lateral deviation and the heading angle deviation between the center of mass of the vehicle and the reference waypoint, respectively.$\sum F_x$, $\sum F_y$, and $\sum M_z$ are the lateral force, longitudinal force, and yaw moment received by the center of mass of the vehicle, respectively.

The above vehicle model is nonlinear. Considering that the heading deviation e_φ is generally small, it is assumed that $\cos(e_\varphi) \approx 1$, $\sin(e_\varphi) \approx 0$. At the same time, considering that the road curvature and lateral deviation are generally small, the lateral speed of the vehicle is generally much smaller than the longitudinal speed. Based on this, the following can be obtained:

$$\begin{cases} \dot{s} \approx v_x - v_y e_\varphi \approx v_x \\ \dot{e}_l \approx v_x e_\varphi + v_y \\ \dot{v}_x = \frac{1}{m}\sum F_x \end{cases} \tag{2}$$

The state equation of the vehicle lateral model is as follows:

$$\frac{d}{dt}\begin{bmatrix} v_y \\ w \\ e_\varphi \\ e_l \end{bmatrix} = \begin{bmatrix} 0 & -v_x & 0 & 0 \\ 0 & 0 & 0 & 0 \\ 0 & 1 & 0 & 0 \\ 1 & 0 & v_x & 0 \end{bmatrix}\begin{bmatrix} v_y \\ w \\ e_\varphi \\ e_l \end{bmatrix} + \begin{bmatrix} \frac{1}{m} & 0 \\ 0 & \frac{1}{I_z} \\ 0 & 0 \\ 0 & 0 \end{bmatrix}\begin{bmatrix} \sum F_y \\ \sum M_z \end{bmatrix} + \begin{bmatrix} 0 \\ 0 \\ -v_x \\ 0 \end{bmatrix}\kappa_r \tag{3}$$

The state equation of the vehicle longitudinal model is as follows:

$$\frac{d}{dt}\begin{bmatrix} s \\ v_x \end{bmatrix} = \begin{bmatrix} 0 & 1 \\ 0 & 0 \end{bmatrix}\begin{bmatrix} s \\ v_x \end{bmatrix} + \frac{1}{m}\sum F_x \tag{4}$$

2.2. Tire Model

At present, there are three types of tire models: the physical tire model, empirical–semi-empirical tire model, and finite element tire model. In this paper, the Pacejka tire model is selected, which is a semi-empirical tire model and can be expressed as follows [21]:

$$Y(x) = D\sin\{C\arctan[Bx - E(Bx - \arctan(Bx))]\}$$

where $Y(x)$ is the lateral tire force. x is the slip angle of tire. B is the stiffness factor. C, D, E are curve shape factor, peak factor, and curvature factor, respectively.

3. Trajectory Tracking Controller Based on Hierarchical Architecture

The hierarchical control algorithm structure has the advantages of clear algorithm design and being convenient for subsequent algorithm updating. For the x-by-wire chassis vehicle, which is a research object with high integration and high controllable degrees of freedom, the method of controlling each subsystem separately cannot give full play to its own performance advantages. The control strategy of the hierarchical structure can realize the coordinated control of each controller of the vehicle with an x-by-wire chassis. As shown in Figure 1, in the hierarchical trajectory tracking controller, it is divided into the trajectory tracking layer, tire force distribution layer, and actuator control layer. In the trajectory tracking layer, which is based on the MPC, the expected total force and moment are calculated by taking the longitudinal force $\sum F_x$, lateral force $\sum F_y$, and yaw moment $\sum M_z$ in the vehicle coordinate system as the control variables. The tire force distribution layer solves for the longitudinal force $F_{l,i,j}$ and lateral force $F_{c,i,j}$ in the vehicle coordinate system of each wheel. The actuator control layer obtains the tire steer angle δ_{ij} and driving torque T_{ij} based on the inverse tire force model and controls the controlled vehicle to track the trajectory.

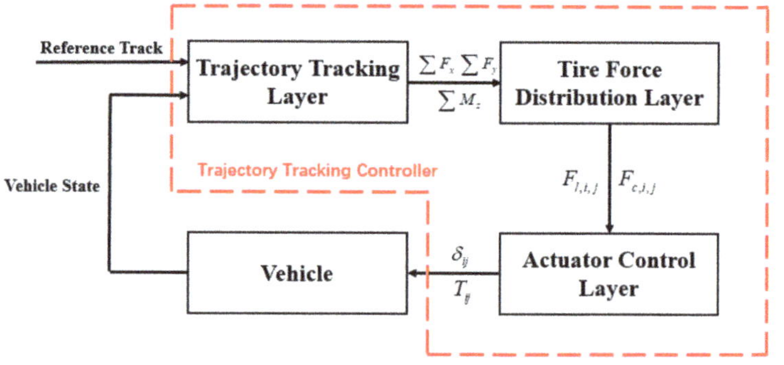

Figure 1. Trajectory tracking controller architecture.

3.1. Trajectory Tracking Layer

3.1.1. Prediction Model

In this paper, the trajectory tracking layer is established based on the MPC algorithm and the state equation established based on Equation (1) is used as the prediction model as follows:

$$\frac{d}{dt}\begin{bmatrix} v_y \\ r \\ e_\varphi \\ e_l \\ s \\ v_x \end{bmatrix} = \begin{bmatrix} 0 & -v_x & 0 & 0 & 0 & 0 \\ 0 & 0 & 0 & 0 & 0 & 0 \\ 0 & 1 & 0 & 0 & 0 & 0 \\ 1 & 0 & v_x & 0 & 0 & 0 \\ 0 & 0 & 0 & 0 & 0 & 1 \\ 0 & 0 & 0 & 0 & 0 & 0 \end{bmatrix} \begin{bmatrix} v_y \\ r \\ e_\varphi \\ e_l \\ s \\ v_x \end{bmatrix} + \begin{bmatrix} \frac{1}{m} & 0 & 0 \\ 0 & \frac{1}{I_z} & 0 \\ 0 & 0 & 0 \\ 0 & 0 & 0 \\ 0 & 0 & 0 \\ 0 & 0 & \frac{1}{m} \end{bmatrix} \begin{bmatrix} \sum F_y \\ \sum M_z \\ \sum F_x \end{bmatrix} + \begin{bmatrix} 0 \\ 0 \\ -v_x \\ 0 \\ 0 \\ 0 \end{bmatrix} \kappa_r \tag{5}$$

Equation (5) is rewritten as follows:

$$\begin{cases} \dot{x}(t) = A_c x(t) + B_{cu} u(t) + B_{cd} d(t) \\ y(t) = C_c x(t) \end{cases} \tag{6}$$

A_c is the state matrix, B_{cu} is the control matrix, B_{cd} is the disturbance matrix, and C_c is the output matrix, respectively:

$$A_c = \begin{bmatrix} 0 & -v_x & 0 & 0 & 0 & 0 \\ 0 & 0 & 0 & 0 & 0 & 0 \\ 0 & 1 & 0 & 0 & 0 & 0 \\ 1 & 0 & v_x & 0 & 0 & 0 \\ 0 & 0 & 0 & 0 & 0 & 1 \\ 0 & 0 & 0 & 0 & 0 & 0 \end{bmatrix}, B_{cu} = \begin{bmatrix} \frac{1}{m} & 0 & 0 \\ 0 & \frac{1}{I_z} & 0 \\ 0 & 0 & 0 \\ 0 & 0 & 0 \\ 0 & 0 & 0 \\ 0 & 0 & \frac{1}{m} \end{bmatrix}, C_c = I_{6\times 6}$$

Discretization of Equation (5):

$$\begin{cases} x(k+1) = A_d x(k) + B_{du} u(k) + B_{dd} d(k) \\ y(t) = C_c x(t) \end{cases} \tag{7}$$

The discrete sampling time is defined as T_s, then the discretized coefficient matrix is as follows:

$$\begin{aligned} A_d &= e^{A_c T_s} \\ B_{du} &= B_{cu} \int_0^{T_s} e^{A_c \tau} d\tau \\ B_{dd} &= B_{cd} \int_0^{T_s} e^{A_c \tau} d\tau \\ C_d &= C_c \end{aligned} \tag{8}$$

In order to reduce the vibration of the control amount, the control amount is rewritten as an incremental type:

$$\Delta u(k) = u(k) - u(k-1) \tag{9}$$

The state equation is as follows:

$$x(k+1) = A_d x(k) + B_{du}(u(k-1) + \Delta u(k)) + B_{dd} d(k) \tag{10}$$

The augmented state is as follows:

$$\xi(k) = \begin{bmatrix} x(k) \\ u(k-1) \end{bmatrix} \tag{11}$$

Then, the equation of state can be rewritten as follows:

$$\begin{cases} \xi(k+1) = \overline{A}_d \xi(k) + \overline{B}_{du} \Delta u(k) + \overline{B}_{dd} d(k) \\ y(k) = \overline{C}_d \xi(k) \end{cases} \tag{12}$$

Among:

$$\overline{A}_d = \begin{bmatrix} A_d & B_{du} \\ O & I \end{bmatrix}, \overline{B}_{du} = \begin{bmatrix} B_{du} \\ I \end{bmatrix}, \overline{B}_{dd} = \begin{bmatrix} B_{dd} \\ O \end{bmatrix}, \overline{C}_d = \begin{bmatrix} C_d & O \end{bmatrix}$$

After applying the control variables, the model prediction process is as follows:

$$\xi(k+1) = \overline{A}_d \xi(k) + \overline{B}_{du} \Delta u(k) + \overline{B}_{dd} d(k) \tag{13}$$

$$\begin{aligned} \xi(k+2) &= \overline{A}_d \xi(k+1) + \overline{B}_{du} \Delta u(k+1) + \overline{B}_{dd} d(k+1) \\ &= \overline{A}_d^2 \xi(k) + \overline{A}_d \overline{B}_{du} \Delta u(k) + \overline{B}_{du} \Delta u(k+1) + \overline{A}_d \overline{B}_{dd} d(k) + \overline{B}_{dd} d(k+1) \end{aligned} \tag{14}$$

$$\begin{aligned}\xi(k+N_c) &= \overline{A}_d\xi(k+N_c-1)+\overline{B}_{du}\Delta u(k+N_c-1)+\overline{B}_{dd}d(k+N_c-1)\\ &= \overline{A}_d^{N_c}\xi(k)+\sum_{i=0}^{N_c-1}\overline{A}_d^{i}\overline{B}_{du}\Delta u(k+N_c-1-i)+\sum_{i=0}^{N_c-1}\overline{A}_d^{i}\overline{B}_{dd}d(k+N_c-1-i)\end{aligned} \tag{15}$$

$$\begin{aligned}\xi(k+N_p) &= \overline{A}_d\xi(k+N_p-1)+\overline{B}_{du}\Delta u(k+N_c-1)+\overline{B}_{dd}d(k+N_p-1)\\ &= \overline{A}_d^{N_p}\xi(k)+\sum_{i=0}^{N_c-1}\overline{A}_d^{i+N_p-N_c}\overline{B}_{du}\Delta u(k+N_c-1-i)+\sum_{i=0}^{N_p-1}\overline{A}_d^{i}\overline{B}_{dd}d(k+N_p-1-i)\end{aligned} \tag{16}$$

Definition:

$$Y(k+1)=\begin{bmatrix}y(k+1)\\ y(k+2)\\ \vdots\\ y(k+N_c)\\ \vdots\\ y(k+N_p)\end{bmatrix}=\begin{bmatrix}\overline{C}_d\xi(k+1)\\ \overline{C}_d\xi(k+2)\\ \vdots\\ \overline{C}_d\xi(k+N_c)\\ \vdots\\ \overline{C}_d\xi(k+N_p)\end{bmatrix}, X(k)=\begin{bmatrix}\xi(k+1)\\ \xi(k+2)\\ \vdots\\ \xi(k+N_c)\\ \vdots\\ \xi(k+N_p)\end{bmatrix}, \Delta D(k)=\begin{bmatrix}d(k)\\ d(k+1)\\ \vdots\\ d(k+N_c)\\ \vdots\\ d(k+N_p)\end{bmatrix}$$

where $Y(k+1)$ is the output vector, $X(k)$ is the state vector, and $\Delta D(k)$ is the disturbance vector.

$$Y(k+1)=S_x\xi(k)+S_u\Delta U(k)+S_dD(k) \tag{17}$$

The state matrix S_x, control matrix S_u, and disturbance matrix S_d are, in respective order, as follows:

$$S_x=\begin{bmatrix}\overline{C}_d\overline{A}_d\\ \overline{C}_d\overline{A}_d^2\\ \vdots\\ \overline{C}_d\overline{A}_d^{N_c}\\ \vdots\\ \overline{C}_d\overline{A}_d^{N_p}\end{bmatrix} S_u=\begin{bmatrix}\overline{C}_d\overline{B}_{du} & O & \cdots & O\\ \overline{C}_d\overline{A}_d\overline{B}_{du} & \overline{C}_d\overline{B}_{du} & \cdots & O\\ \vdots & \vdots & \ddots & \vdots\\ \overline{C}_d\overline{A}_d^{N_c-1}\overline{B}_{du} & \overline{C}_d\overline{A}_d^{N_c-2}\overline{B}_{du} & \cdots & \overline{C}_d\overline{B}_{du}\\ \vdots & \vdots & \ddots & \vdots\\ C_dA_d^{N_p-1}\overline{B}_{du} & C_dA_d^{N_p-2}\overline{B}_{du} & \cdots & \overline{C}_d\overline{A}_d^{N_p-N_c}\overline{B}_{du}\end{bmatrix} S_d=\begin{bmatrix}\overline{C}_d\overline{B}_{dd} & O & \cdots & O\\ \overline{C}_d\overline{A}_d\overline{B}_{dd} & \overline{C}_d\overline{B}_{dd} & \cdots & O\\ \vdots & \vdots & \ddots & \vdots\\ \overline{C}_d\overline{A}_d^{N_c-1}\overline{B}_{dd} & \overline{C}_d\overline{A}_d^{N_c-2}\overline{B}_{dd} & \cdots & O\\ \vdots & \vdots & \ddots & \vdots\\ \overline{C}_d\overline{A}_d^{N_p-1}\overline{B}_{dd} & \overline{C}_d\overline{A}_d^{N_p-2}\overline{B}_{dd} & \cdots & \overline{C}_d\overline{B}_{dd}\end{bmatrix}$$

3.1.2. Cost Function

In order to ensure the safe and accurate driving of the x-by-wire electric vehicle according to the predetermined trajectory, the cost function is as follows:

$$\begin{aligned}J &= \sum_{i=1}^{N_p}\left[\Gamma_{xi}\left(y(k+i)-y_{ref}\right)\right]^2+\sum_{i=0}^{N_c-1}\left[\Gamma_{ui}(\Delta u(k+i))\right]^2\\ &= \left(Y(k+i)-Y_{ref}\right)^T\Gamma_x\left(Y(k+i)-Y_{ref}\right)+\Delta U(k)^T\Gamma_u\Delta U(k)\\ &= \left(S_x\xi(k)+S_u\Delta U(k)+S_dD(k)-Y_{ref}\right)^T\Gamma_x\left(S_x\xi(k)+S_u\Delta U(k)+S_dD(k)-Y_{ref}\right)+\Delta U(k)^T\Gamma_u\Delta U(k)\end{aligned} \tag{18}$$

The error E is defined as follows:

$$E=S_x\xi(k)+S_dD(k)-Y_{ref} \tag{19}$$

Equation (18) be rewritten as follows:

$$\begin{aligned}J &= (S_u\Delta U(k)+E(k))^T\Gamma_x(S_u\Delta U(k)+E(k))+\Delta U(k)^T\Gamma_u\Delta U(k)\\ &= \Delta U(k)^T\left(\Gamma_u+S_u{}^T\Gamma_xS_u\right)\Delta U(k)+2\Delta U(k)^TS_u{}^T\Gamma_xE(k)+E(k)^T\Gamma_xE(k)\end{aligned} \tag{20}$$

The symmetric matrix is as follows:

$$\Gamma_x = diag\left(\Gamma_{x1}, \cdots \Gamma_{xN_p}\right), \Gamma_u = diag(\Gamma_{u1}, \cdots \Gamma_{uN_c})$$

To simplify further,

$$\begin{aligned} H &= \Gamma_u + {S_u}^T \Gamma_x S_u \\ f &= 2{S_u}^T \Gamma_x E(k) + E(k)^T \Gamma_x E(k) \end{aligned} \tag{21}$$

Substituting Equation (21) into Equation (18), the cost function is as follows:

$$J = \Delta U(k)^T H \Delta U(k) + \Delta U(k)^T f + E(k)^T \Gamma_x E(k) \tag{22}$$

3.1.3. Security Constraints

When the vehicle is driving, it is necessary to ensure the stability and safety of the vehicle. Vehicle stability characteristics can be captured by nonlinear tire state models. Bobier [22] uses the center of mass slip angle-yaw rate to handle the vehicle's stability constraint envelope.

When the additional yaw moment generated by the longitudinal force is not considered, in the steady state, the relationship between the yaw rate of the vehicle and the lateral force is as follows:

$$\dot{r} = \frac{\sum F_y}{m v_x} \tag{23}$$

As the road friction coefficient limit is $\sum F_y \leq \mu m g$, the limit of the yaw rate at which the vehicle is stable is as follows:

$$\dot{r} \leq \frac{\mu g}{v_x} \tag{24}$$

The safe driving environment of the vehicle is the safe envelope area considering road boundaries, obstacles, and traffic vehicles. The obstacle avoidance and stability control framework proposed by Erlien et al. [23] uses the safe driving envelope to delineate a collision-free area for the vehicle to exercise and the environmental safety constraints are constrained by road boundaries and traffic vehicles. This paper is based on the road boundary of the tracked trajectory as a constraint, which can be expressed as follows:

$$H_{env}\xi \leq \begin{bmatrix} e_{\max} - \frac{B}{2} - l_{buffer} \\ -e_{\min} + \frac{B}{2} + l_{buffer} \end{bmatrix} \tag{25}$$

$H_{env} = \begin{bmatrix} 0 & 0 & 0 & 1 & 0 & 0 \\ 0 & 0 & 0 & -1 & 0 & 0 \end{bmatrix}$. $e_{\max}$ and $e_{\min}$ represent the upper and lower boundaries of the road, respectively. l_{buffer} is the boundary safety margin.

3.2. Tire Force Distribution Layer

As the four wheels of an x-by-wire electric vehicle can be independently controlled, the control variables obtained based on the trajectory tracking layer are the longitudinal force $\sum F_x$, lateral force $\sum F_y$, and yaw moment $\sum M_z$ at the center of mass. Therefore, the resultant force and torque need to be distributed to each wheel. The tire force distribution layer adopts an optimized method to distribute the force/moment to each wheel under the condition of satisfying the tire force constraint.

3.2.1. Tire Force Constraints

During the driving of the vehicle, the resultant force generated by each tire must satisfy the constraints of the trajectory tracking layered force equation:

$$\begin{cases} F_{x,fl} + F_{x,fr} + F_{x,rl} + F_{x,rr} = \sum F_x \\ F_{y,fl} + F_{y,fr} + F_{y,rl} + F_{y,rr} = \sum F_y \\ \left(F_{y,fl} + F_{y,fr}\right) l_f - \left(F_{y,rl} + F_{y,rr}\right) l_r + \frac{B}{2}\left(-F_{x,fl} + F_{x,fr} - F_{x,rl} + F_{x,rr}\right) = \sum M_z \end{cases} \tag{26}$$

When distributing the tire force, in order to ensure the stability of the vehicle, it is necessary to consider the adhesion capacity of the wheel and tire. Assuming that the road friction coefficient is μ, the maximum resultant force that the tire can generate is $F \leq \mu F_z$, where F_z is the vertical load.

$$F_x^2 + F_y^2 \leq (\mu F_z)^2 \tag{27}$$

As shown in Figure 2, the longitudinal force and lateral force of the tire must satisfy the friction circle constraint. When the wheel is in the extreme condition, the tire force will reach the limit of the friction circle and the wheel will slip, resulting in the loss of vehicle stability. The constraint of tire adhesion is nonlinear and the tire force problem can be defined as an optimization problem with quadratic constraints, but it is often time-consuming. In order to improve the real-time performance of the algorithm, the tire force constraint is simplified on the premise of ensuring the driving stability of the vehicle. Therefore, this paper adopts the circumscribed regular octagon of the friction circle to approximately describe the tire friction constraint and converts the quadratic constraint into a linear constraint.

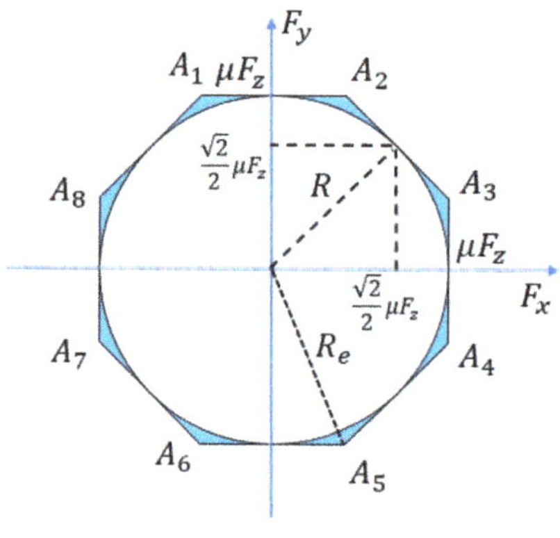

Figure 2. Tire friction constraint.

As shown in Figure 2, R is the radius of friction circle and R_e is the radius of the simplified octagon.

$$R_e = R \cdot \sec 22.5^\circ \approx 1.08R$$

$$\frac{S_{octagon}}{S_{circle}} = \frac{8}{\pi\left(1 + \sqrt{2}\right)} \approx 1.0548$$

According to the geometric calculation, the radius of the regular octagon is 1.08 times the radius of the friction circle and the area of the octagon is about 5.5% larger than that of the friction circle. This linearization treatment has little effect on tire force and ensures the driving stability of an x-by-wire chassis vehicle. The regular octagon is replaced by the friction circle and the octagon linear inequality constraint of the tire friction circle is shown in Equation (28). (F_x, F_y) must be inside the regular octagon $A_1A_2A_3A_4A_5A_6A_7A_8$ and the constraints of the attached ellipse can be expressed as follows:

$$\begin{cases} -\mu F_z \le F_x \le \mu F_z \\ -\mu F_z \le F_y \le \mu F_z \\ -\sqrt{2}\mu F_z \le F_x + F_y \le \sqrt{2}\mu F_z \\ -\sqrt{2}\mu F_z \le F_x - F_y \le \sqrt{2}\mu F_z \end{cases} \tag{28}$$

For the four tires of the vehicle, the matrix form is expressed as follows:

$$F = \begin{bmatrix} F_{x,fl} & F_{x,fr} & F_{x,rl} & F_{x,rr} & F_{y,fl} & F_{y,fr} & F_{y,rl} & F_{y,rr} \end{bmatrix}^T \tag{29}$$

3.2.2. Objective Function

The tire adhesion margin is defined as follows [24]:

$$\varepsilon = 1 - \frac{F}{\mu F_z} \tag{30}$$

The adhesion margin of the tire represents the ratio of the remaining utilization force of the tire to the maximum force provided by the tire. The value range is $0 \sim 1$ and $\varepsilon = 0$ represents the tire reaching the adhesion limit. Thus, the objective function is defined to maximize the attachment margin. The tire usage rate is defined as follows:

$$\eta = 1 - \varepsilon = \frac{F}{\mu F_z} \tag{31}$$

Based on Equation (31), the objective function is defined as the minimum sum of the tire usage rates of the four tires, which is as follows:

$$J = \sum_{i=1}^{4} \frac{F_{xi}^2 + F_{yi}^2}{(\mu F_{zi})^2} \tag{32}$$

In this article, the main object of coordination control is the lateral and longitudinal tire forces, which are related to vertical force. However, the vertical forces are not the control object of the control strategy, so the vertical force is regarded as directly available through sensors or other means.

3.2.3. Tire Force Distribution Algorithm

From Equation (32), the tire force distribution problem is regarded as an optimization problem with constraints:

$$\begin{aligned} &\min_{\{F_{x,ij}, F_{y,ij}\}} J \\ &s.t. \quad Ax \le b \end{aligned} \tag{33}$$

Based on the objective function in Equation (32) and the optimization variable in Equation (29), the optimization problem is a quadratic programming problem, which can be solved quickly.

3.3. Actuator Control Layer

X-by-wire electric vehicles follow a desired trajectory by having independent drive/brake and steering control of each wheel. The actuator control layer converts the longitudinal force and lateral force of each wheel under the vehicle coordinate system obtained by the tire force distribution layer into the tire coordinate system. Figure 3 presents the tire force diagram.

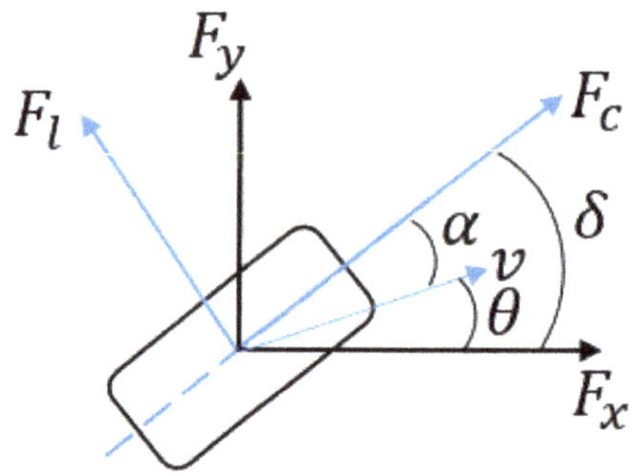

Figure 3. Tire force of a single wheel.

The tire force relationship is as follows:

$$\begin{bmatrix} F_{c,ij} \\ F_{l,ij} \end{bmatrix} = \begin{bmatrix} \cos\delta_{ij} & \sin\delta_{ij} \\ -\sin\delta_{ij} & \cos\delta_{ij} \end{bmatrix} \begin{bmatrix} F_{x,ij} \\ F_{y,ij} \end{bmatrix} \tag{34}$$

δ_{ij} is the wheel steer angle, α_{ij} is the tire slip angle, and θ_{ij} is the angle between the driving direction of the wheel and the longitudinal axis of the wheel coordinate system.

$$\theta_{ij} = \delta_{ij} + \alpha_{ij} \tag{35}$$

The tire side slip angle α_{ij} is generally small. Equation (35) is approximated as follows:

$$\theta_{ij} \approx \delta_{ij}$$

The force in the tire coordinate system can be obtained as follows:

$$\begin{bmatrix} \hat{F}_{c,ij} \\ \hat{F}_{l,ij} \end{bmatrix} = \begin{bmatrix} \cos\theta_{ij} & \sin\theta_{ij} \\ -\sin\theta_{ij} & \cos\theta_{ij} \end{bmatrix} \begin{bmatrix} F_{x,ij} \\ F_{y,ij} \end{bmatrix} \tag{36}$$

The lateral and longitudinal velocity of each wheel are as follows:

$$\begin{cases} v_{y,fl} = v_{y,fr} = v_y + l_f\dot{\varphi} \\ v_{y,rl} = v_{y,rr} = v_y - l_f\dot{\varphi} \\ v_{x,fl} = v_{x,fr} = v_x - \frac{B}{2}\dot{\varphi} \\ v_{x,fr} = v_{x,rr} = v_x + \frac{B}{2}\dot{\varphi} \end{cases} \tag{37}$$

It can be obtained from Equation (37) that the angle between the motion direction of each wheel and the longitudinal direction of the body coordinate system is as follows:

$$\begin{cases} \theta_{fl} = \tan^{-1}\frac{v_{y,fl}}{v_{x,fl}} = \tan^{-1}\frac{v_y + l_f\dot{\varphi}}{v_x - \frac{B}{2}\dot{\varphi}} \\ \theta_{fr} = \tan^{-1}\frac{v_{y,fr}}{v_{x,fr}} = \tan^{-1}\frac{v_y + l_f\dot{\varphi}}{v_x + \frac{B}{2}\dot{\varphi}} \\ \theta_{rl} = \tan^{-1}\frac{v_{y,rl}}{v_{x,rl}} = \tan^{-1}\frac{v_y - l_f\dot{\varphi}}{v_x - \frac{B}{2}\dot{\varphi}} \\ \theta_{rr} = \tan^{-1}\frac{v_{y,rr}}{v_{x,rr}} = \tan^{-1}\frac{v_y + l_f\dot{\varphi}}{v_x + \frac{B}{2}\dot{\varphi}} \end{cases} \tag{38}$$

The actuator control layer solves the relationship between tire slip angle and lateral force based on the arctangent model proposed by Sakai et al. [25], the function of which could fit the magic tire formula. The tire side slip model can be expressed as follows:

$$\begin{cases} F_y = -CG_x\frac{\mu}{k}\tan^{-1}\left(\frac{k}{\mu}\alpha\right) \\ G_x = \sqrt{1-\left(\frac{F_x}{\mu F_z}\right)^2} \\ k = \frac{C\pi}{2F_z} \end{cases} \tag{39}$$

k and G_x are factors and C is the tire side slip stiffness.

The desired $\hat{F}_{l,ij}$ is brought into Equation (39) and the tire slip angle $\hat{\alpha}_{ij}$ can be obtained, then the wheel angle is as follows:

$$\delta_{ij} = \theta_{ij} - \hat{\alpha}_{ij} \tag{40}$$

The longitudinal moment of the wheel is as follows:

$$T_{ij} = \frac{1}{1+\tau s}F_{ij}r_{ij} \tag{41}$$

τ is the time constant and r_{ij} is the radius of rotation of the wheel.

4. Simulation Test

On the one hand, real vehicle experiments have a certain risk owing to a variety of conditions. On the other hand, in order to ensure the vehicle in the experiment has the expected trajectory and speed and to reduce the influence of the experimenter's subjective control on the experimental results, focus is placed upon the influence of the control strategy on the vehicle driving stability in the trajectory tracking control. In this paper, the effectiveness of the control strategy is verified by simulation experiment.

Through MATLAB/Simulink and CarSim simulation, the hierarchical trajectory tracking control algorithm is verified by the double line change (DLC) test. The vehicle parameters are shown in Table 1.

Table 1. Vehicle parameters.

Parameters	Symbol	Unit	Value
Vehicle mass	m	kg	1120
Distance from center of mass to front axle	l_f	m	1.165
Distance from center of mass to rear axle	l_r	m	1.165
Wheelbase	B	m	1.75
Moment of inertia	I_z	$kg \cdot m^2$	1020

4.1. Variable Velocity DLC Condition

On a road with a friction coefficient of 0.85, the trajectory tracking of an x-by-wire electric vehicle under variable speed is simulated. The initial velocity of the vehicle is 20 km/h and the simulation time is set to 20 s.

As shown in Figure 4, the control algorithm can follow the changing vehicle velocity well and accurately track the desired trajectory; the lateral deviation and heading angle deviation of the trajectory tracking are kept within a small range.

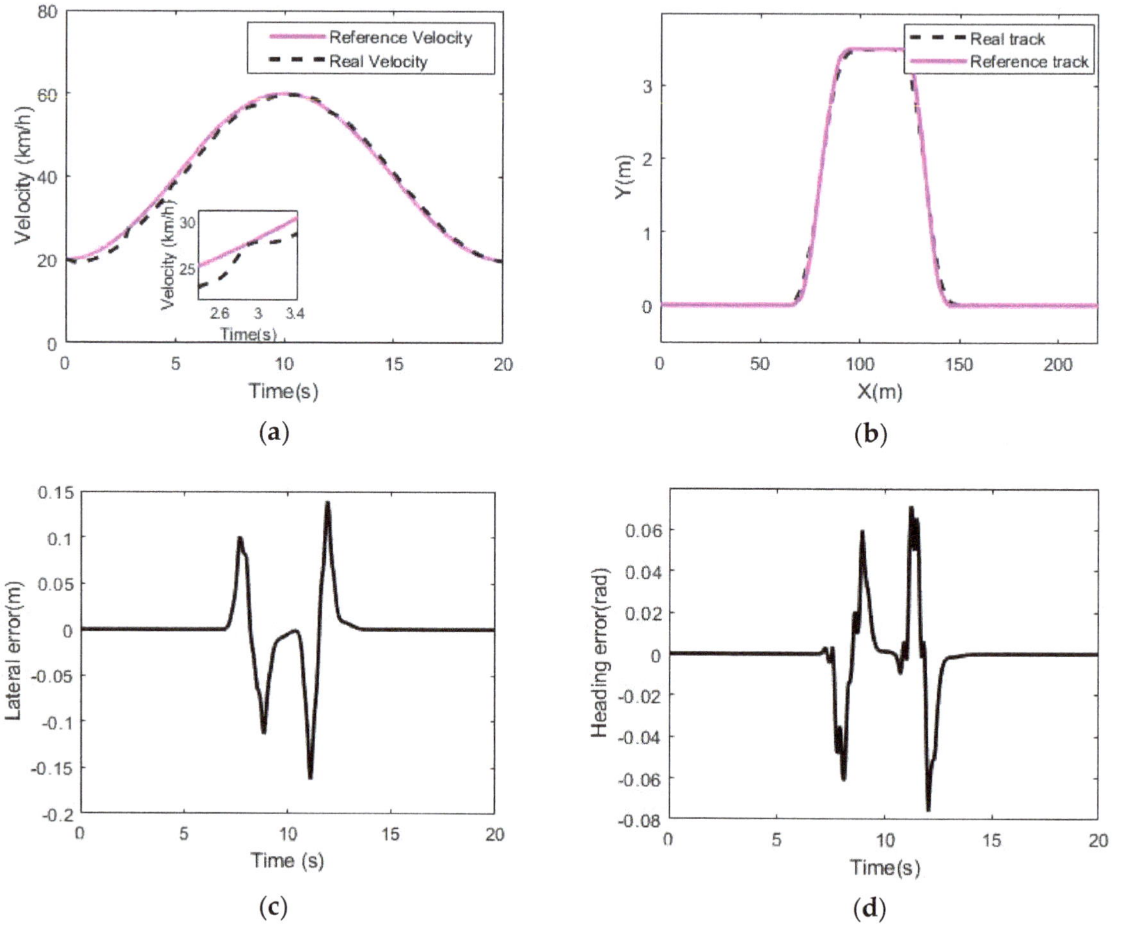

Figure 4. Vehicle state and tracking error: (**a**) velocity; (**b**) vehicle location; (**c**) lateral error; (**d**) heading error.

As shown in Figure 5, the values of the vehicle's center of mass slip angle and yaw angular velocity all change within a small range and the changes are relatively stable; the change range in the vehicle's lateral speed and lateral acceleration is also small, which proves the vehicle has lateral stability.

As shown in Figure 5a, the driving torque has a jitter before 5 s, which is due to the large deviation between the real vehicle velocity and the reference velocity at this time. In order to reduce this deviation, the driving torque is increased. When the deviation is reduced to a reasonable range, the driving torque returns to the normal value. The change in velocity is shown in Figure 4a.

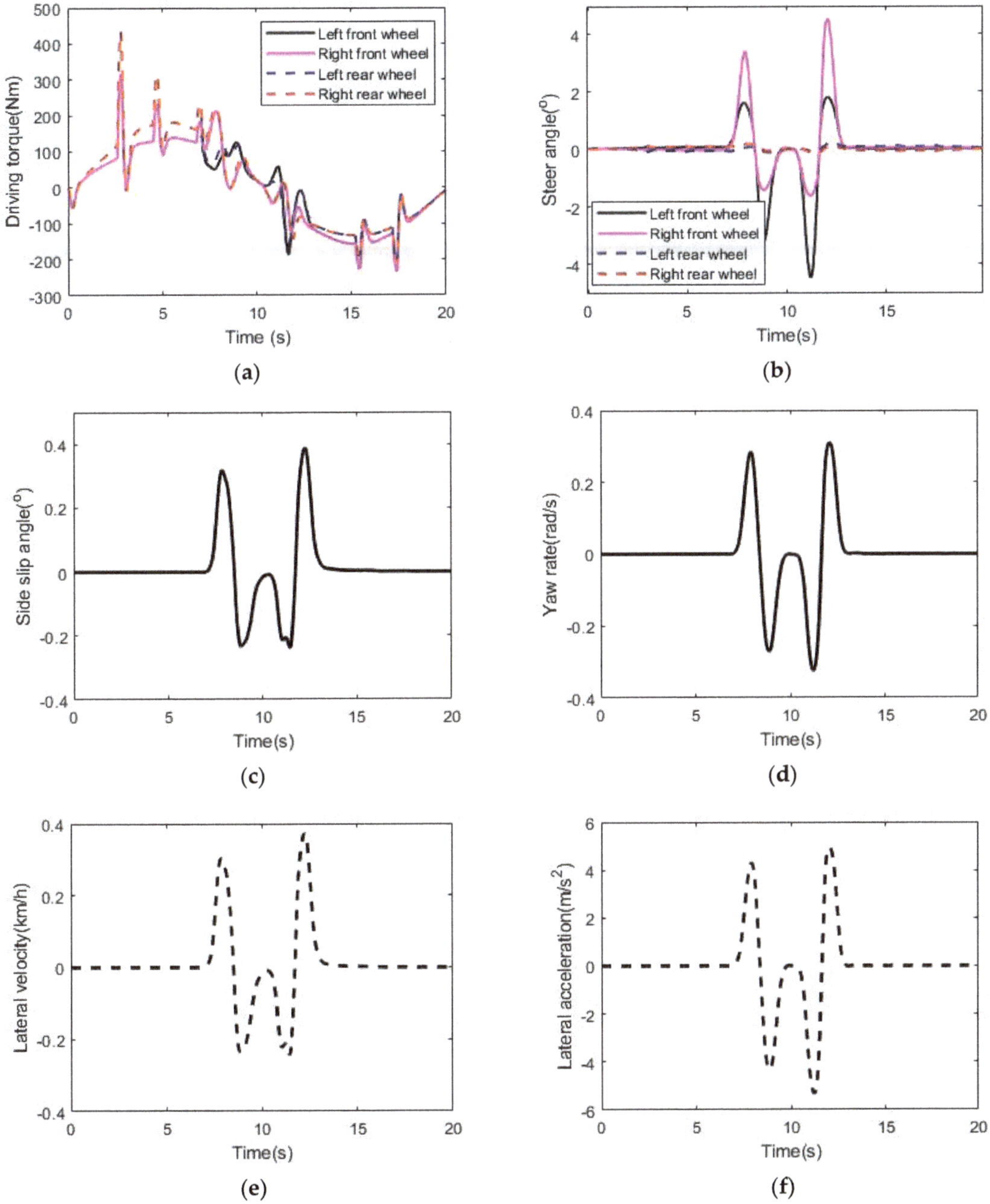

Figure 5. Vehicle state: (**a**) driving torque; (**b**) steer angle; (**c**) side slip angle; (**d**) yaw rate; (**e**) lateral velocity; (**f**) lateral acceleration.

As shown in Figure 6, the trajectory tracking control algorithm can reasonably distribute the load of the four tires while following the desired trajectory to ensure the stability of the vehicle.

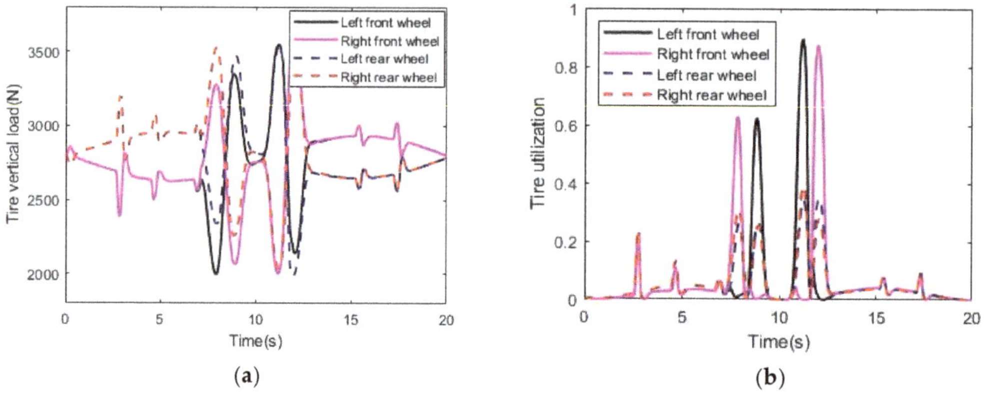

Figure 6. Tire state: (**a**) tire vertical load; (**b**)tire utilization.

4.2. Low Road Friction Coefficient DLC Condition

In order to further verify that the control algorithm has good following ability and stability, a pavement with friction coefficient of 0.35 was selected for the DLC test. The initial speed is 40 km/h.

As shown in Figure 7, the control algorithm can also follow the desired vehicle speed well and accurately track the desired trajectory in the low friction coefficient road environment.

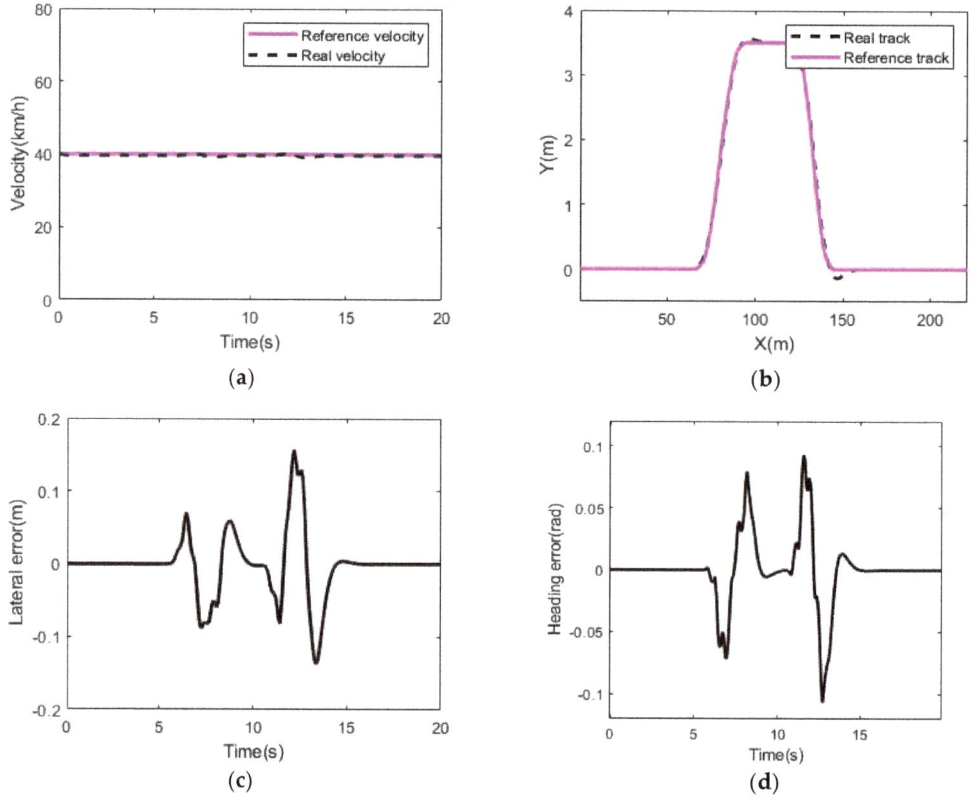

Figure 7. Vehicle state and tracking error: (**a**) velocity; (**b**) vehicle location; (**c**) lateral error; (**d**) heading error.

As shown in Figure 8, the lateral deviation and heading angle deviation of trajectory tracking are kept within a small range. The change in the control amount is relatively stable and there is no major fluctuation. The side slip angle and yaw rate also kept fluctuating within a relatively stable range.

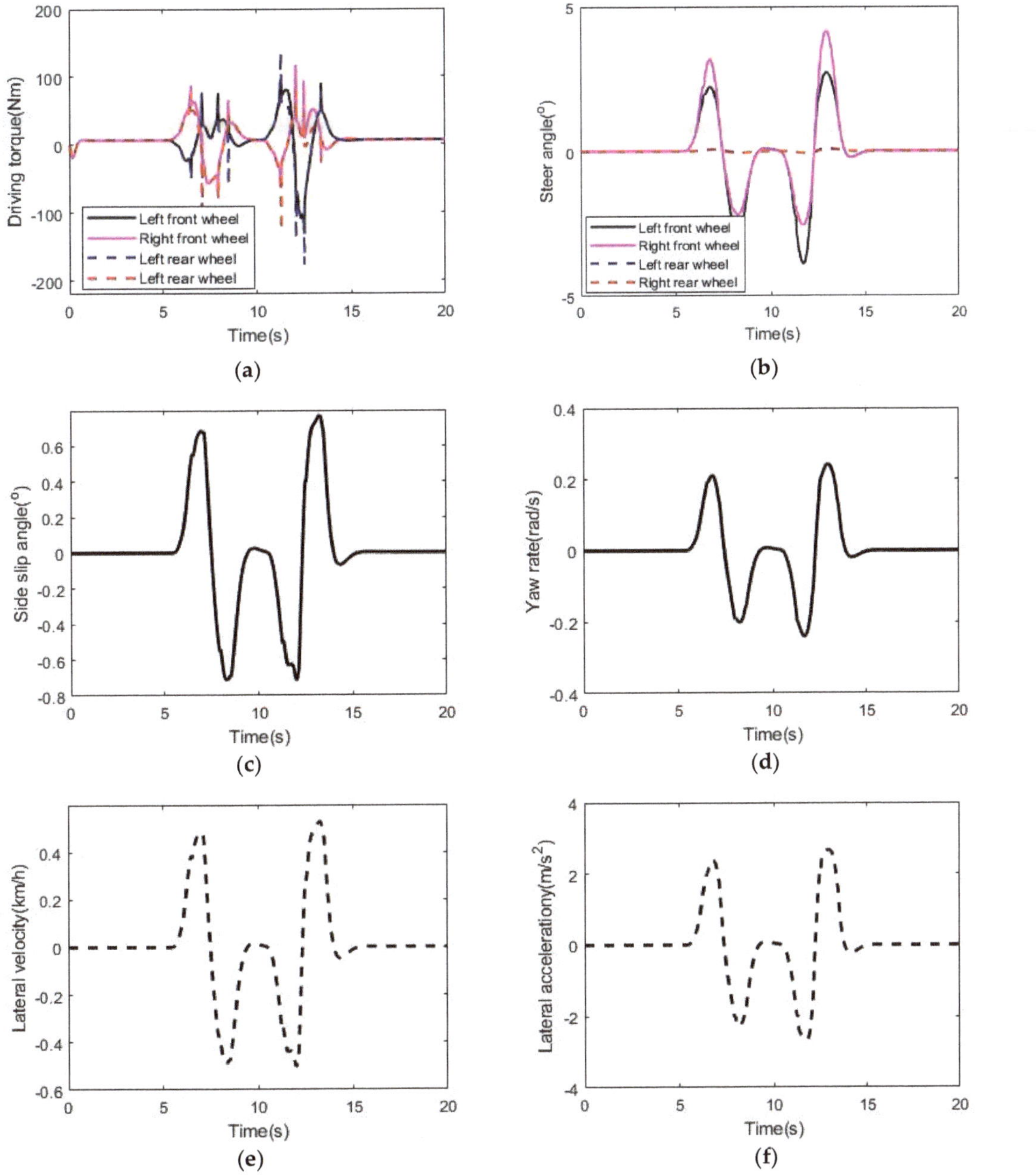

Figure 8. Vehicle state: (**a**) driving torque; (**b**) steer angle; (**c**) side slip angle; (**d**) yaw rate; (**e**) lateral velocity; (**f**) lateral acceleration.

As shown in Figure 9, it can be seen from the tire utilization coefficient of each wheel that the vehicle runs stably and safely on a low-adhesion road surface. It is proved that the algorithm proposed in this paper is suitable for various working conditions. Based on the curve of tire force and tire utilization coefficient in the simulation experiment, it conforms to the change in tire performance of each wheel in the real driving process of the vehicle. On the other hand, the vehicle state parameters also conform to the actual driving performance of the vehicle. Therefore, the results of the simulation experiment are considered to be reliable.

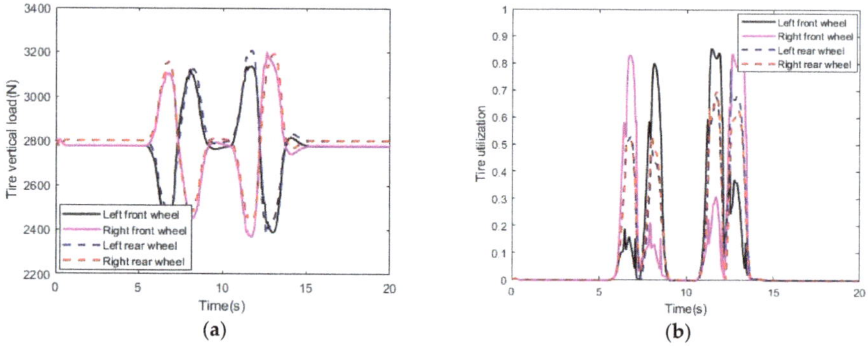

Figure 9. Tire state: (**a**) tire vertical load; (**b**) tire utilization.

5. Conclusions

In this paper, the trajectory tracking control module of an x-by-wire electric vehicle is designed. A trajectory tracking control algorithm based on a hierarchical control architecture is designed to perform effective trajectory tracking control for an x-by-wire electric vehicle, which facilitates the subsequent further development of each layer. The control algorithm includes the trajectory tracking layer, tire force distribution layer, and actuator control layer. The trajectory tracking layer uses longitudinal force, lateral force, and yaw moment as control variables and designs an MPC algorithm to control the vehicle to follow the desired trajectory. The tire force distribution layer distributes the desired resultant force/moment to each vehicle tire. In this paper, the tire force distribution problem is transformed into a quadratic programming problem with constraints to solve such that the longitudinal force and lateral force of each tire in the vehicle coordinate system are obtained. The actuator control layer can obtain the longitudinal force and lateral force in the tire coordinate system through coordinate system transformation in order to obtain the vehicle turning angle and vehicle driving/braking torque control amount. Through the arctangent function tire model, the expected tire slip angle is obtained, and then the vehicle steer angle and driving torque are obtained. In order to verify the effectiveness of the algorithm, the effectiveness of the proposed control algorithm is verified based on simulation experiments.

Author Contributions: Conceptualization, Z.W. and H.Z.; methodology, Z.W.; software, H.Z.; validation, Y.L., C.K., and H.Z.; formal analysis, Z.W.; investigation, H.Z.; data curation, C.K.; writing—original draft preparation, Z.W., Y.L., and H.Z.; writing—review and editing, Z.W. and H.Z.; visualization, Z.W. and H.Z.; supervision, H.Z. All authors have read and agreed to the published version of the manuscript.

Funding: This research was funded by the National Key Research and Development Program of China under Grant, grant number 2021YFB2500703.

Data Availability Statement: Not applicable.

Conflicts of Interest: Chuyo Kaku is an employee of Jiangsu Chaoli Electric Co., Ltd., Danyang 212321, China. The paper reflects the views of the scientists and not the company. Dr Kaku is a part-time professor in our research group; however, his formal affiliation is a company.

References

1. Ni, J.; Hu, J.; Xiang, C. A review for design and dynamics control of unmanned ground vehicle. *Proc. Inst. Mech. Eng. Part D J. Automob. Eng.* **2021**, *235*, 1084–1100. [CrossRef]
2. Zhang, L.; Zhang, Z.; Wang, Z.; Deng, J.; Dorrel, D. Chassis Coordinated Control for Full X-by-Wire Vehicles—A Review. *Chin. J. Mech. Eng.* **2021**, *34*, 42. [CrossRef]
3. Song, P.; Zong, C.; Zheng, H.; He, L. Rapid prototyping for the vehicle control unit of a full drive-by-wire electric vehicle. *Adv. Mater. Res.* **2013**, *694–697*, 1573–1581. [CrossRef]
4. Zhang, D.; Zong, C.; Chen, G.; Song, P.; Zhang, Z. *Study on Dynamic Characteristics and Control Methods for Drive-by-Wire Electric Vehicle*; SAE Technical Paper 2014-01-2291; SAE International: Warrendale, PA, USA, 2014.
5. Ding, W.; Wang, K.; Li, Y.; Liu, Z.; Wang, H. Multidirectional-motion-coupling-based Extreme Motion Control of Distributed Drive Autonomous Vehicle. *Sci. Rep.* 2022; *under review*.
6. Chen, T.; Xu, X.; Li, Y.; Wang, W.; Chen, L. Speed-dependent coordinated control of differential and assisted steering for in-wheel motor driven electric vehicles. *Proc. Inst. Mech. Eng. Part D J. Automob. Eng.* **2018**, *232*, 1206–1220. [CrossRef]
7. Li, B.; Du, H.; Li, W. Trajectory control for autonomous electric vehicles with in-wheel motors based on a dynamics model approach. *IET Intell. Transp. Syst.* **2016**, *10*, 318–330. [CrossRef]
8. Hu, C.; Jing, H.; Wang, R.; Yan, F.; Chadli, M. Robust H∞ output-feedback control for path following of autonomous ground vehicles. *Mech. Syst. Signal Process.* **2016**, *70–71*, 414–427. [CrossRef]
9. Borase, R.; Maghade, D.; Sondkar, S.; Pawar, S. A review of PID control, tuning methods and applications. *Int. J. Dyn. Control* **2021**, *9*, 818–827. [CrossRef]
10. Wang, Y.; Ding, H.; Yuan, J.; Chen, H. Output-feedback triple-step coordinated control for path following of autonomous ground vehicles. *Mech. Syst. Signal Process.* **2019**, *116*, 146–159. [CrossRef]
11. Al-Mayyahi, A.; Wang, W.; Birch, P. Path tracking of autonomous ground vehicle based on fractional order PID controller optimized by PSO. In Proceedings of the 2015 IEEE 13th International Symposium on Applied Machine Intelligence and Informatics (SAMI), Herl'any, Slovakia, 22–24 January 2015.
12. Kapania, N.; Gerdes, J. Design of a feedback-feedforward steering controller for accurate path tracking and stability at the limits of handling. *Veh. Syst. Dyn.* **2015**, *53*, 1687–1704. [CrossRef]
13. Wu, X.; Jin, P.; Zou, T.; Qi, Z.; Xiao, H.; Lou, P. Backstepping trajectory tracking based on fuzzy sliding mode control for differential mobile robots. *J. Intell. Robot. Syst.* **2019**, *96*, 109–121. [CrossRef]
14. Mallem, A.; Nourredine, S.; Benaziza, W. Mobile robot trajectory tracking using PID fast terminal sliding mode inverse dynamic Control. In Proceedings of the 2016 4th International Conference on Control Engineering & Information Technology (CEIT), Hammamet, Tunisia, 16–18 December 2016.
15. Rokonuzzaman, M.; Mohajer, N.; Nahavandi, S.; Mohamed, S. Model predictive control with learned vehicle dynamics for autonomous vehicle path tracking. *IEEE Access* **2021**, *9*, 128233–128249. [CrossRef]
16. Funke, J.; Brown, M.; Erlien, S.; Gerdes, J. Collision avoidance and stabilization for autonomous vehicles in emergency scenarios. *IEEE Trans. Control Syst. Technol.* **2016**, *25*, 1204–1216. [CrossRef]
17. Ahn, T.; Lee, Y.; Park, K. Design of integrated autonomous driving control system that incorporates chassis controllers for improving path tracking performance and vehicle stability. *Electronics* **2021**, *10*, 144. [CrossRef]
18. Hiraoka, T.; Nishihara, O.; Kumamoto, H. Automatic path-tracking controller of a four-wheel steering vehicle. *Veh. Syst. Dyn.* **2009**, *47*, 1205–1227. [CrossRef]
19. Zheng, H.; Yang, S. A trajectory tracking control strategy of 4WIS/4WID electric vehicle with adaptation of driving conditions. *Appl. Sci.* **2019**, *9*, 168. [CrossRef]
20. Chen, X.; Han, Y.; Hang, P. Researches on 4WIS-4WID Stability with LQR Coordinated 4WS and DYC. In Proceedings of the IAVSD International Symposium on Dynamics of Vehicles on Roads and Tracks, Gothenburg, Sweden, 12–16 August 2019.
21. Pacejka, H.; Bakker, E. The magic formula tyre model. *Veh. Syst. Dyn.* **1992**, *21*, 1–18. [CrossRef]
22. Bobier, C. *A Phase Portrait Approach to Vehicle Stabilization and Envelope Control*; Stanford University: Stanford, CA, USA, 2012.
23. Erlien, S.; Fujita, S.; Gerdes, J. Safe driving envelopes for shared control of ground vehicles. In Proceedings of the 7th IFAC Symposium on Advances in Automotive Control, Tokyo, Japan, 4–7 September 2013.
24. Ono, E.; Hattori, Y.; Muragishi, Y.; Koibuchi, K. Vehicle dynamics integrated control for four-wheel-distributed steering and four-wheel-distributed traction/braking systems. *Veh. Syst. Dyn.* **2006**, *44*, 139–151. [CrossRef]
25. Sakai, J.; Sado, H.; Hori, Y. Dynamic driving/braking force distribution in electric vehicles with independently driven four wheels. *Electr. Eng. Jpn.* **2002**, *138*, 79–89. [CrossRef]

World Electric Vehicle Journal

MDPI

Article

MPC-Based Obstacle Avoidance Path Tracking Control for Distributed Drive Electric Vehicles

Hongchao Wu, Huanhuan Zhang * and Yixuan Feng

School of Mechanical and Automotive Engineering, Shanghai University of Engineering Science, Shanghai 201620, China

* Correspondence: zhanghh@sues.edu.cn

Abstract: A path tracking controller based on front wheel steering angle and additional yaw moment control is designed to achieve safe obstacle avoidance of distributed drive electric vehicles. Using sixth-degree polynomial at a given time with anti-collision and anti-rollover conditions, the path planning of obstacle avoidance is proposed. The front wheel steering angle and additional yaw moment are output by the Model Predictive Control (MPC) controller. The wheel torque is distributed by the torque distribution controller. Through additional yaw moment and the vertical force ratio of the wheel, the obstacle avoidance path tracking control is realized. The co-simulation platform is established with Carsim/Simulink. The obstacle avoidance path, model predictive controller and torque distribution controller designed in this paper are simulated. The results show that obstacle avoidance path and tracking controller for the distributed drive electric vehicles effectively meet the requirements of safe obstacle avoidance.

Keywords: distributed driving electric vehicles; polynomial path planning; model predictive control; torque allocation; obstacle avoidance path tracking

Citation: Wu, H.; Zhang, H.; Feng, Y. MPC-Based Obstacle Avoidance Path Tracking Control for Distributed Drive Electric Vehicles. *World Electr. Veh. J.* **2022**, *13*, 221. https://doi.org/10.3390/wevj13120221

Academic Editor: Yong Li

Received: 6 October 2022
Accepted: 16 November 2022
Published: 22 November 2022

1. Introduction

With the continuous increase in car ownership, the modern intelligent transportation system has increasingly regarded vehicle safety as its key factor. Between 2002 and 2012, the lack of proper obstacle avoidance contributed to the deaths of millions of people in traffic accidents around the world, and the economic cost of these accidents amounted to $0.48 trillion [1]. More than 90% of these accidents are caused by human factors [2]. Therefore, in order to avoid vehicle collisions and minimize the impact of accidents, a method of adding an increasing proportion of active safety systems is proposed [3].

The obstacle avoidance trajectory tracking control of intelligent vehicles is one of the key functions. According to the trajectory planned by the upper controller and the real-time state information, real-time vehicle control variables, such as front wheel angle and driving force/braking force, are generated [4]. Compared with traditional vehicles, the drive motors are directly installed in the drive wheels for the distributed drive electric vehicles. So they have outstanding advantages, such as short drive transmission chain, high transmission efficiency and compact structure. It can be more beneficial to realize the tracking control of the obstacle avoidance path by optimizing the distribution of the driving torque of each wheel [5]. Although distributed hub motors can lead to greater unsprung mass, a small amount of unsprung mass increase is negligible, given their advantages in dynamic control.

After the continuous development of path tracking control, different tracking control algorithms have been proposed and verified, such as sliding mode control, fuzzy control, and model predictive control. Reference [6] proposed a distributed drive unmanned vehicle path tracking and stability coordinated control strategy based on the layered control theory. In order to reduce the heading deviation and lateral deviation during the path-tracking process, the sliding mode control method was used. Moreover, it can ensure the stability of the vehicle. On the basis of the traditional dynamic model and preview model, Zhang Bingli [7] designed a new trajectory-tracking controller based on neural networks and fuzzy control theory. Wang Yao [8] proposed a model prediction-fuzzy (MPC-Fuzzy) joint control strategy, which alleviated the conflict between the vehicle tracking accuracy and the controller's computational pressure under a single model prediction control and better solved the distributed driving vehicle path, tracking accuracy and stability issues in tracking. Hu [9] designed a fuzzy linear quadratic regulator (LQR) with preview PID angle compensation and verified that it can maintain good accuracy and stability at different vehicle speeds. However, it also had the problem of low tracking accuracy when the vehicle was running at a low speed. Kou [10] used a state expansion MPC and corner compensation fuzzy controller, designed a dual feedback MPC controller based on state expansion and verified that it had better path tracking performance at medium and low vehicle speeds. Wu [11] proposed a front-wheel steering controller and optimized the controller parameters so that the vehicle had better path-tracking performance at low and medium speeds and ensured that the vehicle had certain steering accuracy and driving stability when driving at high speed on ice and snow roads. In order to improve the stability of distributed drive electric vehicles in extreme working conditions, Li [12] proposed a nonlinear model predictive control (NMPC) based torque coordination control strategy. In order to coordinate the trajectory tracking accuracy and stability of distributed drive electric vehicles and improve the adaptive ability of the control algorithm to different working conditions, Zhuang [13] proposed a trajectory tracking control strategy based on Takagi–Sugeno fuzzy model predictive control (T-S FMPC).

The total required torque is calculated through the yaw moment obtained by the upper controller. The torque distributed to each drive wheel is obtained through different algorithms. There are many distribution methods for torque distribution. A fixed proportional distribution method was proposed in the literature [14,15]. Although this method was simple, it cannot implement different proportional distributions according to different road conditions. There was no optimization in the energy utilization efficiency of the motor, and the best distribution result could not be achieved. Ono E [16] proposed to realize the torque distribution to the four wheels according to the usage of the tire force of four tires. This method made the wear degree of the four tires more balanced, and the life of the tires and the motor were improved, but the performance of vehicle handling was greatly reduced. Mokhiamar O [17] also proposed to analyze the torque distribution problem by establishing a multi-objective optimization function, but this method may show different effects in different models and different operating conditions, so the consistency and robustness were poor. Reference [18] considered the dynamic characteristics of the vehicle when it is unstable and proposed a control allocation scheme based on feedback linearization. This scheme was relatively dependent on the accuracy of the model and had poor stability. Reference [19] distributed torque according to the vertical load distribution ratio of each wheel. Since the friction ellipse of the tire increases with the increase of the vertical load, the torque distribution according to the vertical load ratio of a single tire can avoid the saturation of the tire force and can simply and effectively improve the driving stability of the vehicle.

Based on the above analysis, the whole vehicle control process designed in this paper is shown in Figure 1. In Section 1, using the obstacle information obtained by the environment awareness layer and the vehicle pose information detected by the vehicle sensor, the obstacle avoidance path is planned. Anti-collision and anti-rollover constraints adding to a six-degree polynomial are used for path planning. The front wheel angle δ_f and additional yaw moment ΔM_z are output by the MPC controller in Section 3.2. By processing the information of the path planning layer, the torque distribution controller distributes the torque through the additional yaw moment and the wheel vertical force ratio in Section 3.3. The MPC controller and the torque distribution controller together form a path-tracking controller. During the obstacle avoidance process, the vehicle avoids obstacles through the front wheel angle and four-wheel torque output by the controller. In Section 4, the simulation platform is established, and the effect of the controllers is verified.

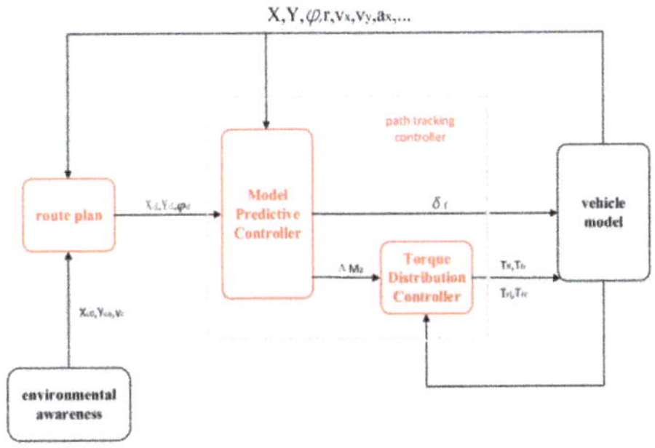

Figure 1. Vehicle control flow chart. Where, X_{c0} and Y_{c0} are the lateral distance and longitudinal distance from the obstacle; v_c is the vehicle speed of the obstacle; X_d and Y_d are the expected lateral and longitudinal distance of the vehicle; φ_d is the expected yaw angle of the vehicle; ΔM_z is the additional yaw moment; T_{fl}, T_{rl}, T_{rr} and T_{fr} are the torques of the left front wheel, the left rear wheel, the right rear wheel and the right front wheel, respectively; X and Y are the transverse and vertical coordinates under the inertial coordinate system; φ is the yaw angle under the body coordinate system; r is the turning radius of the vehicle; v_x and v_y are the lateral and speed longitudinal of the vehicle; a_x is the lateral acceleration of the vehicle.

2. Path Planning

In this paper, the six-degree polynomial path planning method [20] for a given time is used. In the sixth-degree polynomial path planning method, the vehicle's gentle arrival from the starting point to the target point is considered, the collision between the vehicle and the obstacle in the process of running is considered, and the smooth running according to the planned path is also considered. With the addition of anti-collision and anti-rollover conditions, the safety of obstacle avoidance path planning is improved.

Based on the longitudinal and lateral displacements at a given time, the initial expression for the sixth-degree polynomial is:

$$\begin{cases} X(t) = c_0 + c_1 t + c_2 t^2 + c_3 t^3 + c_4 t^4 + c_5 t^5 + c_6 t^6 \\ Y(t) = b_0 + b_1 t + b_2 t^2 + b_3 t^3 + b_4 t^4 + b_5 t^5 + b_6 t^6 \end{cases} \quad (1)$$

Adding anti-collision conditions [20]: set the coordinate center of the obstacle as (X_{ob}, Y_{ob}), and the radius of the circumscribed circle of the vehicle and the obstacle as R_{car} and R_{ob}, this is shown in Figure 2,the constraint condition of anti-collision can be expressed as:

$$(R_{car} + R_{ob})^2 < [X(t) - X_{ob}]^2 + [Y(t) - Y_{ob}]^2 \quad (2)$$

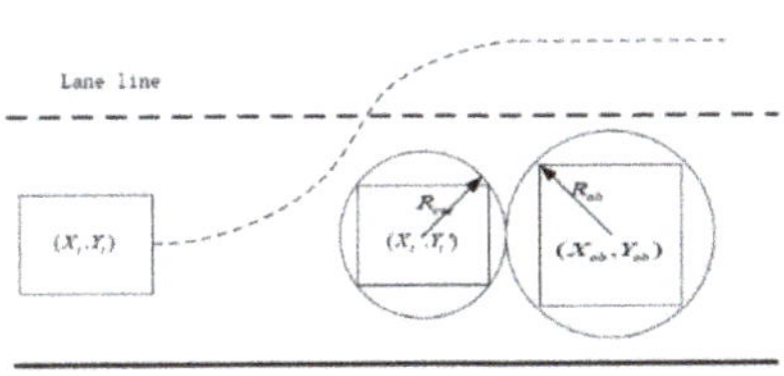

Figure 2. Schematic diagram of collision avoidance.

Adding the anti-rollover condition constraint [20].

According to the simplified model diagram of vehicle rollover in Figure 3, the moment balance equation can be expressed as:

$$m_s a_y h + m_s g \Delta y = (F_{zl} - F_{zr})\frac{s}{2} \tag{3}$$

where, a_y is the lateral acceleration, m_s is the sprung mass, h is the distance from the center of mass to the ground, F_{zl} represents the sum of forces on the left front and rear wheels in the vertical direction, F_{zr} represents the sum of forces on the right front and rear wheels, s represents the track width, g is the gravitational acceleration, $\Delta y = h \cdot \sin\phi$.

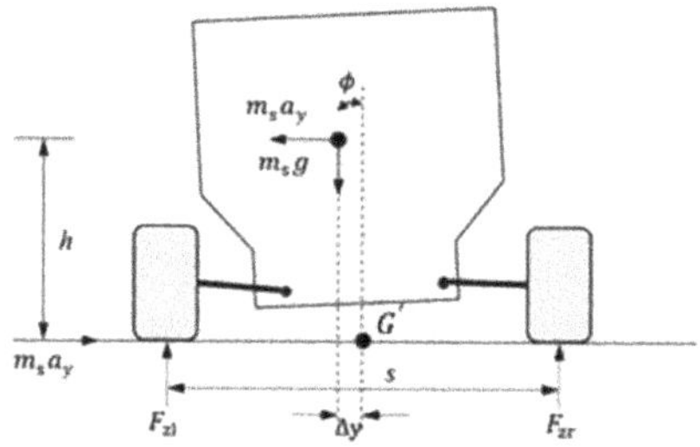

Figure 3. Simplified model of automobile rollover.

The lateral transfer rate of the vehicle is:

$$\text{LTR} = \frac{F_{\text{zl}} - F_{\text{zr}}}{F_{\text{zl}} + F_{\text{zr}}} \tag{4}$$

Combining Equations (3) and (4), LTR can be expressed as:

$$LTR = \frac{2R}{s}\left[\frac{a_y}{g} + \sin\phi\right] \tag{5}$$

where, ϕ is the vehicle roll angle.

The larger the absolute value of LTR indicates, the greater the rollover risk of the car, so the size of a_y should be fully taken into account in path planning. Given the maximum roll risk LTR_{max},

$$a_y = \ddot{Y}(t) \le \left(\frac{s \cdot \text{LTR}_{\text{max}}}{2h} - \sin\phi\right)g \tag{6}$$

$$a_{y\text{max}} = \left(\frac{s \cdot \text{LTR}_{\text{max}}}{2h} - \sin\phi\right)g \tag{7}$$

Constructing the evaluation index function of anti-rollover:

$$J_{anti-r} = \int_{t_0}^{t_d} -\left(a_{y\text{max}} - |a_y|\right)dt \tag{8}$$

The polynomial optimization path planning problem is expressed as:

$$\begin{cases} \min J_{anti-r} \\ \text{s.t.}(R_{car} + R_{ob})^2 < [X(t) - X_{ob}]^2 + [Y(t) - Y_{ob}]^2 \\ a_y = \ddot{Y}(t) \le \left(\frac{s \cdot \text{LTR}_{\text{max}}}{2h} - \sin\phi\right)g \\ b_3{t_d}^3 + b_4{t_d}^4 + b_5{t_d}^5 + b_6{t_d}^6 = \boldsymbol{H} \\ 3b_3{t_d}^2 + 4b_4{t_d}^3 + 5b_5{t_d}^4 + 6b_6{t_d}^5 = 0 \\ 6b_3t_d + 12b_4{t_d}^2 + 20b_5{t_d}^3 + 30b_6{t_d}^4 = 0 \end{cases} \tag{9}$$

where, t_0 and t_d are the initial time and the end time, respectively, a_y is the lateral acceleration, X_{ob} and Y_{ob} are the coordinate centers of the obstacles, the coordinate center of the obstacle:(X_{ob}, Y_{ob}) = (100, 1), R_{car} and R_{ob} are the radiuses of the circumscribed circles between the vehicle and the obstacle, ∅ is the body roll angle, and w is the wheelbase, h is the distance from the center of mass to the ground, g is the acceleration of gravity, H is the self-conjugate matrix, $J_{anti\text{-}r}$ is the anti-rollover evaluation index function, and LTR is the roll risk coefficient.

In Equation (9), the coefficients b_3, b_4, b_5 and b_6 can be obtained by the optimization algorithm.

The yaw angle can be expressed as:

$$\varphi(t) = \arctan\frac{dY(t)}{dX(t)} = \arctan\frac{\dot{Y}(t)}{\dot{X}(t)} \tag{10}$$

3. Path Tracking Control

3.1. Establishment of Vehicle Dynamics Model

The three-degree-of-freedom vehicle dynamics model is shown in Figure 4, including lateral motion, yaw motion and longitudinal motion [21,22]. The aerodynamics, suspension system and steering system are ignored. Considering that the sideslip angle of the center of mass is small when the lateral acceleration is not high, the control accuracy will not be affected if it is ignored, so the sideslip angle of the center of mass is ignored [23]. According to the characteristics of the distributed drive vehicle, the additional yaw moment is considered.

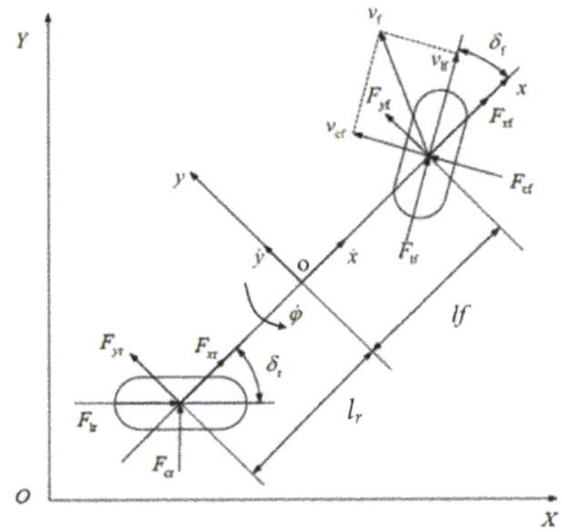

Figure 4. Three-degree-of-freedom vehicle dynamics model.

In Figure 4, the coordinate axis $oxyz$ is the fixed coordinate system of the vehicle, xoz is in the left-right symmetrical plane of the vehicle, the coordinate origin of the center of mass of the vehicle is o, the x-axis is the coordinate axis along the longitudinal direction of the vehicle, and the y-axis is the coordinate perpendicular to the longitudinal direction of the vehicle axis. The z-axis is the coordinate axis perpendicular to the xoy plane, and the direction satisfies the right-hand rule. XOY is the geodetic coordinate system. Considering that F_{xf} and F_{yf} are the force exerted on a single tire, it needs to be multiplied by 2. According to Newton's second law:

$$\begin{cases} m\ddot{x} = m\dot{y}\dot{\varphi} + 2F_{xf} + 2F_{xr} \\ m\ddot{y} = -m\dot{x}\dot{\varphi} + 2F_{yf} + 2F_{yr} \\ I_z\ddot{\varphi} = 2l_fF_{yf} - 2l_rF_{yr} \end{cases} \tag{11}$$

where l_f and l_r are the distances from the center of mass to the front and rear axles, respectively. m is the overall mass of the vehicle, and I_z is the central moment of inertia of the vehicle around the z axis.

The conversion relationship between the resultant force of the tire in the x-direction and y-direction, the longitudinal force and the lateral force is expressed as follows:

$$\begin{cases} F_{xf} = F_{lf}\cos\delta_f - F_{cf}\sin\delta_f \\ F_{xr} = F_{lr}\cos\delta_f + F_{cr}\sin\delta_r \\ F_{yf} = F_{lf}\sin\delta_f + F_{cf}\cos\delta_f \\ F_{yr} = F_{cr}\cos\delta_r - F_{lr}\sin\delta_f \end{cases} \tag{12}$$

where F_{xf} and F_{yf} are the resultant forces on the front wheels in the x and y directions, respectively. F_{xr} and F_{yr} are the resultant forces on the front wheels in the x and y directions, respectively, and F_{lf} and F_{lr} are the longitudinal forces on the front and rear wheels, respectively. F_{cf} and F_{cr} are the lateral forces on the front and rear wheels, respectively. δ_f is the front wheel angle, δ_r is the rear wheel angle.

The lateral and longitudinal forces of vehicle tires can be expressed as complex functions of parameters such as wheel slip angle, slip rate, road friction coefficient and vertical load:

$$F_c = f_c(\alpha, s, \mu, F_z) \tag{13}$$

$$F_l = f_l(\alpha, s, \mu, F_z) \tag{14}$$

where, F_z is the vertical load on the tire, μ is the road friction coefficient, s is the slip ratio, and α is the wheel side angle.

In order to simplify the model, the tire force is usually represented by a linear function under the condition of small tire longitudinal slip rate, lateral acceleration and slip angle. The force is expressed as follows:

$$\begin{cases} F_l = C_l s \\ F_c = C_c \alpha \end{cases} \tag{15}$$

where C_l is the longitudinal stiffness of the vehicle tire, and C_c is the cornering stiffness of the vehicle tire.

The following approximate relationship is satisfied:

$$\begin{cases} \cos\theta \approx 1 \\ \sin\theta \approx \theta \\ \tan\theta \approx \theta \end{cases} \tag{16}$$

where θ can be expressed as front wheel side angle and rear wheel side slip angle, etc.

The wheel side slip angle α can be obtained according to the geometric relationship:

$$\alpha = \tan^{-1}\frac{v_c}{v_l} \tag{17}$$

where v_c and v_l represent the lateral and longitudinal speeds of the tire, respectively, which can be represented by the speeds v_x and v_y in the direction of the coordinate system:

$$\begin{cases} v_l = v_y\sin\delta + v_x\cos\delta \\ v_c = v_y\cos\delta - v_x\sin\delta \end{cases} \tag{18}$$

where δ is the tire deflection angle. The tire speed is generally calculated by the speed conversion of the vehicle, and the conversion relationship is shown in Formula (19):

$$\begin{cases} v_{yf} = \dot{y} + l_f\dot{\varphi} \\ v_{yr} = \dot{y} - l_r\dot{\varphi} \end{cases} \tag{19}$$

$$\begin{cases} v_{xf} = \dot{x} \\ v_{xr} = \dot{x} \end{cases} \tag{20}$$

Through the above formulas, the wheel side slip angle can be obtained as:

$$\begin{cases} \alpha_f = \frac{\dot{y}+l_f\dot{\varphi}}{\dot{x}} - \delta_f \\ \alpha_r = \frac{\dot{y}-l_r\dot{\varphi}}{\dot{x}} \end{cases} \tag{21}$$

From Equation (20), the tire lateral force of the front and rear wheels can be obtained as:

$$\begin{cases} F_{cf} = C_{cf}\left(\delta_f - \frac{\dot{y}+l_f\dot{\varphi}}{\dot{x}}\right) \\ F_{cr} = C_{cr}\frac{l_r\dot{\varphi}-\dot{y}}{\dot{x}} \end{cases} \tag{22}$$

According to the above derivation formula, the following nonlinear vehicle dynamics model can be obtained as follows:

$$\begin{cases} m\ddot{y} = -m\dot{x}\dot{\varphi} + 2\left[C_{cf}\left(\delta_f - \frac{\dot{y}+l_f\dot{\varphi}}{\dot{x}}\right) + C_{cr}\frac{l_r\dot{\varphi}-\dot{y}}{\dot{x}}\right] \\ m\ddot{x} = m\dot{y}\dot{\varphi} + 2\left[C_{lf}s_f + C_{cf}\left(\delta_f - \frac{\dot{y}+l_f\dot{\varphi}}{\dot{x}}\right)\delta_f + C_{lf}s_r\right] \\ I_z\ddot{\varphi} = 2\left[l_fC_{cf}\left(\delta_f - \frac{\dot{y}+l_f\dot{\varphi}}{\dot{x}}\right) - l_rC_{cr}\frac{l_r\dot{\varphi}-\dot{y}}{\dot{x}}\right] + \Delta M_z \\ \dot{Y} = \dot{x}\sin\varphi + \dot{y}\cos\varphi \\ \dot{X} = \dot{x}\cos\varphi - \dot{y}\sin\varphi \end{cases} \tag{23}$$

In this paper, due to the use of a linear time-varying model predictive control algorithm, the displacement and velocity in the longitudinal direction are not considered as control variables. Let $\dot{\varphi}$ = r, $\dot{x}$ = v_x, $\dot{y}$ = v_y, Equation (22) can be simplified as:

$$\begin{cases} \dot{v}_y = -\frac{C_{cf}+C_{cr}}{mv_x}v_y - \left(\frac{l_fC_{cf}-l_rC_{cr}}{mv_x} + v_x\right)r + \frac{C_{cf}}{m}\delta_f \\ \dot{r} = -\frac{l_fC_{cf}-l_rC_{cr}}{I_zv_x}v_y - \frac{l_f^2C_{cf}+l_r^2C_{cr}}{I_zv_x}r + \frac{l_fC_{cf}}{I_z}\delta_f + \frac{\Delta M_z}{I_z} \\ \dot{\varphi} = r \\ \dot{Y} = v_x\varphi + v_y \end{cases} \tag{24}$$

According to the above dynamic model, the following state space equation is established:

$$\begin{cases} \dot{\boldsymbol{\xi}} = \boldsymbol{A\xi} + \boldsymbol{Bu} \\ \boldsymbol{y} = \boldsymbol{C\xi} \end{cases} \tag{25}$$

where the state quantity $\boldsymbol{\xi} = \begin{bmatrix} v_y & r & \varphi & y \end{bmatrix}^{\mathrm{T}}$, the control quantity $\boldsymbol{u} = \begin{bmatrix} \delta_f & \Delta M_z \end{bmatrix}^{\mathrm{T}}$, **A**, **B**, and **C** are coefficient matrices, respectively:

$$\boldsymbol{A} = \begin{bmatrix} -\frac{C_{cf}+C_{cr}}{mv_x} & -\left(\frac{l_fC_{cf}-l_rC_{cr}}{mv_x} + v_x\right) & 0 & 0 \\ -\frac{l_fC_{cf}-l_rC_{cr}}{I_zv_x} & -\frac{l_f^2C_{cf}+l_r^2C_{cr}}{I_zv_x} & 0 & 0 \\ 0 & 1 & 0 & 0 \\ 1 & 0 & v_x & 0 \end{bmatrix} \tag{26}$$

$$\boldsymbol{B} = \begin{bmatrix} \frac{C_{cf}}{m} & 0 \\ \frac{l_fC_{cf}}{I_z} & \frac{1}{I_z} \\ 0 & 0 \\ 0 & 0 \end{bmatrix} \tag{27}$$

$$\boldsymbol{C} = \begin{bmatrix} 0 & 0 & 1 & 0 \\ 0 & 0 & 0 & 1 \end{bmatrix} \tag{28}$$

3.2. MPC Controller Design

The linear time-varying model is shown in Figure 5. In the model, the control value is initial velocity v_0 = 18 m/s; The actual measured value is v_y, r, φ, y; The control input value is α_f, ΔM_z.

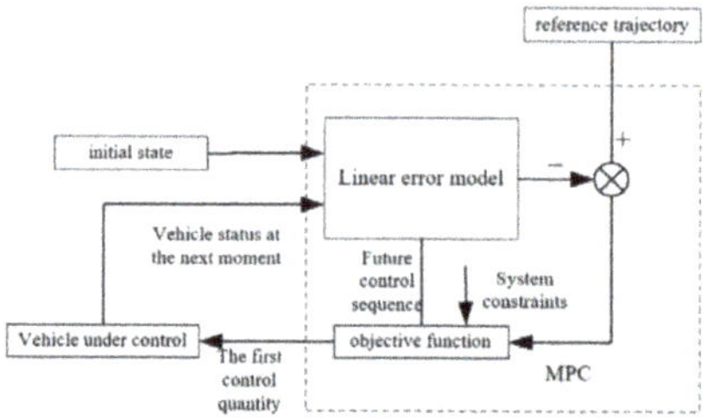

Figure 5. Flow chart of linear time-varying model control.

Since the linear time-varying model [13] is used, it is necessary to perform linear time-varying processing and discretization processing on Equation (24).

Let $\mathbf{A_k}$ = I + A_T, B_k = BT, C_k = C.

where T is the sampling time, I is the identity matrix,

$$A_k = \begin{bmatrix} 1 - \frac{C_{cf}+C_{cr}}{mv_x}T & -\left(\frac{l_fC_{cf}-l_rC_{cr}}{mv_x} + v_x\right)T & 0 & 0 \\ -\frac{l_fC_{cf}-l_rC_{cr}}{I_zv_x}T & 1 - \frac{l_f^2C_{cf}+l_r^2C_{cr}}{I_zv_x}T & 0 & 0 \\ 0 & T & 1 & 0 \\ T & 0 & v_xT & 1 \end{bmatrix} \tag{29}$$

$$B_k = \begin{bmatrix} \frac{C_{cf}}{m}T & 0 \\ \frac{l_fC_{cf}}{I_z}T & \frac{T}{I_z} \\ 0 & 0 \\ 0 & 0 \end{bmatrix} \tag{30}$$

$$C_k = \begin{bmatrix} 0 & 0 & 1 & 0 \\ 0 & 0 & 0 & 1 \end{bmatrix} \tag{31}$$

So the following discrete state space equation is obtained:

$$\begin{cases} \dot{\xi}(k+1) = A\xi(k) + Bu(k) \\ y(k) = C_k\xi(k) \end{cases} \tag{32}$$

In order to avoid the phenomenon of a sudden change in the control quantity of the system, it is necessary to restrict the control increment in each sampling period, so the above formula of this paper is converted and deformed:

Note: ζ =, substitute into the above formula to obtain a new state space equation:

$$\begin{cases} \zeta(k+1) = \widetilde{A}_k\zeta(k) + \widetilde{B}_k\Delta u(k) \\ y(k) = \widetilde{C}_k\xi(k) \end{cases} \tag{33}$$

In the formula,

$$\widetilde{A}_k = \begin{bmatrix} A_k & B_k \\ 0_{3\times 4} & I_2 \end{bmatrix} \tag{34}$$

$$\widetilde{B}_k = \begin{bmatrix} B_k \\ I_2 \end{bmatrix} \tag{35}$$

$$\widetilde{C}_k = \begin{bmatrix} 0 & 0 & 1 & 0 & 0 & 0 \\ 0 & 0 & 0 & 1 & 0 & 0 \end{bmatrix} \tag{36}$$

The optimization objective function of the path tracking and torque distribution controller is expressed as:

$$J = \sum_{i=1}^{N_p} \|y(k+i) - y_{ref}(k+i)\|^2{}_Q + \sum_{i=1}^{N_c-1} \|\Delta u(k+i)\|^2{}_R \tag{37}$$

where y_{ref} is obtained by the path planning layer; **Q** and **R** are the trajectory tracking error weight matrix and the control amount increment weight matrix, respectively; N_p and N_c are the prediction time domain and the control time domain, respectively. Some parameter settings in model predictive control are shown in Table 1.

Table 1. Part of the parameter settings of the controller.

Parameter Name	Value
Nc	8
Np	25
Δu_{min}	−0.02
Δu_{max}	0.2
a_{ymin}	−0.25 g
a_{ymax}	0.25 g
Q	[2000 0; 0 2000]
R	[3000 0; 0 1]

3.3. Torque Distribution Controller Design

Torque distribution strategy has a great impact on vehicle lateral stability. Firstly, the additional yaw moment ΔM_z output by the MPC controller is used to calculate the total demand torque T_{all}. Then using the proportion of the vertical load, calculates the torque of the four wheels. On the premise of ensuring lateral stability, the flexible torque distribution is realized.

The additional yaw moment ΔM_z and the four-wheel torque should satisfy the following relationship:

$$\Delta M_z = \frac{\mathrm{b}}{2R}\left[(T_{Fr} - T_{Fl})\cos\delta_f + (T_{Rr} - T_{Rl})\right] \tag{38}$$

where R is the wheel radius, b is the wheel track, and T_{Fl}, T_{Fr}, T_{Rl} and T_{Rr} are the torques of the left front wheel, the right front wheel, the left rear wheel and the right rear wheel, respectively.

$$F_z = F_{z,Fl} + F_{z,Fr} + F_{z,Rl} + F_{z,Rr} \tag{39}$$

while:

$$\begin{cases} k_1 = \frac{F_{z,Fl}}{F_z} \\ k_2 = \frac{F_{z,Fr}}{F_z} \\ k_3 = \frac{F_{z,Rl}}{F_z} \\ k_4 = \frac{F_{z,Rr}}{F_z} \end{cases} \tag{40}$$

where $F_{z,Fl}$, $F_{z,Fr}$, $F_{z,Rl}$, $F_{z,Rr}$ are the vertical loads of the left front wheel, the right front wheel, the left rear wheel and the right rear wheel, respectively. Force, k_1, k_2, k_3, k_4 are the proportional coefficients of the four-wheel vertical force and the total vertical force.

According to Equations (37)–(39), the total demand torque T_{all} and the distributed four-wheel torques T_{Fl}, T_{Fr}, T_{Rl}, T_{Rr} can be obtained:

$$T_{all} = \frac{\Delta M_z \cdot 2R}{d \cdot \left[(k_2 - k_1)\cos\delta_f + (k_4 - k_3)\right]} \tag{41}$$

$$\begin{cases} T_{Fl} = k_1 T_{all} \\ T_{Fr} = k_2 T_{all} \\ T_{Rl} = k_3 T_{all} \\ T_{Rr} = k_4 T_{all} \end{cases} \tag{42}$$

The solution of the final path tracking and torque distribution controller can be expressed as:

$$\begin{cases} \min Z = \sum_{i=1}^{N_p} \|y(k+i) - y_{ref}(k+i)\|^2{}_Q + \sum_{i=1}^{N_c-1} \|\Delta u(k+i)\|^2{}_R \\ \text{s.t. } \zeta(k+1) = \widetilde{A}_k \zeta(k) + \widetilde{B}_k \Delta u(k) \\ \Delta u_{\min} \le \Delta u(k+i) \le \Delta u_{\max} \\ u_{\min} \le u(k+i) \le u_{\max} \\ y_{\min} \le y(k+i) \le y_{\max} \\ \varphi_{\min} \le \varphi(k+i) \le \varphi_{\max} \end{cases} \tag{43}$$

The control quantity at time $k + 1$ can be expressed by the control quantity at time k plus the control increment at time k:

$$u(k+1) = u(k) + \Delta u(k) \tag{44}$$

4. Simulation Analysis and Verification

The C-class vehicle model is adopted in Carsim. PD18 in-wheel motor is adopted in the motor model. According to the characteristics of the distributed drive vehicle, the vehicle parameters are shown in Table 2.

Table 2. The vehicle model parameters.

Parameters	Value
Vehicle sprung mass m_s/kg	1743
Vehicle mass m/kg	1907
Moment of inertia $I_z/(\text{kg}\cdot\text{m}^2)$	3246.9
front wheelbase l_f/m	1.33
rear wheelbase l_r/m	1.81
The height of the center of mass above the ground h/m	0.781
Wheeltrack b/m	2.029
Lateral stiffness of front wheels $C_{cf}/\left(\text{N}\cdot\text{rad}^{-1}\right)$	116,050
Lateral stiffness of rear wheels $C_{cr}/\left(\text{N}\cdot\text{rad}^{-1}\right)$	104,590

4.1. Verification of Co-Simulation Platform

The initial speed of the vehicle is 60 km/h, the road adhesion coefficient μ = 0.8 and the maximum value of LTR is set at 0.1. The simulation results are shown in Figures 6 and 7.

It can be seen from Figure 5 that the controller can accurately track the reference path. The tracking error of the yaw angle is small, and the lateral acceleration appears slightly jittered between 0 and 1 s, but the value of the lateral acceleration is maintained throughout the process. Within a reasonable range, the front wheel corners shake slightly at 0 s to 1 s and 6 s to 7 s, but the changes in the corners are relatively gentle, and the corner values do not exceed the specified value. It can be concluded that the control effect of the controller is good and can meet the needs of vehicle obstacle avoidance.

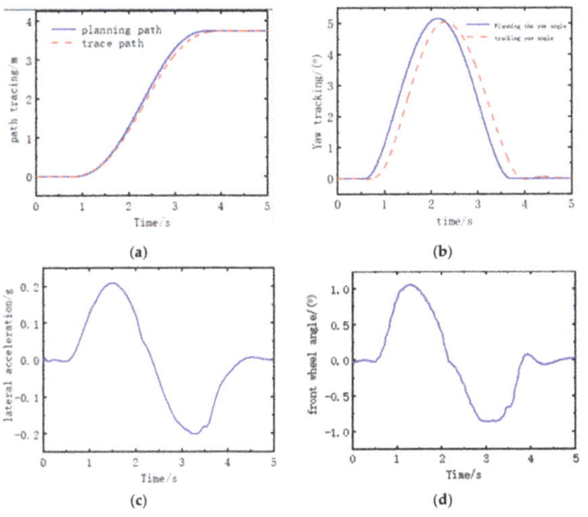

Figure 6. Co-simulation results. (**a**) Path tracking. (**b**) Yaw angle tracking. (**c**) Lateral acceleration. (**d**) Front-wheel turning angle.

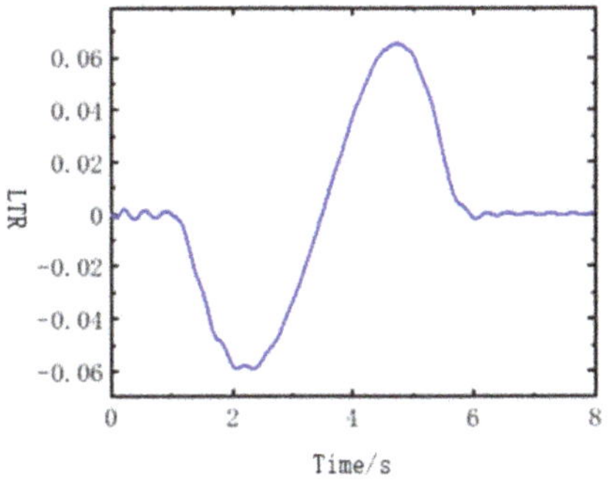

Figure 7. Changes in LTR value.

It can be seen from Figure 6 that the value of LTR is always in a small range during the entire obstacle avoidance process, which verifies the effectiveness of the set obstacle avoidance path and meets the requirements of anti-rollover in the obstacle avoidance process.

4.2. Simulation of Tracking Control Performance

The stability and robustness of the designed path tracking and torque distribution controller are verified by the simulation of different road adhesion coefficient conditions. Select high adhesion pavement coefficient μ = 0.8, low adhesion pavement coefficient μ = 0.2, μ = 0.1. The specific simulation results are shown in Figures 7 and 8.

It can be seen from Figure 8a–d that the tracking control effect of the controller is excellent when the high adhesion coefficient road surface μ = 0.8. The effect is slightly decreased when the low adhesion coefficient road surface μ = 0.2, and the tracking control error is slightly increased. The changes in lateral acceleration and front wheel angle are relatively gentle under both conditions. When μ = 0.1, the vehicle deviates greatly from the planned path and the planned yaw angle. The lateral acceleration and front wheel angle also produce a lot of jitter. This is because the road adhesion coefficient is too low to the adhesion limit. Under this condition, the controller cannot perform tracking control.

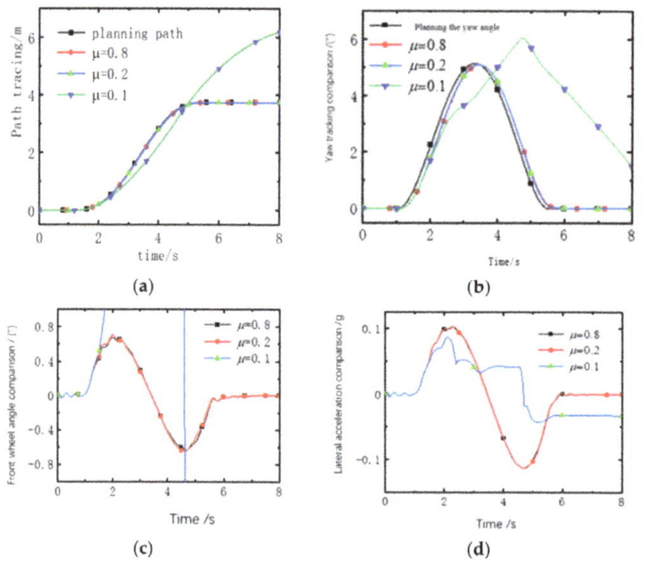

Figure 8. Simulation comparison under different road adhesion coefficient conditions. (**a**) Path tracking. (**b**) Yaw tracking. (**c**) Front-wheel angle. (**d**) Lateral acceleration.

It can be seen from Figure 9 that when μ = 0.8 and μ = 0.2, the vertical force changes of the wheels are relatively gentle, which meets the requirements of stable driving of the vehicle in the process of obstacle avoidance. When μ = 0.1, the tires are not sufficient to provide the vehicle with steady longitudinal and lateral force. At 2.5 s and 4.7 s, there is a large jump, which can no longer meet the stability requirements of the vehicle during driving.

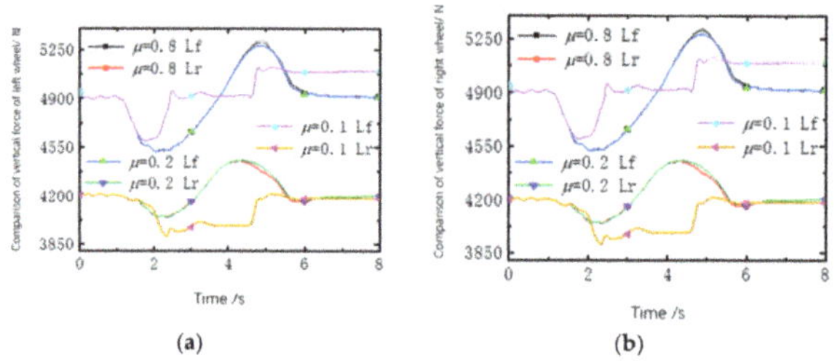

Figure 9. Comparison of vertical forces on wheels. (**a**) vertical force of left wheel; (**b**) vertical force of right wheel.

Figure 10a–d shows that when μ = 0.8, the torque distribution of each wheel meets the requirements of this paper. When μ = 0.2, the torque of the left front wheel begins to increase after 1 s and reaches a peak at 1.5~3.5 s, the maximum value is 107 N·m. The torque starts to decrease continuously at 3.5~5 s and 5~5.5 s. It reaches the reverse maximum value of 116 N·m and then changes to 0 N·m. The torque changes of the remaining wheels are roughly the same as the torque changes of the left front wheel, so this conclusion description will not be repeated. It can be seen that under this working condition, the controller has a good torque distribution effect. When μ = 0.1, the torque of each wheel drops rapidly to about −100 N·m in 2 s, and the controller cannot continue to distribute the torque normally.

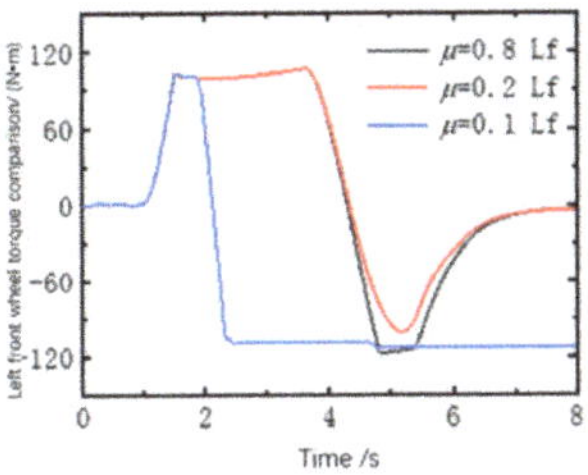

(a)

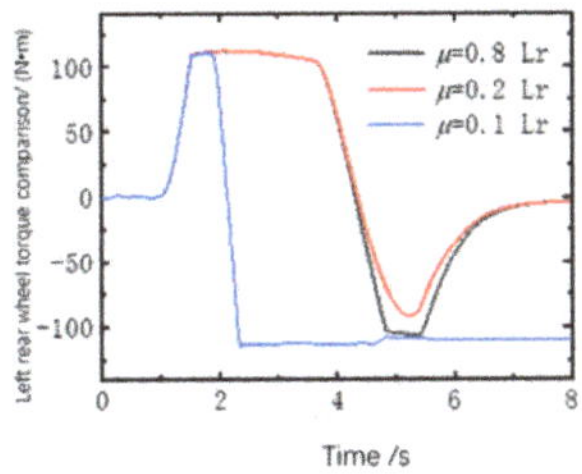

(b)

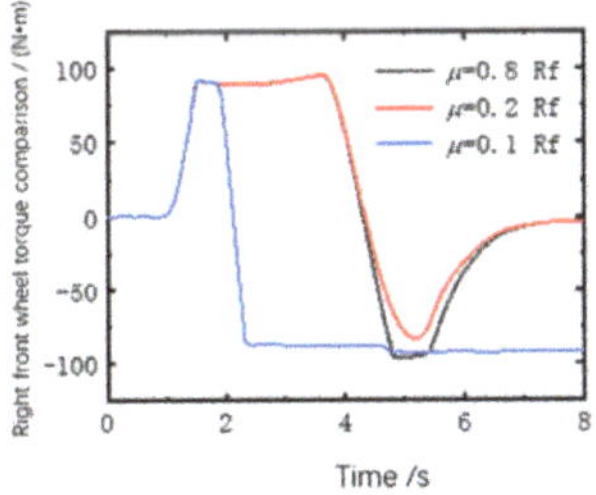

(c)

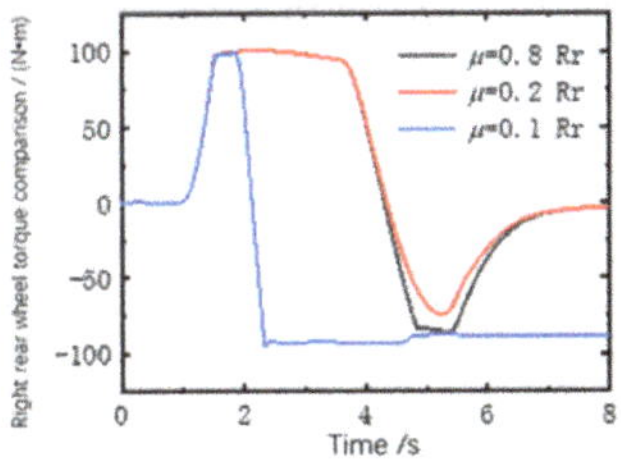

(d)

Figure 10. Comparison of torque distribution among wheels. (**a**) Torque comparison of left front wheel. (**b**) Torque comparison of left rear wheel. (**c**) Torque comparison of right front wheel. (**d**) Torque comparison of right rear wheel.

From the analysis of Figures 9 and 10, it can be seen that the controller designed in this paper can still control the vehicle to drive according to the correct planned path under the condition of low road adhesion coefficient $\mu = 0.2$ and carry out reasonable torque distribution, which verifies the path. The tracking controller has good robustness. When $\mu = 0.1$, the vehicle has already slipped, and the path tracking control can not be realized.

5. Conclusions

(1) Sixth-order polynomial for obstacle avoidance path planning is presented. Through the simulation results, it is verified that the planned path can be accurately tracked, and the *LTR* values are always within the safe range. The vehicle has no risk of rollover. The obstacle avoidance path and tracking controller in this paper can effectively meet the requirements of safe obstacle avoidance.

(2) In this paper, path planning, path tracking and torque distribution are combined to achieve safe obstacle avoidance through the tracking control of the obstacle avoidance path. The path-tracking controller not only realizes the intelligent obstacle avoidance process of unmanned vehicles but also combines it with distributed vehicles. The safety and stability are improved by the torque distribution strategy in the obstacle avoidance process compared with traditional vehicles. Simultaneously, simulations are carried out under different road adhesion coefficient conditions, and the simulation results show that the vehicle can still perform safe and stable automatic obstacle avoidance under the conditions of road adhesion coefficient $\mu = 0.2$, indicating that the controller has good robustness.

This paper provides an idea for the obstacle avoidance path tracking control of distributed drive electric vehicles. In the follow-up research work, the torque distribution controller can be improved, and better torque distribution and anti-skid control can be added to the original design scheme. The control strategy can be realized under the condition of a lower road adhesion coefficient. The real vehicle test link is added to verify the effectiveness of the control strategy in this paper.

Author Contributions: Writing—original draft, H.W.; Writing—review & editing, H.Z. and Y.F. All authors have read and agreed to the published version of the manuscript.

Funding: This research was funded by National Natural Science Foundation of China grant number 51705306.

Data Availability Statement: Not applicable.

Conflicts of Interest: The authors declare no conflict of interest.

References

1. World Health Organization. *Global Status Report on Road Safety 2013: Supporting a Decade of Action*; World Health Organization: Geneva, Switzerland, 2013.
2. Toroyan, T. Global status report on road safety. *Inj. Prev.* **2009**, *15*, 286. [CrossRef] [PubMed]
3. Chang, S.; Gordon, T. A flexible hierarchical model-based control methodology for vehicle active safety systems. *Veh. Syst. Dyn.* **2008**, *46*, 63–75. [CrossRef]
4. Guo, J.; Hu, P.; Wang, R. Nonlinear coordinated steering and braking control of vision-based autonomous vehicles in emergency obstacle avoidance. *IEEE Trans. Intell. Transp. Syst.* **2016**, *17*, 3230–3240. [CrossRef]
5. Yu, Z.; Jiang, W.; Zhang, L. Torque distribution control of electric vehicle driven by four wheel hub motor. *J. Tongji Univ. Nat. Sci. Ed.* **2008**, *36*, 5.
6. Chen, T.; Chen, L.; Xu, X. Coordinated Control of Path Tracking and Stability of Distributed Driving Unmanned Vehicles. *Automot. Eng.* **2019**, *41*, 1109–1116.
7. Zhang, B.; Li, Z.; Shen, G. Research on intelligent vehicle track tracking based on fuzzy neural network. *Automot. Eng.* **2019**, *41*, 7.
8. Wang, Y.; Chen, Q.; Gao, L. A distributed driving vehicle path tracking method based on prediction-fuzzy joint control. *Sci. Technol. Eng.* **2020**, *20*, 15100–15108.
9. Hu, J.; Zhong, X.; Chen, R. Intelligent Vehicle Path Tracking Control Based on Fuzzy LQR. *Automot. Eng.* **2022**, *44*, 17–25.
10. Kou, F.; Zheng, W.; Zhang, X.; Yang, H.; He, J. Path following control of unmanned vehicle using state extended MPC and corner compensation. *Mech. Sci. Technol. Aerosp. Eng.* **2022**, *4*, 1–8.
11. Wu, F.; Guo, S. Intelligent vehicle path tracking algorithm based on nonlinear model predictive control. *Automot. Technol.* **2020**, *5*, 1–7.
12. Li, S.; Yang, Z.; Wang, X. Trajectory Tracking Control of Intelligent Vehicle Based on T-S Fuzzy Variable Weight MPC. *J. Mech. Eng.* **2022**, *15*, 1–14.
13. Zhuang, Y.; Wu, Y.; Zhang, B. Coordinated Torque Control of Distributed Drive Electric Vehicle Based on Nonlinear MPC. *J. Vib. Shock.* **2021**, *40*, 9.
14. Shino, M.; Nahai, M. Yaw-moment control of electric vehicle for improving handling and stability. *JSAE Rev.* **2001**, *22*, 473–480. [CrossRef]
15. Zou, G.; Luo, Y.; Li, K. Optimal distribution method of all-wheel longitudinal force for four-wheel independent electric vehicle. *J. Tsinghua Univ. Nat. Sci. Ed.* **2009**, *49*, 719–722.
16. One, E.; Hattori, Y.; Muragishi, Y. Vehicle dynamics integrated control for four-wheel-distributed steering and four-wheel-distributed traction/braking systems. *Veh. Syst. Dyn.* **2006**, *44*, 139–151. [CrossRef]
17. Mokhiamar, O.; Abe, M. How the four wheels should share forces in an optimum cooperative chassis control. *Control Eng. Pract.* **2006**, *14*, 295–304. [CrossRef]
18. Yu, Z.; Yang, P.; Xiong, L. Application of Control Distribution Theory in Vehicle Dynamics Control. *Chin. J. Mech. Eng.* **2014**, *50*, 99–107. [CrossRef]
19. Mutoh, N. Driving and braking torque distribution methods for front-and rear-wheel-independent drive-type electric vehicles on roads wit low friction coefficient. *IEEE Trans. Ind. Electron.* **2012**, *59*, 3919–3933. [CrossRef]
20. Yang, B.; Zhang, W.; Jiang, Z. Simulation analysis of obstacle avoidance path planning and tracking control for intelligent vehicles. *China Test* **2021**, *47*, 71–78.
21. Ren, Y.; Zheng, L.; Zhang, W. Research on Active Collision Avoidance Control of Intelligent Vehicles Based on Model Predictive Control. *Automot. Eng.* **2019**, *41*, 405–410.
22. Zhang, F.; Wei, M.; Huang, L. Research on vehicle emergency lane change control based on model predictive control. *Mod. Manuf. Eng.* **2017**, *3*, 57–64.
23. Gong, J.; Jiang, W.; Xu, W. *Model Predictive Control for Self-Driving Vehicles*, 1st ed.; Beijing Institute of Technology Press: Beijing, China, 2014; pp. 22–25.

World Electric Vehicle Journal

MDPI

Article

Adaptive Robust Path Tracking Control for Autonomous Vehicles Considering Multi-Dimensional System Uncertainty

Mengyuan Chen [1], Yue Ren [2,*] and Minghui Ou [1]

[1] Big Data and Internet of Things School, Chongqing Vocational Institute of Engineering, Chongqing 402260, China
[2] College of Engineering and Technology, Southwest University, Chongqing 400715, China
* Correspondence: renyueok@hotmail.com; Tel.: +86-15223081321

Abstract: As the bottom layer of the autonomous vehicle, path tracking control is a crucial element that provides accurate control command to the X-by-wire chassis and guarantees the vehicle safety. To overcome the deterioration of control performance for autonomous vehicle path-tracking controllers caused by modeling errors and parameter perturbation, an adaptive robust control framework is proposed in this paper. Firstly, the 2-DOF vehicle dynamic model is established and the non-singular fast terminal sliding mode control algorithm is adopted to formulate the control law. The unmeasured model disturbance and parameter perturbation is regarded as the system uncertainty. To enhance the control accuracy, the radial basis forward neural network is introduced to estimate such uncertainty in real time. Then, the dynamic model of an active front steering system is established. The model reference control algorithm is applied for the steering torque control considering model uncertainty brought by the dissipation of manufacturing and mechanical wear. Finally, the Simulink–CarSim co-simulation platform is used and the proposed control framework is validated in two test scenarios. The simulation results demonstrate the proposed adaptive robust control algorithm has satisfactory control performance and good robustness against the system uncertainty.

Keywords: autonomous vehicle; path tracking control; non-singular fast terminal sliding mode control; model reference control

Citation: Chen, M.; Ren, Y.; Ou, M. Adaptive Robust Path Tracking Control for Autonomous Vehicles Considering Multi-Dimensional System Uncertainty. *World Electr. Veh. J.* **2023**, *14*, 11. https://doi.org/10.3390/wevj14010011

Academic Editors: Yong Li, Xing Xu, Lin Zhang, Yechen Qin and Yang Lu

Received: 28 November 2022
Revised: 20 December 2022
Accepted: 27 December 2022
Published: 2 January 2023

1. Introduction

Promoted by growing demands on safety, efficiency, and low carbon emissions, the intelligent transportation system (ITS) has become one of the hottest fields of research in recent years. As the most important part, autonomous vehicles play a crucial role for the development of the ITS. Owing to the gradual reduction in sensor cost and the popularity of drive-by-wire chassis technology, advanced driver assistant systems are equipped by more and more vehicle manufacturers, such as autonomous emergency braking (AEB), lane-keeping assistance (LKA), and adaptive cruise control (ACC) [1–3]. With the continuous development of autonomous vehicles, it is estimated that one in ten cars will be fully autonomous-driving by the year 2030.

The key technology of autonomous vehicles includes environment perception, decision making, trajectory planning, and chassis control. As the bottom layer of the typical hierarchical architecture, vehicle chassis control is a vital part of autonomous vehicles, which implements accurate, stable, and safe tracking of desired trajectory by applying precise control instructions to actuators. As the fundamental function achieved by an autonomous vehicle, vehicle chassis control has been extensively researched, especially for vehicle trajectory tracking. Due to the complexity of the vehicle dynamic model with the multi degrees of freedom, non-linear, and multi-dimensional coupling characteristics, the simplicity of a vehicle dynamic model is widely adopted for model-based controller design, including linear quadratic regulators (LQRs) and model predictive control (MPC) to guarantee the real-time performance of the controller [4–6].

Salepour et al. proposed an integrated controller via LQR, which established a tracking optimization function and regulated the yaw moment [7]. Xu et al. considered the variation of road curvature and introduced the multiple point preview into the LQR controller that reduced the overshooting of the tracking error and smoothened the steering wheel angle [8]. Ji et al. adopted the linear tire model and formulated the MPC-based path-tracking controller with multiple constraints to guarantee the vehicle collision and stability safety [9]. Based on the prediction of the cut path of the obstacle vehicle, Chen et al. designed the model predictive tracking control with the dynamic prediction time domain, which avoided the collision effectively while tracking the desired path [10]. Guo et al. presented a dual-constraint vehicle collision avoidance control algorithm. The variable step MPC is utilized for path tracking. In addition, road boundaries and obstacles were integrated into the controller as constraints, which realized the vehicle obstacle avoidance with high tracking accuracy [11]. Funk et al. proposed a cooperative control mode in which two safety space models are defined in the process of vehicle collision avoidance to avoid collision and ensure vehicle stability. One is the vehicle stability limit, the other is the boundary of vehicle collision space. The variable predictive step MPC controller is used to take over the driver's operation in case of danger [12,13].

However, the tracking performance of the model-based controller directly depends on the model accuracy. Most of the above control strategies adopted linear tire model based on the fixed lateral cornering stiffness. Although they have achieved considerable tracking effect in simulations and experimental tests, the deterioration of the tracking error under extreme conditions such as low friction coefficient and emergency lane-changing is inevitable. More seriously, the instability of the control system caused by model mismatch may lead to critical traffic accidents. To alleviate the negative impact of the varying cornering stiffness and acquire better system robustness, robust control is also widely considered in vehicle trajectory tracking control.

Hu et al. considered the variation in the vehicle speed and tire-cornering stiffness uncertainty during the trajectory tracking process, and established the time-varying parameter model and designed the H∞ output feedback robust controller. The gain-scheduling matrix is obtained by off-line optimization via genetic algorithm [14]. Guo et al. expanded the dimension of the vehicle dynamic model according to the upper and lower bounds of the uncertain parameters and designed the robust tracking controller to ensure the convergence of tracking errors even in the severe working conditions in which tire-cornering stiffness deviated far from its nominal value [15,16]. Chen et al. established the human–vehicle–road closed-loop model based on TS fuzzy theory for vehicle lane-keeping. The H_∞ robust steering torque compensation controller was designed to reduce the lateral offset [17]. Though H_∞ robust control guarantees the system robustness under a certain parameter perturbation range, the control performance is conservative. Adaptive control is considered as another suitable candidate. Akermi et al. proposed a path-tracking architecture with the combination of sliding mode control, fuzzy logic, and perturbations observer. The SMC gain is automatically adjusted by fuzzy organ [18]. Ao et al. developed the super twisting sliding model control algorithm based on Lyapunov theory and applied back-stepping technology. The system robustness is enhanced and the chattering phenomenon is attenuated [19]. Sun et al. proposed the adaptive non-singular fast terminal sliding mode (NFTSM) control for yaw stability control of a bus. Meanwhile, the robust least-squares allocation method is adopted for braking force distribution of each tire, which significantly improves the vehicle lateral stability under special driving conditions [20]. Most research focuses on the variation in the vehicle tire-cornering stiffness under different working conditions. Except for the vehicle state parameters, the system uncertainty of the vehicle chassis subsystem also has an important impact on the control performance.

In this paper, a novel adaptive autonomous vehicle path-tracking controller is proposed. The control framework is shown in Figure 1. The main contributions are listed as follows:

(1) The NFTSM controller is adopted for the vehicle lateral tracking control considering the vehicle dynamic model uncertainty and parameter disturbance, which has a

faster convergence rate and transient response than a linear sliding model controller. In addition, the RBFNN is introduced to estimate and compensate the nonlinear uncertainty terms in real time, which enhances the control performance;

(2) The steering system dynamics are established and the model reference adaptive control (MRAC) is utilized for the steering torque control to overcome the model uncertainty caused by the dissipation of production and system degradation

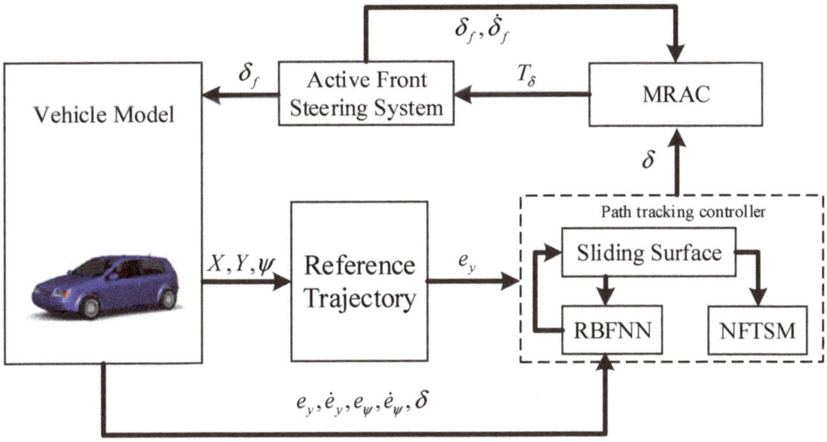

Figure 1. Control framework for vehicle path tracking.

The rest of the paper is outlined as follows. In Section 2, the vehicle dynamic model is introduced for controller design. In Section 3, the path-tracking controller based on the NFTSM is established with the radial basis forward neural network (RBFNN) estimator. Then, the system stability and finite time convergence are analyzed. Section 4 provides the MRAC for steering torque control. The CarSim–Simulink co-simulation results are presented in Section 5 to verify the effectiveness and advantages of the proposed robust path-tracking control algorithm. Finally, the conclusions are drawn in Section 6.

2. Vehicle Dynamic Modeling

In this paper, the 2-DOF bicycle tracking model is utilized for controller design [21,22], which is shown in Figure 2.

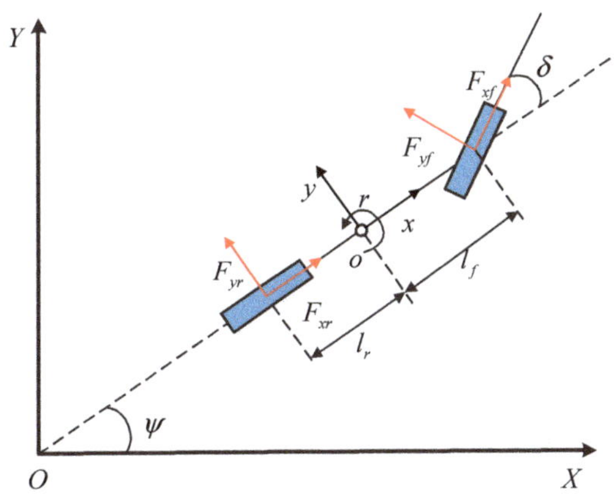

Figure 2. The 2-DOP bicycle-tracking model.

To follow the reference trajectory, both lateral position error e_y and orientation error e_ψ should be eliminated. The relationship of e_y and e_ψ can be expressed as Equation (1) based on vehicle kinematic model:

$$\begin{cases} \dot{e}_y = u \sin e_\psi + v \cos e_\psi \\ \dot{e}_\psi = \dot{\psi} - \dot{\psi}_d \end{cases}, \tag{1}$$

where u, v, ψ represent the vehicle longitudinal speed, lateral speed, and heading angle, respectively. ψ_d is the desired heading angle. With the small angle assumption and simplified linear tire model, the vehicle dynamic model is described as:

$$\begin{cases} \dot{Y} = u \sin \psi + v \cos \psi \\ \dot{v} = \frac{1}{m}\left(F_{yf} + F_{yr}\right) - ur \\ \dot{\psi} = r \\ \dot{r} = \frac{1}{I_z}\left(l_f F_{yf} - l_r F_{yr}\right) \\ F_{yf} = C_f \alpha_f = C_f\left(\delta - \frac{v + l_f r}{u}\right) \\ F_{yr} = C_r \alpha_r = C_r\left(-\frac{v - l_r r}{u}\right) \end{cases}, \tag{2}$$

where $\dot{Y}$ is change rate of the vehicle lateral position in the global coordinate. l_f, l_r represent the distance from the center of gravity to the front and rear axle, respectively. r is the vehicle yaw rate. F_{yf}, F_{yr} denote the lateral force of the front and rear axle, respectively, which can be calculated by the cornering stiffness C_f, C_r and steering wheel angle δ.

Combining Equations (1) and (2), and defining the state vectors as $X = \left[e_y, \dot{e}_y, e_\psi, \dot{e}_\psi\right]^T$, the system input as $U = \delta$ and reference as $D = \dot{\psi}_d$, the nominal state equation of tracking model can be written as:

$$\dot{X} = AX + BU + GD, \tag{3}$$

where

$$A = \begin{bmatrix} 0 & 1 & 0 & 0 \\ 0 & -\frac{C_f + C_r}{um} & \frac{C_f + C_r}{m} & -\frac{C_f l_f - C_r l_r}{um} \\ 0 & 0 & 0 & 1 \\ 0 & -\frac{C_f l_f - C_r l_r}{I_z u} & \frac{C_f l_f - C_r l_r}{I_z} & -\frac{C_f l_f^2 + C_r l_r^2}{I_z} \end{bmatrix}, B = \begin{bmatrix} 0 \\ \frac{C_f}{m} \\ 0 \\ \frac{C_f l_f}{I_z} \end{bmatrix}, G = \begin{bmatrix} 0 \\ -\frac{C_f l_f - C_r l_r}{um} - u \\ 0 \\ -\frac{C_f l_f^2 + C_r l_r^2}{I_z u} \end{bmatrix}$$

3. Path-Tracking Control Algorithm

In this section, the NFTSM is designed to eliminate the lateral position error that guarantees the vehicle accurately tracking the reference path. Meanwhile, the RBFNN is adopted to estimate the model uncertainty terms.

3.1. Control Law Design

Based on the nominal tracking error model presented in Equation (3), the second-time derivative of lateral position error could be written as:

$$\ddot{e}_y = L + g\delta, \tag{4}$$

where

$$L = -\left(\frac{C_f + C_r}{um}\right)\dot{Y} - \left(u + \frac{C_f l_f - C_r l_r}{um}\right)\dot{\psi} + u\left(\dot{\psi} - \dot{\psi}_d\right),$$

However, the simplicity model cannot accurately describe the dynamic characteristics of the vehicle in a complex environment. Moreover, the parametric perturbation and disturbances also deteriorate model accuracy. Hence, Equation (4) is rewritten as:

$$\ddot{e}_y = L + g\delta + \varsigma, \tag{5}$$

where ς is the system uncertainty that cannot be directly measured.

To eliminate lateral position error, the sliding surface is defined as:

$$s = e_y + p^{-1}e_y^{\alpha} + q^{-1}\dot{e}_y^{\beta}, \tag{6}$$

which satisfies $p > 0$, $q > 0$, $1 < \beta < 2$, $\beta < \alpha$ [20].

It can be seen that if $|e_y| \leq 1$, Equation (6) can be approximated by ignoring the high order terms as:

$$s = e_y + q^{-1}\dot{e}_y^{\beta}, \tag{7}$$

When system maintains on the sliding surface, Equation (7) is equivalent to:

$$\dot{e}_y = -q^{\frac{1}{\beta}}e^{\frac{1}{\beta}}, \tag{8}$$

which obviously reveals that the convergence rate is better than the linear sliding surface with the same parameters. If $|e_y| \leq 1$, the $\dot{e}_y$ can be approximated by:

$$\dot{e}_y = -\left(qp^{-1}\right)^{\frac{1}{\beta}}e_y^{\frac{\alpha}{\beta}}, \tag{9}$$

Based on the sliding surface in Equation (6), the control law can be conceived as:

$$\delta = \delta_{eq} + \delta_{sw} - \varsigma, \tag{10}$$

where δ_{eq} is equivalent control term, δ_{sw} is switching control term, and ς is system uncertainty. The equivalent control term is designed as:

$$\delta_{eq} = -\frac{1}{g}\left(L + \left(1 + p^{-1}\alpha e_y^{\alpha-1}\right)q\beta^{-1}\dot{e}_y^{2-\beta}\right), \tag{11}$$

To improve the approach speed and system robustness, the fast terminal-switching control term is proposed as:

$$\delta_{sw} = -q\beta^{-1}\left(\lambda_1 s + \lambda_2 s^{\theta_1} + \lambda_3 s^{\theta_2}\right), \tag{12}$$

where $\lambda_1, \lambda_2, \lambda_3$ are positive constants.

3.2. RBFNN-Based System Uncertainty Estimator

Note that Equation (10) contains unmeasurable system uncertainty ς. Owing to the simple structure, fast learning, and fine approximation ability, RBFNN is much more widely used for complex dynamical system control than other neural networks such as multilayer perception, which not only has less network parameters for tuning, but can also avoid local minimum problems [23,24]. Here, the RBFNN is adopted to approximate the system uncertainty ς.

Define the ideal output and the actual output of RBFNN as f^* and f, which are expressed as:

$$f^* = w^{*T}\sigma(x) + \varepsilon^*, \tag{13}$$

$$f = w^T\sigma(x) + \varepsilon, \tag{14}$$

where $w = [w_1, w_2, \ldots, w_n]^T$, $\varepsilon = [\varepsilon_1, \varepsilon_2, \ldots, \varepsilon_n]^T$ represent the estimation of optimal network weight w^* and bias ε^*, respectively, n is number of nodes in the hidden layer of network, and $\sigma = [\sigma_1, \sigma_2, \ldots, \sigma_n]^T$ is Gaussian radial basis function, expressed as:

$$\sigma_i(x) = \exp\left(-\frac{\|x - c_i\|^2}{2b_i^2}\right), i = 1, 2, \ldots, n, \tag{15}$$

where x is input vectors of the network. c_i and b_i are network parameters of the ith radial basis function. In this paper, the steering wheel angle δ and state vectors $e_y, \dot{e}_y, e_\psi, \dot{e}_\psi$ are chosen as the network input.

Usually, the switching gain is set as larger to suppress the approximation error of the neural network. However, such a method is conservative, which leads to control chattering. Here, an adaptive compensation term is introduced, reducing network estimation error. The control law in Equation (10) can be redefined as:

$$\delta = -f - \frac{1}{g}\left(L + \left(1 + p^{-1}\alpha e^{\alpha-1}\right)q\beta^{-1}\dot{e}^{2-\beta}\right) - p\beta^{-1}\left(\lambda_1 s + \lambda_2 s^{\theta_1} + \lambda_3 s^{\theta_2}\right) - \varsigma, \tag{16}$$

To guarantee that the actual output of network f adaptively approximates the real one f^* and weight of the neural network is bounded, the update law of weight and error compensation term can be designed as:

$$\dot{w} = \Gamma_w \sigma s q^{-1}\beta\tau - \Gamma_w \eta_w \psi s^2 w, \tag{17}$$

$$\dot{\varepsilon} = \Gamma_\varepsilon s q^{-1}\beta\tau - \Gamma_\varepsilon \eta_\varepsilon \psi s^2 \varepsilon, \tag{18}$$

where Γ_w, η_w, Γ_ε, η_ε are positive constants and $\tau = \dot{e}^{\alpha-1} > 0$.

Note that the network weight w and bias ε are updated iteratively online at each sampling time via Equations (17) and (18). The detailed structure of the RBFNN-based NFTSM control algorithm is shown in Figure 3.

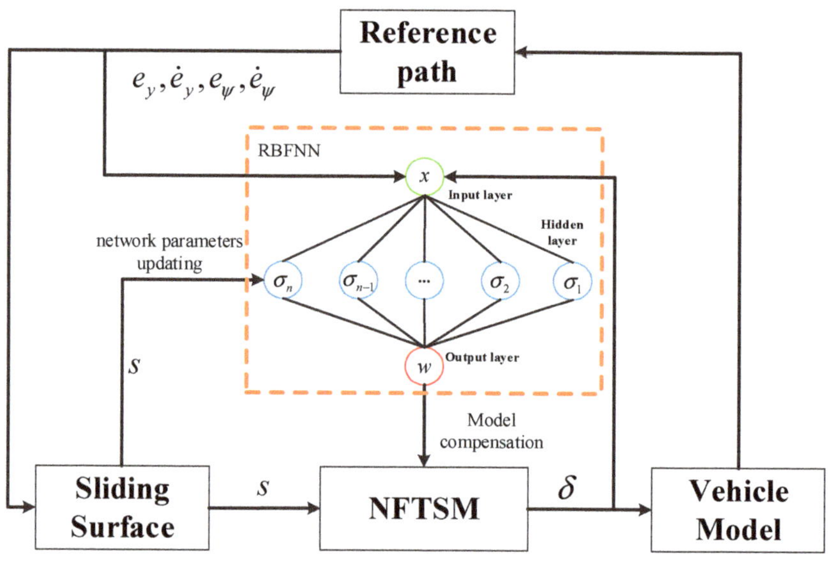

Figure 3. The structure of RBFNN-based NFTSM control algorithm.

3.3. System Performance Analysis

Next, the system stability and finite time control convergence are proven to illustrate the satisfactory system performance of the proposed control algorithm.

3.3.1. System Stability

Theorem 1. *For the path-tracking system expressed as Equation (3), the closed-loop system is global asymptotically stable and the weight and error compensation term of the network are bounded by choosing the control law as Equation (16) and adaptive update law as Equations (17) and (18).*

Proof of Theorem 1. Define the Lyapunov function as:

$$V_1 = \frac{1}{2}\left(s^T s + \mathrm{tr}\left(\hat{w}^T \Gamma_w^{-1} \hat{w}\right) + \hat{\varepsilon}^T \Gamma_w^{-1} \hat{\varepsilon}\right), \tag{19}$$

where $\hat{w}, \hat{\varepsilon}$ denote the error between estimation and actual value of network weight and approximation error, represented as $\hat{w} = w^* - w$ and $\hat{\varepsilon} = \varepsilon^* - \varepsilon$, respectively. Differentiating Equation (19), we obtain:

$$\dot{V}_1 = s^T \dot{s} - \mathrm{tr}\left(\hat{w}^T \Gamma_w^{-1} \dot{w}\right) - \hat{\varepsilon}^T \Gamma_w^{-1} \dot{\varepsilon}, \tag{20}$$

Substitute Equations (6) and (16)–(18) into (20):

$$\begin{aligned} \dot{V}_1 &= s^T\left\{q^{-1}\beta \dot{e}^{\beta-1}\left[\hat{f} - q\beta^{-1}\left(\lambda_1 s + \lambda_2 s^{\theta_1} + \lambda_3 s^{\theta_2}\right) - \varepsilon\right]\right\} - \mathrm{tr}\left(\hat{w}^T \Gamma_w^{-1} \dot{w}\right) - \hat{\varepsilon}^T \Gamma_w^{-1} \dot{\varepsilon} \\ &= sq^{-1}\beta\psi\hat{\varepsilon} - s\psi\lambda_1 s - s\psi\lambda_2 s^{\theta_1} - s\psi\lambda_3 s^{\theta_2} - \mathrm{tr}\left(\hat{w}^T \sigma s q^{-1}\beta\psi\right) + \mathrm{tr}\left(\eta_w \psi s^2 \hat{w}^T w\right) - \hat{\varepsilon}^T \varepsilon q^{-1}\beta\psi + \eta_\varepsilon \psi s^2 \hat{\varepsilon}^T \varepsilon \\ &= -\psi\lambda_1 s^2 - \psi\lambda_2 s^{\theta_1+1} - \psi\lambda_3 s^{\theta_2+1} + \mathrm{tr}\left(\eta_w \psi s^2 \hat{w}^T w\right) + \eta_\varepsilon \psi s^2 \hat{\varepsilon}^T \varepsilon \end{aligned} \tag{21}$$

where $\hat{f} = f^* - f$. Since f is bounded, w^*, ε^* satisfies $\|w^*\| \le w_{\max}$ and $|\varepsilon^*| \le \varepsilon_{\max}$. Based on the F-norm properties [25], we obtain:

$$\dot{V}_1 \le -\psi\lambda_1 s^2 - \psi\lambda_2 s^{\theta_1+1} - \psi\lambda_3 s^{\theta_2+1} - \eta\psi s^2\left(\|\hat{w}\| - \frac{w_{\max}}{2}\right)^2 + \eta\psi s^2 \frac{w_{\max}}{4} - \eta\psi s^2\left(\|\hat{\varepsilon}\| - \frac{\varepsilon_{\max}}{2}\right)^2 + \eta\psi s^2 \frac{\varepsilon_{\max}}{4}, \tag{22}$$

For Equation (22), if $\|\hat{w}\| \ge w_{\max}$ and $|\hat{\varepsilon}| \ge \varepsilon_{\max}$, it is obvious that $\dot{V} \le 0$. The closed system is asymptotically stable and $\|\hat{w}\|, |\hat{\varepsilon}|$ finally converge to $w_{\max}$ and $\varepsilon_{\max}$, respectively.

If $|s| \ge 1$, Equation (22) can be further expressed as follows based on $s^{\theta_1+1} \ge s^2$:

$$\dot{V}_1 \le -\left(\lambda_1 + \lambda_2 - \frac{\eta w_{\max}^2 + \eta \varepsilon_{\max}^2}{4}\right)\psi s^2, \tag{23}$$

Here, the $\dot{V}_1 \le 0$ can be guaranteed only if the following inequality is satisfied:

$$\lambda_1 + \lambda_2 \ge \left(\eta w_{\max}^2 + \eta \varepsilon_{\max}^2\right)/4, \tag{24}$$

Similarly, if $|s| < 1$, $\dot{V}_1 \le 0$ can be guaranteed under the condition that $\lambda_1 + \lambda_3 \ge \left(\eta w_{\max}^2 + \eta \varepsilon_{\max}^2\right)/4$.

In conclusion, by choosing the reasonable $\lambda_1, \lambda_2, \lambda_3$, Equation (22) demonstrates that the closed system retains asymptotic stability. When $\|s\| \to \infty$, $\|\hat{w}\| \to \infty$, or $\|\hat{\varepsilon}\| \to \infty$, $V_1 \to \infty$, the Lyapunov function is positive definite, which demonstrates that the closed-loop system is globally asymptotically stable. □

3.3.2. Finite Time Convergence

Set another Lyapunov function as:

$$V_2 = \frac{1}{2} s^T s, \tag{25}$$

Differentiating Equation (25), we obtain:

$$\begin{aligned} \dot{V}_2 &= s^T\left\{q^{-1}\beta \dot{e}^{\beta-1}\left[\hat{f} - q\beta^{-1}(\lambda_1 s + \lambda_2 s^{\theta_1} + \lambda_3 s^{\theta_2}) - \varepsilon\right]\right\} \\ &= sq^{-1}\beta\psi\hat{w}^T\sigma + sq^{-1}\beta\psi\hat{\varepsilon} - s\psi\lambda_1 s - s\psi\lambda_2 s^{\theta_1} - s\psi\lambda_3 s^{\theta_2} \end{aligned} \tag{26}$$

Note that the range of the Gaussian function is [0,1], so $\|\sigma\| \le \sqrt{L}$, and $\hat{w}^T\sigma$ is bounded due to $\|\hat{w}^T\sigma\| \le \|\hat{w}^T\|\|\sigma\|$. For the following inequalities:

$$\left|\hat{w}^T\sigma\right| \le q\beta^{-1}K_\sigma|s|, \tag{27}$$

$$|\varepsilon| \le q\beta^{-1}K_\varepsilon|s|, \tag{28}$$

they are tenable if the following inequalities hold:

$$\|s\| \ge \frac{\|\hat{w}^T\sigma\|}{q\beta^{-1}K_\sigma} = \Delta_1, \tag{29}$$

$$\|s\| \ge \frac{\varepsilon}{q\beta^{-1}K_\varepsilon} = \Delta_2, \tag{30}$$

where K_σ, K_ε are positive constants. Based on Equations (29) and (30), we can obtain the following inequality by setting $\|s\| = \Delta_3 \ge \max(\Delta_1, \Delta_2)$ and substituting Equations (27) and (28) into (26):

$$\begin{aligned} \dot{V} &\le s\psi K_\sigma s + s\psi K_\varepsilon s - s\psi\lambda_1 s - s\psi\lambda_2 s^{\theta_1} - s\psi\lambda_3 s^{\theta_2} \\ &\le -\psi(\lambda_1 - K_\sigma - K_\varepsilon)s^2 - \psi\lambda_3 s^{\theta_2+1} \end{aligned} \tag{31}$$

For simplicity, set $\Theta_1 = \lambda_1 - K_\sigma - K_\varepsilon$, $\Theta_2 = \psi\lambda_3$. Once $\lambda_1 \ge K_\sigma + K_\varepsilon$ is guaranteed, the finite convergence time T_s satisfies:

$$T_s \le \frac{1}{2\Theta_1(1-v)}\ln\frac{2\Theta_1 V_2(0)^{1-v} + 2^v(1-v)\Theta_2}{2^v\Theta_2}, \tag{32}$$

where $v = (\theta_2 + 1)/2$ and $V_2(0)$ is the initial value of the Lyapunov function. Equation (32) demonstrates that the sliding function s eventually converges in the field of Δ_3. Meanwhile, it can be perceived that the tracking performance is better with smaller Δ_3, which can be acquired by tuning parameters q and β.

4. MRAC for Active Steering System

This section provides a model reference adaptive control algorithm for steering torque generation that tracks the desired steering wheel angle calculated by the path-tracking controller. The feedback control law is designed via quadratic Lyapunov function to achieve the faster rate of convergence and better tracking performance.

The structure of the active front wheel steering (AFS) system based on the steering-by-wire technology is shown in Figure 4. The dynamic model of AFS can be described as a second-order system:

$$J\ddot{\delta}_f + b\dot{\delta}_f + K\delta_f = T_\delta, \tag{33}$$

where δ_f is the eventual front steering angle. J, b, K represent the nominal moment of inertia, damping coefficient, and angular stiffness of the steering system, respectively. T_δ is the output torque of the steering motor.

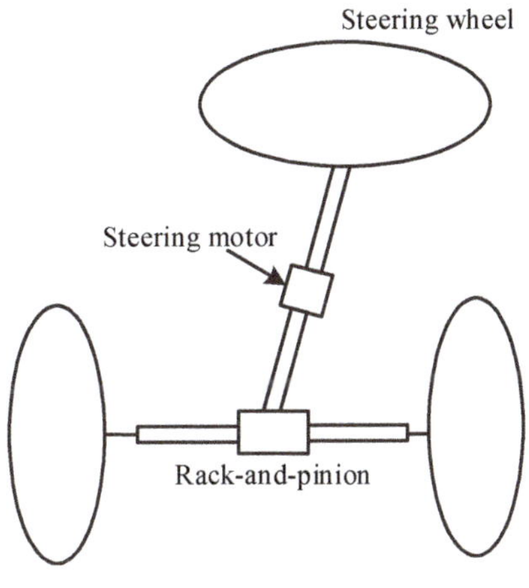

Figure 4. Active front wheel steering system.

By specifying $\omega_n^2 = K/J$ and $2\xi\omega_n = b/J$, the state–space Equation can be established with the measurable state vector $x_p = \left[\delta_f, \dot{\delta}_f\right]^T$ and input $u_p = T_\delta$ as:

$$\dot{x}_p = A_p x_p + B_p u_p, \tag{34}$$

where

$$A_p = \begin{bmatrix} 0 & 1 \\ -\omega_p^2 & -2\xi_p\omega_p \end{bmatrix}, \ B_p = \begin{bmatrix} 0 \\ \frac{1}{J} \end{bmatrix},$$

The object of MRAC is to design the input u_p that the closed–loop system is bounded to and the state vectors x_p that track the reference signals. Here the second-order low-pass filter is adopted to describe the reference model:

$$\dot{x}_m = A_m x_m + B_m \delta, \tag{35}$$

where

$$A_m = \begin{bmatrix} 0 & 1 \\ -\omega_r^2 & -2\xi_r\omega_r \end{bmatrix}, \ B_m = \begin{bmatrix} 0 \\ 1 \end{bmatrix}$$

Here, $x_m = \left[\delta_r, \dot{\delta}_r\right]^T$ is reference steering wheel angle and angular speed, and δ is the steering wheel command calculate by the path-tracking controller. ω_r and ξ_r are the cut-off frequency and damping coefficient of the filter.

It is obvious that if the following control law is applied,

$$u_p = \Omega_1^* x_p + \Omega_2^* \delta, \tag{36}$$

the closed-loop system can be expressed as:

$$\dot{x}_p = \left(A_p + B_p\Omega_1^*\right)x_p + B_p\Omega_2^*\delta, \tag{37}$$

If the Ω_1^* and Ω_2^* can be chosen as $\Omega_1^* = B_p^{-1}\left(A_m - A_p\right), \Omega_2^* = B_p^{-1}B_m$, the closed-loop steering system is the same as the reference model, which implies that the state vector x_p asymptotically tracks x_m for any bounded reference signal δ [26].

$$\lim_{t\to\infty} e_\delta = \lim_{t\to\infty}\left(x_p(t) - x_m(t)\right) = 0, \tag{38}$$

However, due to the gear abrasion, nonlinear characteristics of steering motor, and other factors, the dynamic model of AFS has parameter perturbation including J, b, K, which is difficult to accurately identify in real time. On the other hand, the different longitudinal tire forces between the left and right wheels, differential steering angle caused by scrub radius, and tire self-aligning moment also bring uncertainty to the steering system. The Ω_1^* and Ω_2^* cannot be obtained analytically in practice. Thus, the adaptive estimator has to be carried out for online estimation. Define the $\hat{\Omega}_1$, $\hat{\Omega}_2$ as the estimation values of Ω_1^*, Ω_2^*. The parameter error is represented as $\widetilde{\Omega}_1 = \hat{\Omega}_1 - \Omega_1^*$, $\widetilde{\Omega}_2 = \hat{\Omega}_2 - \Omega_2^*$. Combining Equations (35)–(38), the tracking error satisfies the followed Equation:

$$\dot{e}_\delta = A_m e_\delta + B_P\left(\widetilde{\Omega}_1 x_p + \widetilde{\Omega}_2 \delta\right), \tag{39}$$

Set the Lyapunov candidate as:

$$V_3 = e_\delta^T P e_\delta + \gamma_x^{-1} \widetilde{\Omega}_1^T \widetilde{\Omega}_1 + \gamma_\delta^{-1} \widetilde{\Omega}_2^T \widetilde{\Omega}_2, \tag{40}$$

where γ_x, γ_δ are the parameters learning rate, then the time derivative of Lyapunov function V_3 is:

$$\dot{V}_3 = e_\delta^T P\left(A_m e_\delta + B_P\left(\widetilde{\Omega}_1 x_p + \widetilde{\Omega}_2 \delta\right)\right) + \left(e_\delta^T A_m^T + x_p^T \widetilde{\Omega}_1^T B_p^T + \delta^T \widetilde{\Omega}_2^T B_p^T\right) P e + 2\gamma_x^{-1} \widetilde{\Omega}_1^T \dot{\widetilde{\Omega}}_1 + \gamma_\delta^{-1} \widetilde{\Omega}_2^T \dot{\widetilde{\Omega}}_2, \tag{41}$$

If the adaptive law is chosen as:

$$\dot{\widetilde{\Omega}}_1 = \dot{\hat{\Omega}}_1 = -\gamma_x x_p e_\delta^T P B_P, \tag{42}$$

$$\dot{\widetilde{\Omega}}_2 = \dot{\hat{\Omega}}_2 = -\gamma_\delta \delta e_\delta^T P B_P, \tag{43}$$

The Equation (41) can be written as:

$$\dot{V}_3 = e_\delta^T \left(P A_m + A_m^T P\right) e + 2\widetilde{\Omega}_1^T \left(x_p e_\delta^T P B_p + \gamma_x^{-1} \dot{\widetilde{\Omega}}_1\right) + 2\widetilde{\Omega}_2^T \left(\delta e_\delta^T P B_p + \gamma_\delta^{-1} \dot{\widetilde{\Omega}}_2\right) = -e_\delta^T P e_\delta \leq 0, \tag{44}$$

Therefore, the adaptive control scheme is established in case that $\widetilde{\Omega}_1, \widetilde{\Omega}_2, e_\delta$ are bounded and $e_\delta(t) \to 0$ as $t \to \infty$.

5. Simulation Results and Discussion

Considering the certain risk for real vehicle experiments, especially for some extreme conditions at high speed and a variety of road friction coefficients, the MATLAB/CarSim co-simulation test was conducted in this work to verify the effectiveness and advantage of the proposed control architecture. The simulation scenarios and high fidelity vehicle dynamic model is established in CarSim and the controller is developed via Simulink.

In this section, two typical test scenarios including double-lane change maneuver and slalom-like maneuver are carried out to evaluate the control performance of the adaptive robust control. These two scenarios can obviously reflect the vehicle stability and system response in the process of high-speed obstacle avoidance and continuous steering, which are widely adopted for vehicle lateral stability evaluation.

5.1. Double-Lane Change (DLC) Maneuver

This test scenario simulates the vehicle emergency obstacle avoidance behavior at high speed. The vehicle longitudinal speed is kept at 90 km/h and the road friction coefficient is set as 0.5. The front and rear tire-cornering stiffness variation are regarded as the model uncertainty, which is set as only 60% of the nominal value. The tracking performance of two controllers is compared here. One is the proposed NFTSM controller and the other is the convention sliding mode control with linear sliding surface controller. The simulation results are shown in Figure 5.

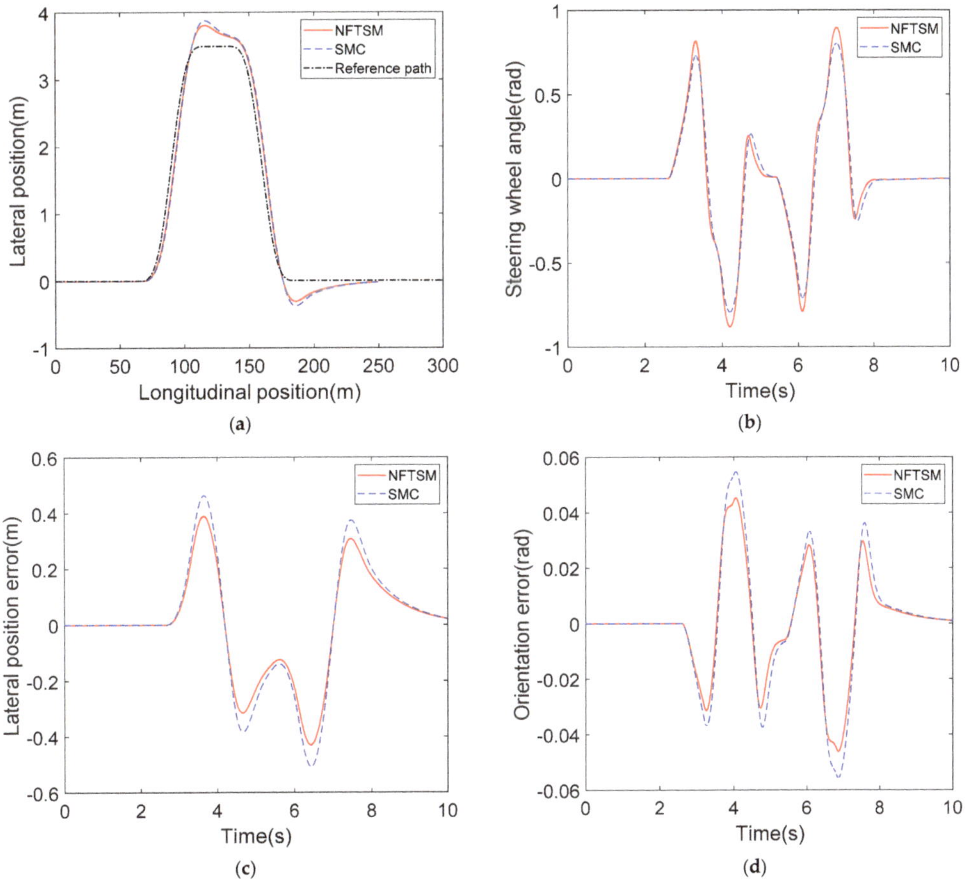

Figure 5. The simulation results for scenario 1: (**a**) Global vehicle trajectory; (**b**)Steering wheel angle; (**c**) Lateral position error; (**d**) Orientation error.

Figure 5a–d illustrate the global vehicle trajectory, steering wheel angle, lateral position tracking error, and orientation error, respectively. From Figure 5a, it can be seen that both controllers accomplish the DLC maneuver. However, due to the high initial speed, poor road friction condition, and the perturbation of the tire-cornering stiffness, they fail to follow the reference path accurately. Due to the attenuation of the tire-cornering stiffness, the steering angle calculated based on the nominal dynamic model cannot generate the corresponding lateral force. Therefore, the conventional sliding mode controller has a relatively large deviation when the vehicle changes to the adjacent lane and switches back to the main lane because of the insufficient system input response. As shown in Figure 5b, the proposed adaptive controller generates larger steering angle during the lane-change process. The adaptive sliding mode controller identifies the parameters perturbation of the system through the RBFNN and gives the controller a certain amount of compensation. Thus, the path of the adaptive sliding mode control strategy is smoother in the tracking process and the offset of the reference path is smaller. Figure 5c,d reveal that the maximum lateral position error and orientation error of the conventional sliding mode control are 0.48 m and 0.053 rad, respectively, which are only 0.43 m and 0.046 rad, respectively, for the RBFNN-based NFTSM controller. The tracking error is reduced by about 10%. Therefore, it can be demonstrated that the adaptive control algorithm proposed in this paper has higher path tracking accuracy. It also has better robustness, especially when system uncertainty or parameter perturbation occur under complex working conditions.

To further illustrate the advantage of the proposed controller, two other commonly used controllers are introduced for comparison as MPC and H_∞ robust control. Here, the tracking performance is compared under two conditions. One is when the control system is healthy (without any parameter perturbation or external disturbance) and the other one has tire-cornering stiffness variation, as stated above. The simulation results are shown in Figure 6.

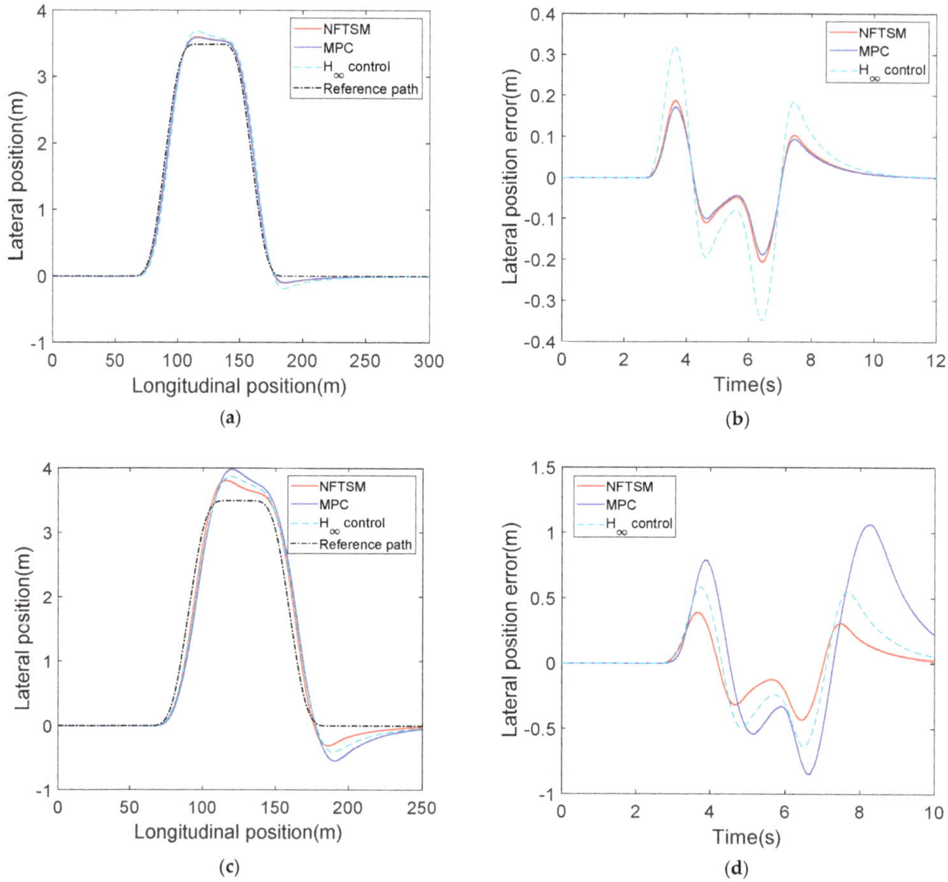

Figure 6. The comparison for three controllers: (**a**) Global vehicle trajectory with healthy system; (**b**) Lateral position error with healthy system; (**c**) Global vehicle trajectory with parameter perturbation; (**d**) Lateral position error with parameter perturbation.

The MPC is designed with a nominal vehicle dynamic model. Thus, it has the best tracking performance when the control system is totally healthy. The maximum lateral position error of MPC is only 0.17 m. On the contrary, the feedback gain matrix of the H_∞ controller is regulated by solving linear matrix inequalities considering the potential system uncertainty, which obtains a wide range of robustness by sacrificing certain control accuracy. The tracking error is largest when system is healthy, which reaches 0.31 m, although the tracking performance of the proposed NFTSM control is inferior to MPC. Its maximum lateral position error is 0.18 m, only about 7% worse than MPC. However, when parameter perturbation occurs, MPC no longer generates the optimal steering command. The tracking performance becomes the worst among these three controllers, and has the largest tracking error. In contrast, the H_∞ controller and NFTSM controller have better robustness against the tire-cornering stiffness variation. The proposed NFTSM controller, in particular, modifies the control law in real time by estimating system uncertainty online, which significantly reduces the tracking error compared with the fixed control gain of the H_∞ controller. Hence, compared with MPC and the H_∞ controller, the proposed control algorithm has better adaptability and a more balanced tracking performance in complex working conditions.

5.2. Slalom-like Maneuver

In this test scenario, the reference path is generated via continuous sinusoidal signal, which simulates the continuous obstacle avoidance maneuver. The initial vehicle speed is 80 km/h and the road friction coefficient is set as 0.85. This scenario mainly tests the system robustness and tracking accuracy under the condition that the AFS performance deteriorates. Thus, the damping coefficient and angular stiffness of the steering system are set as 50% higher than nominal value and the vehicle tire-cornering stiffness is still set as 60% of the nominal value. Similarly, to illustrate the advantage of the proposed control algorithm, another controller without MRAC is introduced here as the comparison. It has the same upper NFTSM controller but the steering torque is directly calculated by Equation (33) based on the nominal moment of inertia, damping coefficient, and angular stiffness of the steering system. The detailed simulation results are shown in Figure 7.

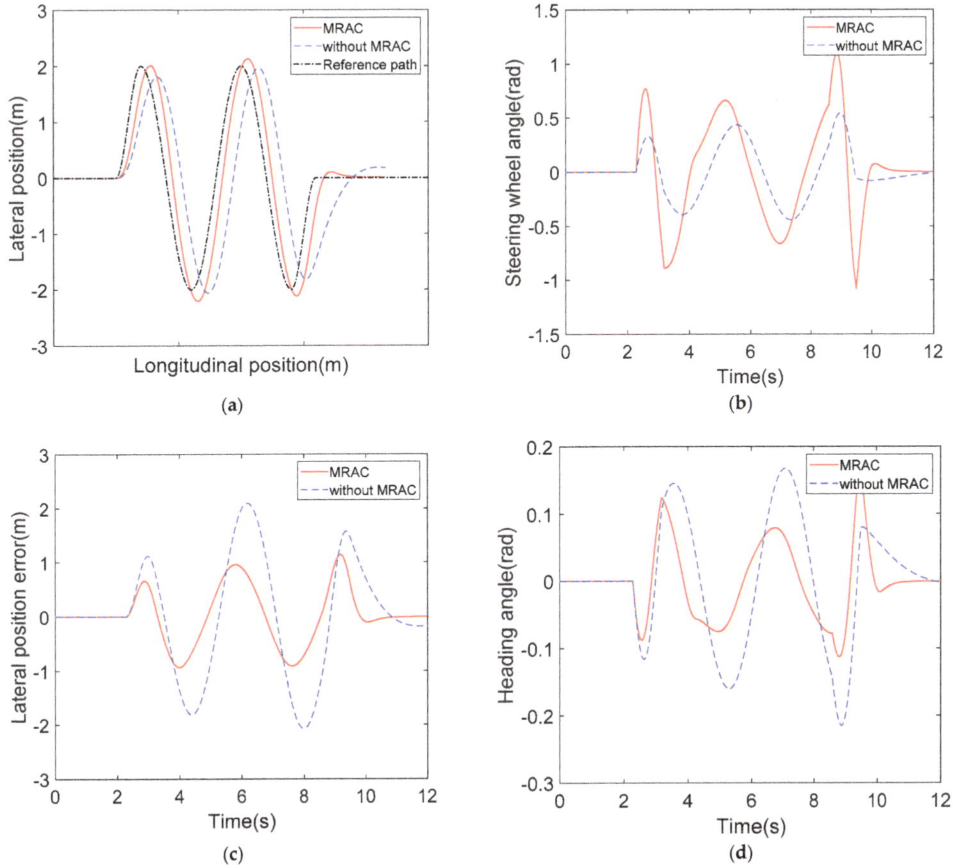

Figure 7. The simulation results for scenario 2: (**a**) Global vehicle trajectory; (**b**) Steering wheel angle; (**c**) Lateral position error; (**d**) Orientation error.

Figure 7a–d illustrate the global vehicle trajectory, steering wheel angle, lateral position tracking error, and orientation error, respectively. Due to the increase in the damping coefficient and angular stiffness of the steering system dynamics model, the system response speed decreases. The steering torque calculated by the controller without MRAC is based on the nominal model parameters, which is unable to generate enough steering wheel angle in time for the real steering system. It can be seen from Figure 7a,b that the response speed of steering wheel angle is slow, and there is a large time delay between the actual vehicle trajectory and the reference path. In addition, during the process of continuous steering for obstacle avoidance, the phenomenon of time delay is further amplified. On the contrary, the proposed controller in the paper adjusts the steering torque by comparing the response error between the actual steering system and reference model, which has a faster response speed and higher tracking accuracy. From Figure 7c,d, it is clear that the maximum lateral position error and orientation error of the control algorithm without MRAC are 2.1 m and 0.22 rad, respectively, which are only 1.2 m and 0.14 rad, respectively, for the proposed controller in this paper.

6. Conclusions

This paper proposed a cascade adaptive robust control architecture that overcomes the influence of multi-dimensional system uncertainty on control performance. The higher-level controller regarded the dynamic model error and parameter disturbance as the system uncertainty and adopted the NFTSM control algorithm with RBFNN estimator to generate the steering wheel angle command. The lower-level controller applied MRAC, considering the uncertainty of stiffness and damping coefficient in AFS to realize the accurate steering torque control. The system stability and finite time convergence are proven by choosing the appropriate Lyapunov function. Two simulation scenarios are carried out to validate the feasibility of the proposed control architecture. The simulation results reveal that the RBFNN-based NFTSM has satisfactory tracking performance under different road friction coefficients and has good robustness against varying tire-cornering stiffness. Meanwhile, it has better control accuracy and a faster convergence rate compared with the conventional sliding mode controller. On the other hand, the model reference adaptive steering torque controller could generate the accurate steering torque that follows the steering wheel angle command with the variation in the steering model parameters or under the circumstance that system performance deteriorates.

Author Contributions: M.C. and Y.R.; methodology, M.C. and Y.R.; software, M.C.; validation, Y.R.; formal analysis, Y.R. and M.O.; investigation, M.O.; data curation, Y.R. and M.O.; writing—original draft preparation, M.C.; writing—review and editing, Y.R. and M.O.; supervision, Y.R.; funding acquisition, Y.R. and M.O. All authors have read and agreed to the published version of the manuscript.

Funding: This work is supported by Natural Science Foundation of Chongqing (cstc2020jcyj-msxmX0496, cstc2021jcjy-msxmX0532), Science and Technology Research Program of Chongqing Municipal Education Commission (KJQN202103401), and The Program for Innovation Research Groups at Institutions of Higher Education in Chongqing (CXQT21032).

Data Availability Statement: Not applicable.

Conflicts of Interest: The authors declare no conflict of interest.

References

1. Gray, A.; Ali, M.; Gao, Y.; Hedrick, J.K.; Borrelli, F. A unified approach to threat assessment and control for automotive active safety. *IEEE Trans. Intell. Transp. Syst.* **2013**, *14*, 1490–1499. [CrossRef]
2. Kim, H.; Kim, D.; Shu, I.; Yi, K. Time-varying parameter adaptive vehicle speed control. *IEEE Trans. Veh. Technol.* **2015**, *65*, 581–588. [CrossRef]
3. Lian, Y.; Zhao, Y.; Hu, L.; Tian, Y. Longitudinal collision avoidance control of electric vehicles based on a new safety distance model and constrained-regenerative-braking-strength-continuity braking force distribution strategy. *IEEE Trans. Veh. Technol.* **2015**, *65*, 4079–4094. [CrossRef]

4. Li, S.; Li, K.; Rajamani, R.; Wang, J. Model predictive multi-objective vehicular adaptive cruise control. *IEEE Trans. Control Syst. Technol.* **2010**, *19*, 556–566. [CrossRef]
5. Amer, N.H.; Hairi, Z. Modelling and Control Strategies in Path Tracking Control for Autonomous Ground Vehicles: A Review of State of the Art and Challenges. *J. Intell. Robot. Syst.* **2017**, *86*, 225–254. [CrossRef]
6. Kapania, N.R.; Gerdes, J.C. Design of a feedback-feedforward steering controller for accurate path tracking and stability at the limits of handling. *Veh. Syst. Dyn.* **2015**, *53*, 1687–1704. [CrossRef]
7. Salehpour, S.; Pourasad, Y.; Taheri, S.H. Vehicle path tracking by integrated chassis control. *J. Central South Univ.* **2015**, *22*, 1378–1388. [CrossRef]
8. Xu, S.; Peng, H. Design, analysis, and experiments of preview path tracking control for autonomous vehicles. *IEEE Trans. Intell. Transp. Syst.* **2019**, *21*, 48–58. [CrossRef]
9. Ji, J.; Khajepour, A.; Melek, W.W.; Huang, Y. Path planning and tracking for vehicle collision avoidance based on model predictive control with multiconstraints. *IEEE Trans. Veh. Technol.* **2016**, *66*, 952–964. [CrossRef]
10. Chen, Y.; Hu, C.; Wang, J. Human-Centered Trajectory Tracking Control for Autonomous Vehicles with Driver Cut-In Behavior Prediction. *IEEE Trans. Veh. Technol.* **2019**, *68*, 8461–8471. [CrossRef]
11. Guo, H.; Liu, J.; Cao, D.; Chen, H.; Yu, R.; Lv, C. Dual-envelop-oriented moving horizon path tracking control for fully automated vehicles. *Mechatronics* **2018**, *50*, 422–433. [CrossRef]
12. Funke, J.; Brown, M.; Erlien, S.M.; Gerdes, J.C. Collision avoidance and stabilization for autonomous vehicles in emergency scenarios. *IEEE Trans. Control. Syst. Technol.* **2016**, *25*, 1204–1216. [CrossRef]
13. Brown, M.; Funke, J.; Erlien, S.; Gerdes, J.C. Safe driving envelopes for path tracking in autonomous vehicles. *Control Eng. Pr.* **2017**, *61*, 307–316. [CrossRef]
14. Jing, H.; Hu, C.; Yan, F.; Chadli, M.; Wang, R.; Chen, N. Robust H∞ output-feedback control for path following of autonomous ground vehicles. *Mech. Syst. Sig. Pro.* **2016**, *70–71*, 414–427.
15. Guo, J.; Luo, Y.; Li, K. Robust gain-scheduling automatic steering control of unmanned ground vehicles under velocity-varying motion. *Veh. Syst. Dyn.* **2018**, *57*, 595–616. [CrossRef]
16. Guo, J.; Wang, J.; Hu, P.; Li, L. Robust guaranteed-cost path-following control for autonomous vehicles on unstructured roads. *Proc. Inst. Mech. Eng. Part D J. Automob. Eng.* **2017**, *232*, 896–908. [CrossRef]
17. Chen, W.; Zhao, L.; Wang, H.; Huang, Y. Parallel Distributed Compensation /H∞ Control of Lane-keeping System Based on the Takagi-Sugeno Fuzzy Model. *Chin. J. Mech. Eng.* **2020**, *33*, 61. [CrossRef]
18. Akermi, K.; Chouraqui, S.; Boudaa, B. Novel SMC control design for path following of autonomous vehicles with uncertainties and mismatched disturbances. *Int. J. Dyn. Control* **2018**, *8*, 254–268. [CrossRef]
19. Ao, D.; Huang, W.; Wong, P.K.; Li, J. Robust backstepping super-twisting sliding mode control for autonomous vehicle path following. *IEEE Access* **2021**, *9*, 123165–123177. [CrossRef]
20. Sun, X.; Wang, Y.; Cai, Y.; Wong, P.K.; Chen, L. An adaptive nonsingular fast terminal sliding mode control for yaw stability control of bus based on STI tire model. *Chin. J. Mech. Eng.* **2021**, *34*, 79. [CrossRef]
21. Norouzi, A.; Masoumi, M.; Barari, A.; Sani, S.F. Lateral control of an autonomous vehicle using integrated backstepping and sliding mode controller. *Pro. Inst. Mech. Eng. Part K J. Mult. Dyn.* **2019**, *233*, 141–151. [CrossRef]
22. Yang, L.; Yue, M.; Ma, T. Path Following Predictive Control for Autonomous Vehicles Subject to Uncertain Tire-ground Adhesion and Varied Road Curvature. *Int. J. Control Autom. Syst.* **2019**, *17*, 193–202. [CrossRef]
23. Aiguo, W.; Liu, H.; Dong, N. Nonsingular Fast Terminal Sliding Mode Control of Robotic Manipulators Based on Neural Networks. *Trans. Chin. Soc. Agri. Mach.* **2018**, *44*, 395–404.
24. Pham, C.V.; Wang, Y.N. Robust Adaptive Trajectory Tracking Sliding mode control based on Neural networks for Cleaning and Detecting Robot Manipulators. *J. Intell. Robot. Syst.* **2015**, *79*, 101–114. [CrossRef]
25. Yong, F.; Yu, X.; Man, Z. Non-singular terminal sliding mode control of rigid manipulators. *Automatica* **2002**, *38*, 2159–2167.
26. Ioannou, P.A.; Sun, J. *Robust Adaptive Control*; Prentice-Hall, Inc.: Hoboken, NJ, USA, 1995.

World Electric Vehicle Journal

MDPI

Article

Research on Active Collision Avoidance and Hysteresis Reduction of Intelligent Vehicle Based on Multi-Agent Coordinated Control System

Chaochun Yuan [1,*], Yongfeng Lin [1], Jie Shen [2], Long Chen [1], Yingfeng Cai [1], Youguo He [1], Shuofeng Weng [3], Xinkai Wu [1], Yuqi Yuan [4], Yuxuan Gong [5] and Qiuye Yu [6]

1 Automotive Engineering Research Institute, Jiangsu University, Zhenjiang 212013, China
2 Department of Computer and Information Science, University of Michigan-Dearborn, Dearborn, MI 48128-1491, USA
3 School of Agricultural Engineering, Jiangsu University, Zhenjiang 212013, China
4 School of Automotive and Traffic Engineering, Jiangsu University, Zhenjiang 212013, China
5 College of Foreign Languages, Northeastern University, Shenyang 110004, China
6 China Automotive Technology & Research Center Co., Ltd., Tianjin 300399, China
* Correspondence: 1000003481@ujs.edu.cn

Abstract: This paper provides a multi-agent coordinated control system to improve the real-time performance of intelligent vehicle active collision avoidance. At first, the functions and characteristics of longitudinal and lateral collision avoidance agents are analyzed, which are the main components of the multi-agent. Then, a coordinated solution mechanism of an intelligent vehicle collision avoidance system is established based on hierarchical control and blackboard model methods to provide a reasonable way to avoid collision in complex situations. The multi-agent coordinated control system can handle the conflict between the decisions of different agents according to the rules. Comparing with existing control strategies, the proposed system can realize multi decisions and planning at the same time; thus, it will reduce the operation time lag during active collision avoidance. Additionally, fuzzy sliding mode control theory is introduced to guarantee accurate path tracking in lateral collision avoidance. Finally, co-simulation of Carsim and Simulink are taken, and the results show that the real-time behavior of intelligent vehicle collision avoidance can be improved by 25% through the system proposed.

Keywords: multi-agent coordinated control system; active collision avoidance; blackboard model; real-time

Citation: Yuan, C.; Lin, Y.; Shen, J.; Chen, L.; Cai, Y.; He, Y.; Weng, S.; Wu, X.; Yuan, Y.; Gong, Y.; et al. Research on Active Collision Avoidance and Hysteresis Reduction of Intelligent Vehicle Based on Multi-Agent Coordinated Control System. *World Electr. Veh. J.* **2023**, *14*, 16. https://doi.org/10.3390/wevj14010016

Academic Editors: Yong Li, Xing Xu, Lin Zhang, Yechen Qin and Yang Lu

Received: 17 October 2022
Revised: 29 November 2022
Accepted: 3 January 2023
Published: 5 January 2023

1. Introduction

With the development of the intelligent vehicle, studies on the active collision avoidance system of intelligent vehicles have attracted more and more attention. Based on the perception of the driving environment, the intelligent vehicle can avoid collision risk by braking or steering.

To improve the performance of intelligent vehicles' active collision avoidance system, researchers have carried out effective research on longitudinal and lateral avoidance. In the aspect of longitudinal collision avoidance, Li Suhua et al. [1] proposed a longitudinal collision avoidance method for electric vehicles, establishing a safe distance model with consideration of road adhesion coefficient and driving intention. Li Shifu et al. [2] made a theoretical derivation of the critical distance of the warning and the critical distance based on the braking process and obtained a safe distance model considering the emergency of the preceding vehicle. Considering the characteristics of vehicle dynamics and synthesizing the influence of road environment and vehicle factors, an RV hierarchical safety distance model was established in the paper [3]. Hou Dezao et al. [4] designed the upper controller based on the optimal tracking theory and driver priority principle.

In the aspect of lateral collision avoidance, Boada et al. [5] designed an emergency steering path tracking controller with a fuzzy control logic method based on vehicle yaw rate. Soudbakhsh et al. [6] constructed the state equation with actual lateral acceleration, ideal lateral acceleration, yaw angle, and ideal yaw angle error. Li Wei et al. [7] proposed a lane change path planning method with an RBF neural network in which the boundary conditions of the path planning algorithm and path change based on polynomials are designed. This method has advantages in complex road conditions. Papers [8,9] studied the braking and steering modes based on the analysis of the vehicle braking process to design a longitudinal safety distance model under various working conditions and braking controllers [8]. In paper [9], the longitudinal and lateral safety distance models were also designed with different collision avoidance methods by analyzing the state of the preceding vehicle. Paper [10] introduced the advantages and shortcomings of the traditional APF method, solving the problem of excessive initial attractive force and the intelligent vehicle cannot reach the target by improving the potential field functions.

It can be seen from the above results that most active collision avoidance systems currently focus on a single collision avoidance method, and there is no reasonable integration of longitudinal and lateral collision avoidance. With the increasingly complex driving conditions of smart cars, the independent active collision avoidance method is difficult to meet the driving requirements of intelligent cars due to its poor flexibility. Therefore, under the premise of ensuring the timeliness of active safety control, it is of great significance to design a comprehensive coordinated control strategy for longitudinal collision avoidance and lateral collision avoidance. At present, most researchers believe that an agent is a computing entity with a life cycle that exists in a specific environment and has the characteristics of real-time perception of the surrounding environment and the ability to operate independently and affect the environment [11,12]. A single agent mainly has four basic characteristics: autonomy, sociality, responsiveness, and initiative. A Multi-Agent System (MAS, Multi-Agent System) is composed of multiple single agents. Through the coordinated control of each agent, its problem-solving ability is far beyond the ability of a single agent, so the multi-agent system is widely used in the coordinated control of complex systems [13,14]. A novel hybrid artificial intelligence-layered multi-agent architecture was presented in the paper [15] to help the digital transformation of energy and the smart grid.

In this paper, an intelligent vehicle active collision avoidance method based on a multi-agent coordinated control system is designed. The longitudinal and the lateral collision avoidance agents are designed based on the blackboard model, to provide reasonable collision avoidance way under different driving conditions. This proposed system can realize the multi-parallel operation of decision and planning at the same time. The lateral and the longitudinal collision avoidance agent can provide collision avoidance planning simultaneously, which can achieve the integration of collision avoidance decisions and planning. It will help reduce the time lag caused by the collision avoidance decision-planning process and improve real-time performance.

The rest of this paper is organized as follows. In Section 2, the main agents of the intelligent vehicle are produced. In Section 3, the blackboard model is introduced to coordinated control of each agent, and the real-time performance of the multi-agent active collision avoidance system is evaluated through simulations in Section 4, followed by some concluding remarks in Section 5.

2. Main Agents

Decision-making agents with lateral collision avoidance and longitudinal collision avoidance are designed in this section to provide decisions for collision avoidance.

2.1. Longitudinal Collision Avoidance Agent

The safety distance model [16] of the longitudinal collision avoidance agent is designed as Equations (1) and (2):

$$S_l = \frac{V_r^2}{2a_{r-max}} + V_r(t_{detect1} + t_{decision1} + t_{excution1}) + d_0 \tag{1}$$

$$a_{r-max} = \begin{cases} \mu g cos\alpha - g sin\alpha & \text{(downhill)} \\ \mu g & \text{(zero slope)} \\ \mu g cos\alpha + g sin\alpha & \text{(uphill)} \end{cases} \tag{2}$$

where $t_{detect1}$ is the environment perception time, $t_{decision1}$ is the decision-planning time of the longitudinal collision avoidance agent, $t_{excution1}$ is the mechanical delay time, V_r is the vehicle speed, a_{r-max} is the maximum braking deceleration of the vehicle, d_0 is the minimum safety threshold between vehicles, μ is the coefficient of road adhesion, g is the acceleration of gravity and α is the vehicle slope angle.

A longitudinal collision avoidance agent is designed to get a safe distance from obstacles in front of the intelligent vehicle. Brake pressure will be calculated if a smaller distance between the obstacle and the intelligent vehicle than the safety distance is detected.

The analysis of vehicle forces during braking is made to get the brake pressure and the diagram is shown in Figure 1.

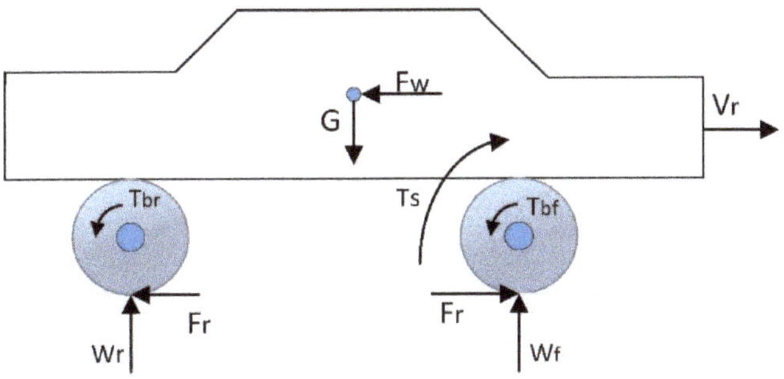

Figure 1. Diagram of vehicle force during braking.

The force balance equation of the vehicle is shown in Equation (3):

$$ma_{des} = F_t - F_{Xb} - \sum F(V_r) \tag{3}$$

F_w is the air resistance force, T_s is the driving torque, T_{bf} and T_{br} are the braking torque of the front and rear wheels, F_f and F_r are the ground friction of the ground acting on the front and rear wheels, W_f and W_r are the vertical force of the front and rear wheels, F_t is the driving force, F_{Xb} is the braking force, and $\sum F(V_r)$ is the total resistance.

Air resistance and ground friction are shown in Equation (4):

$$\sum F(V_r) = \frac{1}{2} C_D A_a \rho v V_r^2 + mgf \tag{4}$$

Desired braking pressure can be calculated based on Equations (3) and (4), which is shown in Equation (5):

$$P_{des} = \frac{|ma_{des} + \frac{1}{2} C_D A_a \rho v V_r^2 + mgf|}{K_b} \tag{5}$$

P_{des} is the desired braking pressure and K_b is the braking pressure ratio.

Most important of all, the braking force is represented approximately as a linear function of oil pressure in the brake system, as shown in Equation (6):

$$\frac{T_{bf} + T_{br}}{r_r} = K_b P_b \tag{6}$$

r_r is the rolling radius of wheels, and P_b is the pressure of the brake pipe.

2.2. Lateral Collision Avoidance Agent

A fifth-order polynomial lane-changing model is used in this paper to present a lateral collision avoidance agent [17]:

$$y(x) = y_e[10(\frac{x}{x_e})^3 - 15(\frac{x}{x_e})^4 + 6(\frac{x}{x_e})^5] \tag{7}$$

y_e is the lateral displacement for the vehicle to avoid a collision.

Longitudinal speed V_r is considered constant during the lane-changing and the relationship between trajectory and time can be shown as Equation (8):

$$y(t) = y_e[10(\frac{t}{t_e})^3 - 15(\frac{t}{t_e})^4 + 6(\frac{t}{t_e})^5] \tag{8}$$

t_e is the lane-changing time, $x_e = V_r t_e$.

The equation of lateral acceleration in the course of vehicle lane change can be obtained based on Equation (8), which is shown in Equation (9):

$$\ddot{y(t)} = a_y(t) = \frac{60y_e}{t_e^5}[2t^3 - 3t_e t^2 + t_e^2 t] \tag{9}$$

The maximum lateral acceleration during the process can be calculated by Equation (10):

$$a_{ymax} = \frac{10\sqrt{3}y_e}{3t_e^2} \tag{10}$$

It can be seen from Equation (10) that the maximum lateral acceleration during the lane-changing process is related to the lane-changing time t_e and the lateral distance y_e. The minimum lane-changing time of the intelligent vehicle on dry asphalt pavement and wet asphalt pavement is set to 1.68 s and 2.1 s, and the maximum lateral accelerations are 7.67 m/s^2 and 4.91 m/s^2.

As shown in Figure 2, to avoid collision with the front obstacle, the lateral displacement of the right-corner vehicle should be greater than the width of the obstacle W_b [18].

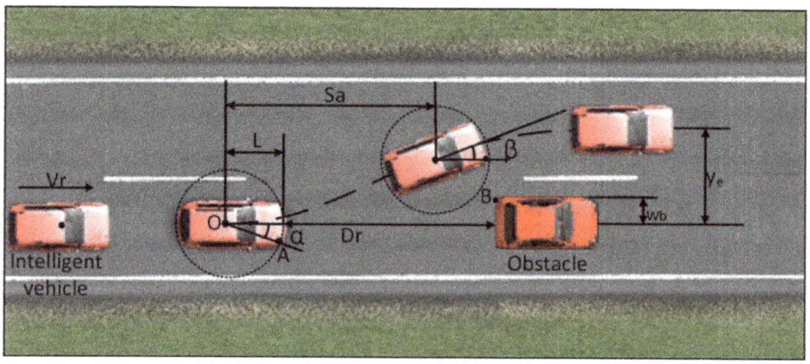

Figure 2. Intelligent vehicle lateral collision avoidance route.

Assuming that there is no vehicle in adjacent lanes. The lateral displacement of point A should meet the requirement of Equation (11) based on the path provided in Equation (11):

$$W_b = y_e\left[10\left(\frac{t_c}{t_e}\right)^3 - 15\left(\frac{t_c}{t_e}\right)^4 + 6\left(\frac{t_c}{t_e}\right)^5\right] - \left(\frac{W_s}{2}\right) \tag{11}$$

where W_s is the vehicle width, t_c is the collision time, W_b is the lateral distance between the obstacle edge and the vehicle center.

The longitudinal displacement S_a of the vehicle can be calculated as Equation (12):

$$S_a = V_r(t_c + t_{detect2} + t_{decision2} + t_{excution2}) \tag{12}$$

where $t_{detect2}$ is the perception delay, $t_{decision2}$ is the decision-planning time and $t_{excution2}$ is the steering mechanical delay.

The minimum longitudinal safety distance to accomplish horizontal change is shown in Equation (13):

$$S_{fmin} = S_a + L_{OA} - L + d_0 \tag{13}$$

The lateral collision avoidance agent is designed to calculate the desired steering wheel angle to follow the preset collision avoidance trajectory. A fuzzy sliding mode control, which has good robustness and real-time performance, is introduced to ensure the accuracy of the path tracking of the vehicle during the lateral collision avoidance process.

The yaw rate and derivative of the sideslip angle based on the vehicle two DOF model are shown in Equation (14) [19–21]:

$$\begin{cases} \dot{\omega} = \frac{a^2C_f + b^2C_r}{I_z v_r}\omega + \frac{aC_f - bC_r}{I_z}\beta - \frac{aC_f}{I_z}\delta \\ \dot{\beta} = \left(\frac{aC_f - bC_r}{Mv_r^2} - 1\right)\omega + \frac{C_f + C_r}{Mv_r}\beta - \frac{C_f}{Mv_r}\delta \end{cases} \tag{14}$$

β is the sideslip angle. v_y is the lateral speed. M is the vehicle mass. ω is the yaw rate. a and b are the front and rear wheelbase. I_z is the vehicle's moment of inertia. C_f and C_r represent the stiffness of the front and rear tires.

The vehicle yaw rate of the vehicle in this paper can be expressed as Equation (15):

$$\begin{gathered} \dot{\omega}_r = a_{11}\omega_r + a_{12}\beta + b_{11}u(t) \\ b_{11} = \frac{aC_f}{I_z} \\ a_{11} = \frac{a^2C_f + b^2C_r}{I_z v_r} \\ a_{21} = \frac{aC_f - bC_r}{Mv_r^2} - 1 \end{gathered} \tag{15}$$

The vehicle yaw rate is chosen as the controlled variable. The tracking error between the yaw rate and the ideal yaw rate can be shown in Equation (16):

$$e = \omega_r - \omega_d \tag{16}$$

The controller switching function is designed as Equation (17):

$$s = e + \gamma\int_0^t e(\tau)d\tau \tag{17}$$

where γ is the sliding surface gain.

The sliding mode control law is designed based on Equations (15)–(17), which is shown in Equations (18) and (19):

$$u = \frac{1}{b_{11}}[-f(\omega_r) + \gamma(\omega_r - \omega_d) + K(t)sgn(s)] \tag{18}$$

$$K(t) = -ksgn(s) \quad k > 0 \tag{19}$$

Control of front wheel angle can be expressed as Equation (20):

$$\delta = \frac{I_z}{aC_f}\left[\frac{aC_f - bC_r}{I_z}\beta + \frac{a^2C_f + b^2C_r}{I_zV_r}\omega_r - \omega_d + \gamma(\dot{\omega}_r - \omega_d)\right] \tag{20}$$

The steering wheel angle can be obtained as Equation (21):

$$\theta_s = \delta * i_{SW} \tag{21}$$

i_{sw} is the steering system ratio.

3. Multi-Agent Coordinated Control System Based on Blackboard Mode

Longitudinal and lateral collision avoidance agents are taken into consideration to ensure the multi-agent coordinated control system deals with the traffic accident risk. Additionally, global path planning agents, path tracking agents, and actuator control agents are taken as fundamental agents in the system.

Each agent of a multi-agent coordinated control system can be carried out in its default mode. Thus, the conflict problem in the driving process is generally classified into three categories according to the cause: resource conflicts, target conflicts, and result conflicts [22].

As can be seen from Figure 3, conflicts between longitudinal and lateral collision avoidance agents are easily issued. When front obstacles are detected by vehicle sensors, safe distance will be calculated by the longitudinal collision avoidance agent and braking force will be transmitted to the active braking agent. Correspondingly, steering control signals will be transmitted to the active steering agent by the lateral collision avoidance agent. Different solutions for front-distance avoidance may lead to the result of conflicts in multi-agent systems.

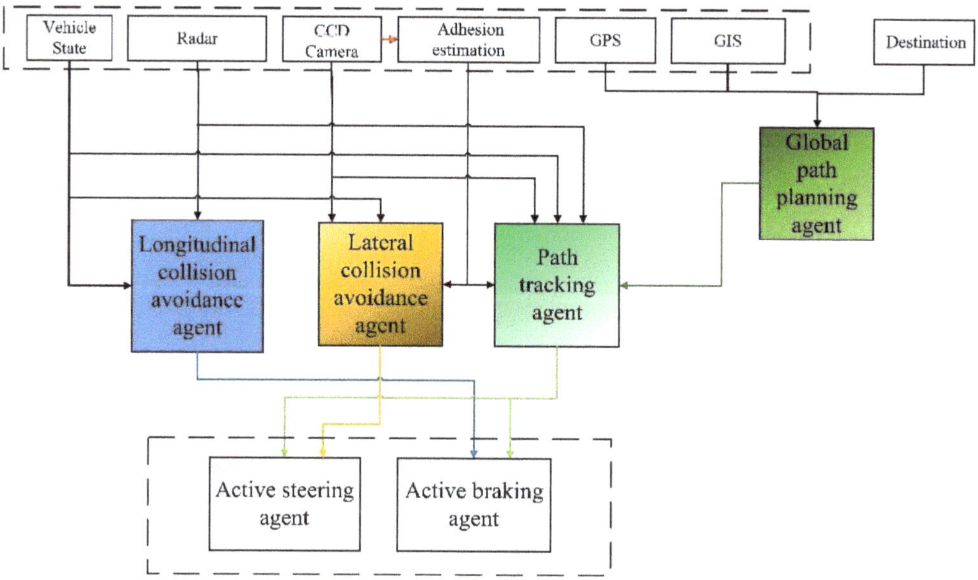

Figure 3. Topology of multi-agent coordinated control system.

The blackboard model is used to solve result conflicts in this section [23], in which each agent exchanges data and writes its solution on the blackboard. The model is composed of three basic components: blackboard, knowledge source, and control mechanism. The internal coordination module manages the data on the blackboard in a unified manner. When there is a collision avoidance decision conflict between the agents, the coordination module can choose a reasonable collision avoidance method according to the internal rule base. A multi-agent coordinated control system based on a blackboard model is shown in Figure 4. As shown in Figure 4, the multi-agent active collision avoidance decision-making system based on the blackboard model is divided into three layers: a planning layer, decision and coordination layer, and execution control layer. Longitudinal and lateral collision avoidance agents constitute the planning layer of the system, which also includes the basic global path planning agent, path tracking agent, and actuator control agents. They can be regarded as different knowledge sources, and each agent can interact with the blackboard model to obtain information and complete a relatively independent and complete problem-solving. In the decision and coordination layer, the blackboard module is used to store the environmental information obtained by the environment perception system of the intelligent vehicle and the solution results of each agent, and the internal coordination module is used to manage the data on the blackboard in a unified way. When encountering collision avoidance decision conflicts between agents, the coordination module needs to choose a reasonable collision avoidance method according to the internal rule base. The executive control layer includes the active braking control agent and the active steering agent, which interacts with the blackboard model and finally, executes the decision of the multi-agent collision avoidance control system.

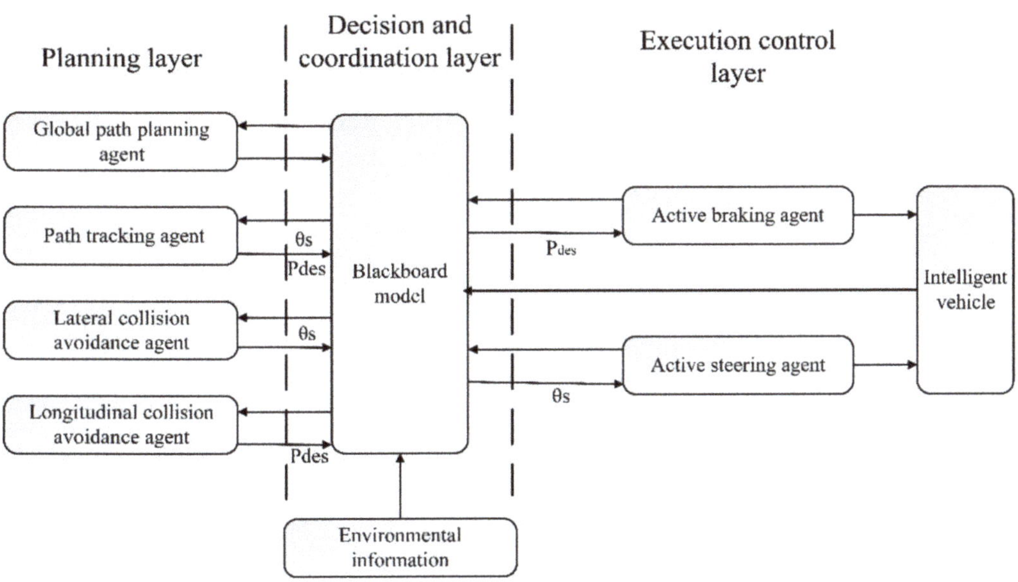

Figure 4. Multi-agent coordinated control system based on blackboard model.

During the driving process, the information acquired by the environment perception layer is uploaded to the blackboard in real time, and each agent obtains information by interacting with the blackboard. When an obstacle appears on the default path, longitudinal and lateral collision avoidance agents are activated. The brake pressure and steering wheel angle are uploaded to the blackboard separately. If the two agents are activated at the same time, a choice must be made by the coordinated control module according to the actual situation.

Figure 5 shows the choice of active collision avoidance during an emergency in different environments.

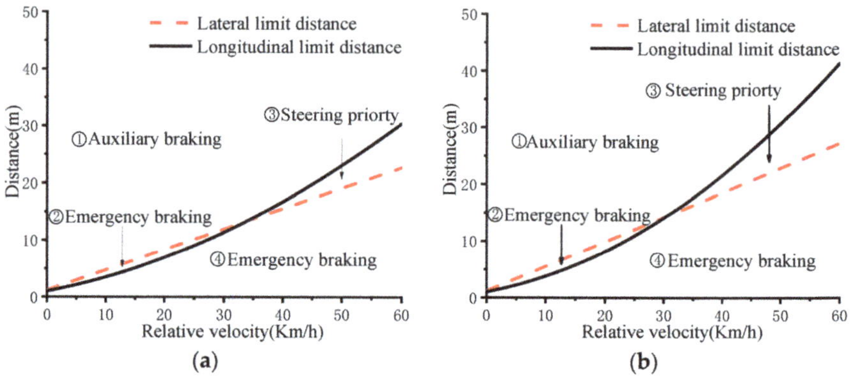

Figure 5. The choice of active collision avoidance during an emergency in different environments: (**a**) Dry asphalt pavement (μ = 0.8); (**b**) Wet asphalt pavement (μ = 0.5).

Longitudinal lateral collision avoidance agents are plotted in Figure 5a,b. The minimum safety distance required for emergency longitudinal collision avoidance is related to the square of the vehicle speed, and the minimum longitudinal safety distance required for emergency lateral collision avoidance is related to the square of the vehicle speed. When the vehicle speed is in the lower range, the collision avoidance distance required for longitudinal collision avoidance is small. With the increase in speed, the collision avoidance distance required for longitudinal collision avoidance increases rapidly, and the longitudinal collision avoidance distance required for lateral collision avoidance begins to be smaller than that for longitudinal collision avoidance. It can be seen that D_b in area ① is greater than the longitudinal and lateral collision limit distance. In this case, the vehicle faces no risk and the auxiliary braking mode is adopted. At this time, more attention should be paid to the traffic efficiency and occupant comfort of the vehicle. The absolute value of the vehicle deceleration is limited to 4 m/s^2 or less.

When located in areas ② and ④, the maximum braking deceleration should be adopted in this emergency condition to reduce the risk of traffic accidents.

When D_b is located in area ③, the speed of the vehicle is higher and the longitudinal distance required for collision avoidance is large. So, a lane-changing strategy will be taken to avoid a collision.

According to the analysis above, the decision and judgment process of the coordinated module is shown as follows, which is designed to make sure the vehicle drives along the planned path. When an obstacle ahead is detected by an intelligent vehicle, the distance between the intelligent vehicle and the obstacle must be compared with the safe distance of the limit collision distance of the longitudinal collision avoidance agent and the lateral collision avoidance agent, firstly. If the distance D_b is longer than the braking limit distance, then it is according to the braking acceleration to judge whether to use auxiliary braking mode or emergency braking mode. However, if the distance is shorter than the braking distance and longer than the steering limit distance, which will use the steering collision avoidance mode.

The active collision avoidance control strategy in this paper is shown in Figure 6a and the collision avoidance control strategy is shown in Figure 6b. T_1 is the planning time of longitudinal collision avoidance, T_2 is the decision time of collision avoidance way, T_3 is the planning time of lateral collision avoidance, T_4 is the execution control time.

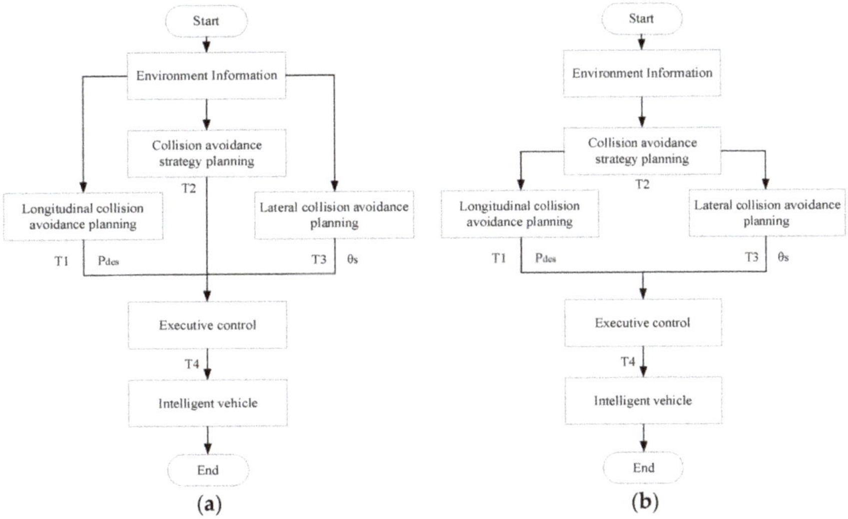

Figure 6. Active collision avoidance system control method: (**a**) Distributed coordinated control system; (**b**) Sequential control system.

As shown in Figure 6a, time consumption $T_{D-Braking}$ and $T_{S-Braking}$used during the entire collision avoidance can be calculated as Equations (22) and (23):

$$T_{D-Braking} = max\{T_1,\ T_2\} + T_4 \, or \, T_{D-Steering} = max\{T_2,\ T_3\} + T_4 \tag{22}$$

$$T_{S-Braking} = T_2 + T_1 + T_4 \, or \, T_{S-Steering} = T_2 + T_3 + T_4 \tag{23}$$

As can be seen from Equations (22) and (23), the separation of the collision avoidance decision and the collision avoidance planning will take more time.

The distributed multi-agent coordinated system based on the blackboard model performs parallel operations. The lateral and longitudinal collision avoidance agents, respectively, solve the steering wheel angle and brake pressure required for collision avoidance. At the same time, the blackboard coordinated module selects the optimal collision avoidance way according to the environmental information and can directly output the control instruction to agents of the execution control layer. By rationally unifying the decision and planning, the running time lag of the active collision avoidance system is reduced, and the real-time behavior of the vehicle collision avoidance is effectively improved.

4. Simulation Analysis

To verify the real-time performance of the distributed coordinated multi-agent system of the active collision avoidance proposed, the longitudinal collision avoidance agent, the lateral collision avoidance agent, and the blackboard model were built in CarSim and Simulink.

Some of the important vehicle parameters are given in Table 1. Most of them are directly measured from a B-class vehicle.

Co-simulation results are compared with the Sequential control system collision avoidance and the co-simulation model is shown in Figure 7.

Two different conditions are adopted to characterize the effectiveness of a multi-agent coordinated control system and superiority in improving the real-time performance of active collision avoidance. The vehicle that used the distributed coordinated multi-agent control system of active collision avoidance based on the blackboard proposed in this paper is recorded as vehicle A and the vehicle that used the sequential control system of active collision avoidance is recorded as vehicle B.

Table 1. Basic parameters of the vehicle.

Parameter	Value	Unit
Mass (m)	1274	kg
Distance from c.g.to front axle (a)	1.8	m
Distance from c.g.to rear axle (b)	1.31	m
Atmospheric density (ρ)	1.206	kg/m^3
Frontal area (A_a)	1.6	m^2
The rolling radius of wheels (r_r)	0.31	m
Intelligent vehicle width (W_s)	1.695	m
Stiffness of the front tire (C_f)	976.24	N/rad
Stiffness of the rear tire (C_r)	980.18	N/rad
Coefficient of air resistance (C_D)	0.3	1
Inertia around the vertical shaft (I_z)	1523	kg·m^2

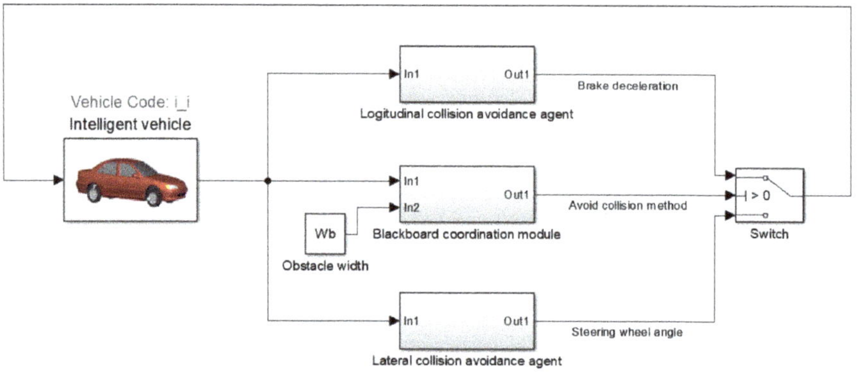

Figure 7. Distributed multi-agent collision avoidance system simulation.

4.1. Simulation of the Longitudinal Collision Avoidance Agent

The vehicle speed is set as 36 km/h, and suddenly, there is an obstacle that falls 10 m in front of the vehicle. The green dotted line in the figure is the moment when the obstacle appears. At this time, D_b is located in the area ④. Simulation results are shown in Figure 8.

Results show that after detecting the obstacle, the vehicle selects braking to avoid a collision. As can be seen from Figure 8c, the brake pressure output time of vehicle A is 0.13 s ahead of vehicle B. What is more, from Figure 8a,b, we can see that vehicle A is 3.12 m away from the obstacle at 3.1 s when it stops, and vehicle B is 1.4 m away from the obstacle at 3.18 s when it stops. In the face of sudden emergency conditions, the coordinated control system adopted by vehicle A can make decisions and simultaneously conduct collision avoidance planning to realize the integration of collision avoidance decision-making and planning, so that vehicle A will provide safer control. So, the data from the simulation experiments show that a multi-agent coordinated control system improves the real-time performance of active collision avoidance, which has a shorter braking distance than the sequential control system.

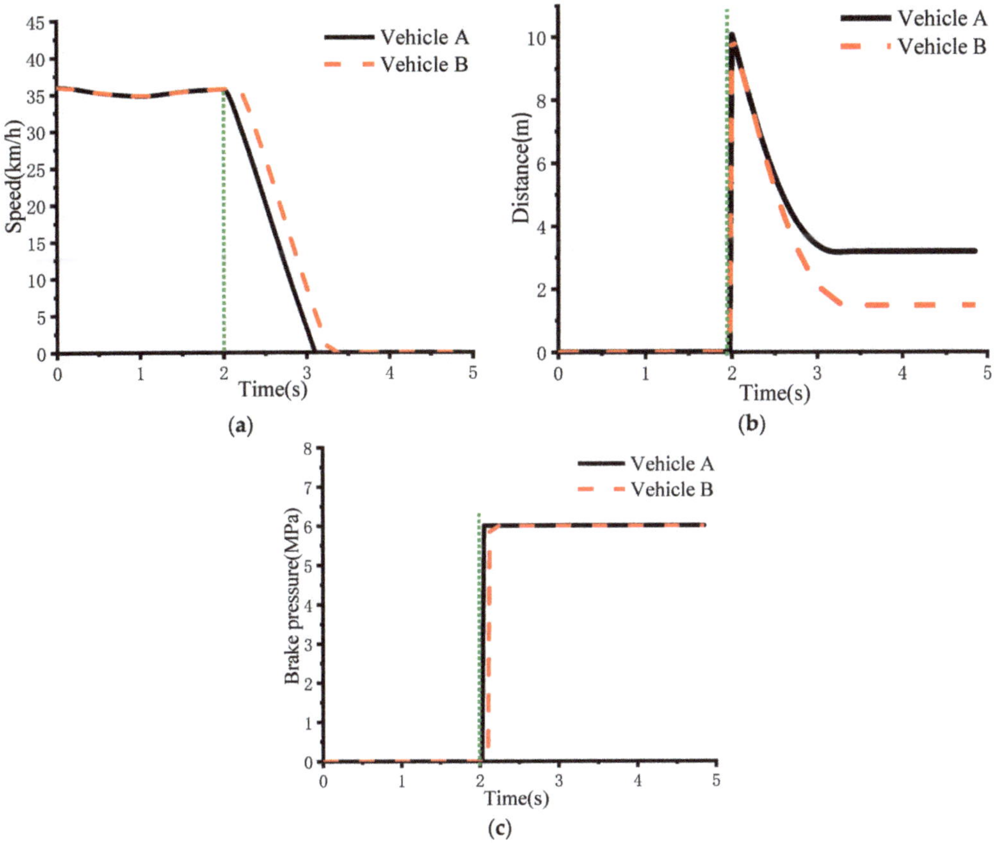

Figure 8. Simulation of longitudinal collision avoidance: (**a**) Vehicle speed; (**b**) Distance between vehicle and obstacle; (**c**) Brake pressure of the vehicles.

4.2. Simulation of the Lateral Collision Avoidance Agent

This simulation will be used to verify the performance of the multi-agent coordinated control system at higher vehicle speeds. The vehicle runs at a constant speed along the road at 80 km/h, and suddenly, there is an obstacle falling 30 m from the front as shown in Figure 9a. The green dotted line in Figure 9a,b is the location and moment when the obstacle appears. The relative distance between the lateral width of the obstacle and the vehicle center of mass is 2 m. At this time, D_b is in the area ③. Simulation results are shown in Figure 9.

Results show that when the obstacle appears, the vehicle selects steering to avoid collision in this situation. The steering wheel angle output time of vehicle A is 0.14 s ahead of vehicle B as shown in Figure 9b. The critical collision time is calculated to be 1.09 s, the longitudinal displacement of vehicle A is 28.43 m, and the longitudinal displacement of vehicle B is 31.54 m. At this time, vehicle B collided with the obstacle, and vehicle A still has a certain distance from the obstacle. Figure 9a shows a close-up view of the collision with vehicle B. The timeliness and effectiveness of vehicle A's lane-changing collision avoidance control have been verified, but its trajectory tracking performance has room for improvement. Therefore, as shown in Figure 9c,d, the fuzzy sliding control is adopted to ensure the good tracking performance of the vehicle on the lane-changing trajectory and the lateral acceleration curve of the vehicle is also more gradual, which meets the stability requirements of the vehicle. The fuzzy sliding control mainly acts on the turn-back stage of the vehicle's emergency steering. As can be seen from Figure 9, the lateral acceleration of the vehicle decreases significantly after the vehicle quickly escapes from the danger of collision, which can ensure the stability of the vehicle and the comfort of the occupants.

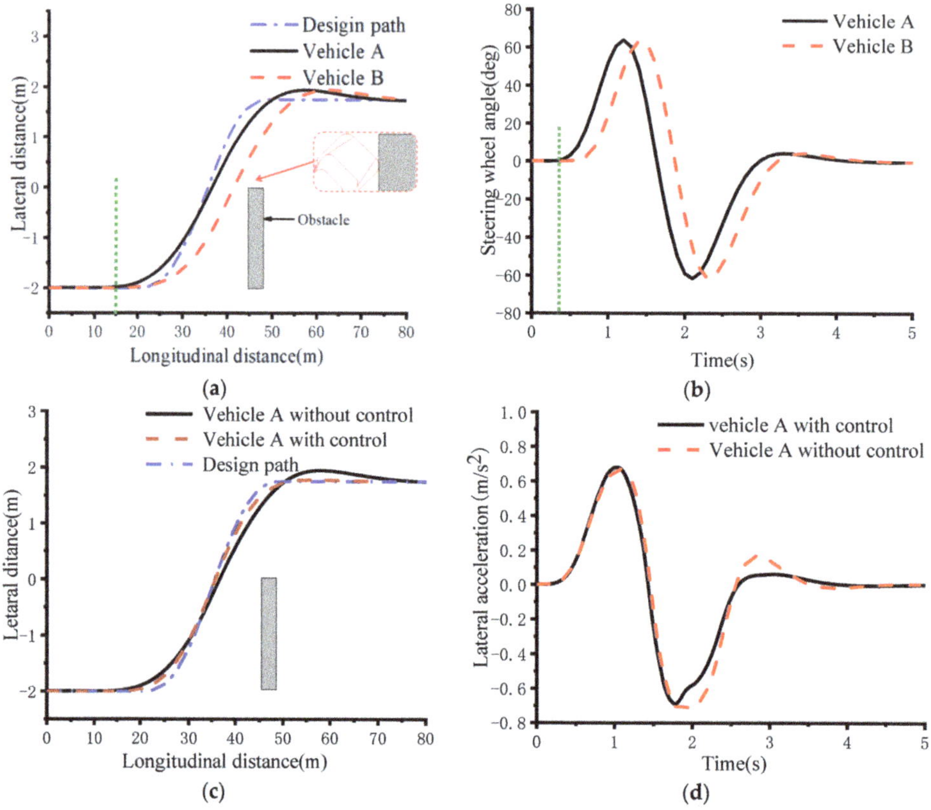

Figure 9. Simulation of lateral collision avoidance: (**a**) Lane-changing track; (**b**) Steering wheel angle; (**c**) Lane-changing track; (**d**) Lateral acceleration.

5. Conclusions

A multi-agent coordinated control system based on a blackboard model is proposed for improving the real-time performance of an active collision avoidance system in this paper. To do this, some agents, including a longitudinal collision avoidance agent and lateral collision avoidance agent, are established, and all of them can support and cooperate under the unified goal to produce reasonable control rules in coordination with the blackboard model, which can select a reasonable collision avoidance method under different driving conditions. In the process of active collision avoidance, the decision and planning are simultaneously operated. At the same time as the decision and planning are completed, underlying control instructions can be executed immediately, and the decision and planning integration of the collision avoidance system is realized, which effectively reduces the time lag during the process of active collision avoidance. The simulation results also indicate that the proposed multi-agent active collision avoidance system can reduce the decision and planning time, improving the real-time behavior of the intelligent vehicle.

The next step will further consider the state of the preceding vehicle. In terms of emergency steering and collision avoidance, the next step will be to consider vehicles in adjacent lanes to ensure that vehicles in adjacent lanes are not affected during the process of changing lanes to avoid collisions. At the same time, a real vehicle test is arranged.

Author Contributions: Conceptualization, C.Y.; methodology, J.S.; software, S.W. and Q.Y.; validation, Y.H., Y.C. and L.C.; formal analysis, C.Y.; investigation, S.W.; resources, Y.G.; data curation, Y.Y.; writing—original draft preparation, Y.L. and X.W.; writing—review and editing, S.W.; supervision, S.W.; project administration, L.C.; funding acquisition, C.Y. All authors have read and agreed to the published version of the manuscript.

Funding: This project is supported by the National Natural Science Foundation of China (52172346, 52225212, U20A20333, U20A20331, 51875255), Key Research and Development Program of Jiangsu Province (BE2020083-3, BE2019010-2), "Research and development project of key technologies of basic simulation software for the automotive industry (ZX20220002)" fund, a key project of the Department of Agricultural Equipment of Jiangsu University (NZXB20210103).

Data Availability Statement: Not applicable.

Conflicts of Interest: Qiuye Yu is employee of the China Automotive Technology & Research Center Co., Ltd. The paper reflects the views of the scientists and not the company.

Nomenclature

Symbol	
$t_{detect1}$ (s)	the environment perception time
$t_{decision1}$ (s)	the decision-planning time of longitudinal collision avoidance agent
$t_{excution1}$ (s)	the mechanical delay time
V_r (km/h)	the vehicle speed
a_{r-max} (m/s^2)	the maximum braking deceleration of the vehicle
d_0 (m)	the minimum safety threshold
μ	the coefficient of road adhesion
g (m/s^2)	the acceleration of gravity
α (rad)	the vehicle slope angle.
F_w (N)	the air resistance force
T_s (N·m)	the driving torque
T_{bf} (N·m)	the braking torque of the front wheels
T_{br} (N·m)	the braking torque of the rear wheels
F_f (N)	the ground friction of the ground acting on the front wheels
F_r (N)	the ground friction of the ground acting on the rear wheels
W_f (N)	the vertical force of the front wheels
W_r (N)	the vertical force of the rear wheels
F_t (N)	the driving force
F_{Xb} (N)	the braking force
$\sum F(V_r)$ (N)	the total resistance.
P_{des} (Pa)	the desired braking pressure
K_b	the braking pressure ratio
r_r (m)	the rolling radius of wheels
P_b (Pa)	the pressure of the brake pipe
y_e (m)	the lateral displacement for the vehicle to avoid collision
t_e (s)	the lane-changing time
W_s (m)	the vehicle width
t_c (s)	the collision time
W_b (m)	the lateral distance
S_a (m)	the longitudinal displacement
$t_{detect2}$ (s)	the perception delay
$t_{decision2}$ (s)	The decision-planning time
$t_{excution2}$ (s)	the steering mechanical delay
β (rad)	sideslip angle
v_y (km/h)	the lateral speed
M (kg)	the vehicle mass
ω (rad/s)	the yaw rate
a and b (m)	the front and rear wheel base

I_z (kg·m^2)	the vehicle's moment of inertia
C_f and C_r (N/rad)	the stiffness of the front and rear tire
γ	the sliding surface gain
i_{SW}	the steering system ratio
θ_s (rad)	the steering wheel angle
T_1 (s)	the planning time of longitudinal collision avoidance
T_2 (s)	the decision time of collision avoidance way
T_3 (s)	the planning time of lateral collision avoidance
T_4 (s)	the execution control time
$T_{D-Breaking}$ (s)	time consumption using distributed coordinated control system
$T_{S-Breaking}$ (s)	time consumption using a sequential control system

References

1. Li, S. Research on longitudinal active collision avoidance system of electric vehicle using dSPACE. *Mech. Des. Manuf.* **2017**, *7*, 111–114.
2. Li, S. Modeling and Simulation of Vehicle Collision Avoidance Control System. Master's Thesis, Hunan University, Changsha, China, 2009.
3. Shuofeng, W. Research on Active Collision Avoidance Algorithm for Intelligent Vehicle Based on Improved Artificial Potential Field. Master's Thesis, Jiangsu University, Zhenjiang, China, 2019.
4. Hou, D.; Liu, G.; Gao, F. A New Vehicle Active Collision Safety Distance Mode. *Automot. Eng.* **2005**, *27*, 186–190.
5. Soudbakhsh, D.; Eskandarian, A. *A Collision Avoidance Steering Controller Using Linear Quadratic Regulator*; SAE Technical Paper 2010-01-0459.
6. Boada, M.J.L.; Boada, B.L.; Munoz, A. Integrated control of front-wheel steering and front braking forces on the basis of fuzzy logic. *Proc. Inst. Mech. Eng. Part D J. Automob. Eng.* **2006**, *220*, 253–267. [CrossRef]
7. Li, W.; Duan, J.; Gong, J. Near optimal solution of intelligent vehicle lane change path planning under obstacles. *J. Cent. South Univ. (Nat.) Sci. Ed.* **2011**, *42* (Suppl. 1), 505–511.
8. Huang, L. Research on Active Collision Avoidance Control System Based on Braking/Steering. Master's Thesis, Nanjing University of Aeronautics and Astronautics, Nanjing, China, 2016.
9. Li, Y. Research on Automobile Active Braking/Steering Obstacle Avoidance Control System. Master's Thesis, Hefei University of Technology, Hefei, China, 2018.
10. Tian, J.; Bei, S.; Li, B.; Hu, H.; Quan, Z.; Zhou, D.; Zhou, X.; Tang, H. Research on Active Obstacle Avoidance of Intelligent Vehicles Based on Improved Artificial Potential Field Method. *World Electr. Veh. J.* **2022**, *13*, 97. [CrossRef]
11. Ferber, J.; Gutknecht, O. A meta-model for the analysis and design of organization in multi-agent systems. In Proceedings of the 3rd International Conference on MAS, Paris, France, 3–7 July 1998; pp. 258–266.
12. Pachcco, O.; Carmo, J. A Role Based Model for the Normative Specification of Organized Collective Agency and Agents Interaction. *Auton. Agents Multi-Agent Syst.* **2003**, *6*, 145–184. [CrossRef]
13. Dongyu, L. Multilayer formation control of multi-agent systems. *Automatica* **2019**, *109*, 108558.
14. Kao-Shing, H.; Jin-Ling, L. A Multi-Layer Architecture for Cooperative MultiAgent Systems. In Proceedings of the 2019 8th International Congress on Advanced Applied Informatics (IIAI-AAI), Toyama, Japan, 7–11 July 2019; pp. 515–520.
15. Laayati, O.; El Hadraoui, H. An AI-Layered with Multi-Agent Systems Architecture for Prognostics Health Management of Smart Transformers: A Novel Approach for Smart Grid-Ready Energy Management Systems. *Energies* **2022**, *15*, 7217. [CrossRef]
16. Zhang, L. Research on Key Technology of Active Collision Avoidance Braking System Based on Pavement Identification. Master's Thesis, Jiangsu University, Zhenjiang, China, 2017.
17. Yuan, W.; Jiang, Z.; Guo, Y. Study on Vehicle Collision Avoidance Control of Braking and Steering Coordination. *China J. Highw. Transp.* **2019**, *32*, 173–181.
18. Liu, Z.; Wang, Y.; Wu, X.; Zhang, C.; Ni, J. Research on Lane Control and Collision Control Strategy Based on Linear Path Tracking Control. *China J. Highw. Transp.* **2019**, *32*, 86–95.
19. Yu, Z. *Automotive Theory*; Mechanical Industry Press: Beijing, China, 2006.
20. Guo, K. Previewing Theory of Follow-up and Simulation of Large Angle Manipulation Motion of Human-Car Closed-Loop System. *Automot. Eng.* **1992**, *14*, 1–11.
21. Han, S.I.; Lee, J.M. Precise positioning of no smooth dynamic systems using fuzzy wavelet echo state networks and dynamic surface sliding mode control. *IEEE Trans Ind. Electron.* **2013**, *60*, 5124–5136. [CrossRef]

22. Niu, L.; Chen, L.; Jiang, H. Application of Multi-Agent Theory in Vehicle Chassis Integrated Control. *Automob. Technol.* **2008**, *8*, 31–35.
23. Zhu, T.; Liu, G.; Jia, L.M. A Cooperative Making Multi-Agent Model on Railway Daily Dispatching Plan Based on Blackboard. In Proceedings of the International Conference on Digital Manufacturing & Automation, Changcha, China, 18–20 December 2010; pp. 18–21.

World Electric Vehicle Journal

MDPI

Article

Analysis and Roll Prevention Control for Distributed Drive Electric Vehicles

Xiaoyu Chang, Huanhuan Zhang *, Shuai Yan, Shengli Hu and Youming Meng

School of Mechanical and Automotive Engineering, Shanghai University of Engineering Science, Shanghai 201620, China
* Correspondence: zhanghh@sues.edu.cn

Abstract: This work presents an approach to improve the roll stability of distributed drive electric vehicles (DDEV). The effect of the reaction torque from the in-wheel motor exerts additional roll moment, which is different from traditional vehicles. The additional roll moment can be achieved by active control of the wheel torque adjustment, which achieves a control effect similar to the active suspension. The anti-roll control strategy of decoupling control of roll motion and yaw motion are proposed. The direct yaw moment is calculated by the linear quadratic regulator (LQR) algorithm while the additional rolling moment is calculated by the sliding mode variable structure. For maneuvering rollover caused by excessive lateral acceleration, an anti-rollover control strategy is designed based on differential braking. A fuzzy control theory is used to decide the yaw moment to be compensated. The distribution method of the braking torque applied to the outer wheel alone, and the lateral load transfer rate is the main evaluation index for simulation verification of typical working conditions. The simulation results show that the proposed control strategy for DDEV is effective.

Keywords: distributed drive electric vehicles; additional roll moment; decoupling control; load transfer rate

Citation: Chang, X.; Zhang, H.; Yan, S.; Hu, S.; Meng, Y. Analysis and Roll Prevention Control for Distributed Drive Electric Vehicles. *World Electr. Veh. J.* **2022**, *13*, 210. https://doi.org/10.3390/wevj13110210

Academic Editors: Yong Li, Xing Xu, Lin Zhang, Yechen Qin and Yang Lu

Received: 7 October 2022
Accepted: 31 October 2022
Published: 7 November 2022

1. Introduction

In recent years, the problem of environmental pollution and energy shortage caused by the massive use of fossil energy has become increasingly serious. The traditional automobile belongs to the industry of high energy consumption and high pollution. Therefore, more and more researchers and enterprises engaged in automobile related work focus on electric vehicles. Distributed drive electric vehicles (DDEV) are one kind of electric vehicles, and the research and development of its key technologies has always been the focus of many automotive and industrial experts. Compared with traditional fuel vehicles, there is no transmission system to transmit power in the distributed drive electric vehicle (DDEV). It not only enables the vehicle to have more controllable degrees of freedom, but also greatly improves its efficiency and response speed, which helps to solve the sustainability problems of energy and vehicles. In [1] a study using bibliometric analysis, analyses sustainable mobility in relation to economic returns, environmental benefits and societal advantages.

The rollover accidents caused by the loss of the stability of vehicles seriously threaten people's lives, property and safety. It has become a safety issue and attracted worldwide attention. The statistics of the US Highway Traffic Safety Administration show that the degree of harm caused by vehicles rollover accidents are second only to vehicles collision accidents in all traffic accidents [2]. Many scholars all over the word have conducted research on the vehicle roll motion control, including controlling the body posture and changing the trajectory of the vehicles. Body posture control can be subdivided into lateral stabilizer control [3–5] and active suspension control [6,7] while the trajectory change strategy consists of active steering control [8–11] and differential brake control [12,13].

The unique structure of DDEV affects the roll and rollover performance which includes the following aspects: (1) in-wheel motors increase the unsprung mass, which will deteriorate the vibration isolation performance of the suspension and cause the lifting effect of the wheels; (2) because of rigid connection between the suspension and the motor stator, the ground driving force and the reaction torque of the motor to be transmitted to the body will form a large roll torque; (3) The cancellation of the differential will cause the coaxial drive wheels to lose the torque self-balancing mechanism. The torque difference between the two sides of the wheels form a large yaw moment. It will cause excessive steering of the car or sharp turn. Therefore, to study the roll stability of the DDEV is the basis and premise for proposing the anti-rollover control [14].

Currently, the safety control of DDEV is mainly focused on the yaw stability. However, many studies on the roll stability control are mainly to control the suspension. Among them, the literature [15] aimed at the roll phenomenon of in-wheel driving vehicles. Based on the control of the suspension technology through the vertical load distribution transfer, the vehicle's front and rear axle lateral stiffness are changed. In [16] an active suspension control algorithm was designed based on the optimal control theory LQG and the robust control theory respectively. According to the driving status, the extra force is applied to the suspension to reduce the dynamic displacement. It made the vertical motion in an optimal state. However, the active suspension is both expensive and complex. Consequently, other control schemes have been developed to maintain the roll stability. A strategy of applying driving/braking torque to different drive motors is proposed in [17,18]. An additional roll moment is generated to the vehicle's body that effectively improves the posture of the vehicle body. A roll stability controller is designed in literature [19], which takes the suppression of the body's roll angle as the control target and its roll torque was generated by the wheel drive torque difference, so it is no longer necessary to design a separate suspension actuator. A joint control system for roll stability and yaw stability is designed in literature [20]. The controller considered the coupling effect of yaw rate and lateral acceleration. The adjustment factor RI is proposed to allocate the proportion of the roll control, which effectively improved the roll stability of the vehicle. Integrated control of the roll, yaw and pitch of DDEV was implemented in literature [21]. It based on the algorithm of optimal allocation of different wheel torques. The strategy did not require the analysis of complex equations to achieve spatial stability of the vehicle. The additional roll moment force of DDEV is analyzed in literature [22]. It is generated by the driving/braking torque. The in-wheel motor drive itself has the ability to self-adjust the body posture, it is found.

It should be noted that different driving torque could realize the control of the rolling moment. Its effect is similar to the active suspension. But the roll and yaw of the vehicle will affect each other to exacerbate the instability of the vehicle. In this paper, the anti-roll control strategy for decoupling control of roll and yaw is proposed. For the roll torque generated by the centrifugal force due to excessive lateral acceleration and centrifugal force, an anti-rollover control strategy for differential braking is proposed.

Section 2 analyzes the generating mechanism of the DDEV roll moment. Section 3.1 discusses the rolling stability control based on active distribution of wheel driving torques. The change of wheel driving torque will affect both the vehicle's roll stability and yaw stability, so the anti-roll control strategy for decoupling control of roll motion and yaw motion is proposed, which is the main contribution of this work. Aiming at the maneuvering rollover caused by excessive roll acceleration, an anti-rollover control strategy based on differential braking is proposed, which is in Section 3.2. Section 4 shows the behavior of the control system in a simulation environment.

2. Generating Mechanism of the Rolling Moment of DDEV

2.1. Rolling Moment of DDEV

The vehicles' rolling moment is mainly composed of three parts. The first term represents the rolling moment caused by the centrifugal force of the sprung mass. The second term is caused by the deviation of the spring center of gravity The third term is produced by the force of the suspension on the body [23].

DDEV also produces the above three kinds of rolling moments when turning. The interaction between the vehicle body and the suspension has a great impact for the rolling moment. After introducing the in-wheel motor, the ground driving force and the reaction torque of the motor are transmitted to the body through the wheels and suspension, thereby generating additional "vertical force", as shown in Figure 1. In the case that the lateral acceleration is not large, the value of the roll moment generated by this "vertical force" is sufficiently large compared to the rolling moment due to lateral acceleration. The additional rolling moment will have a greater impact on the roll attitude of the vehicle body. The roll motion equation of the DDEV is expressed as

$$I_x\ddot{\phi} = m_s h_s a_y + m_s h_s g \sin\phi - (K_\phi \phi + C_\phi \dot{\phi}) - \Delta M_X \tag{1}$$

where I_x is the rotational inertia of the vehicle around the x axis; φ is vehicle roll angle; m_s is vehicle sprung mass; h_s is the distance from the center of the sprung mass to the roll center of the car; a_y is lateral acceleration at the center of mass; g is gravitational acceleration; K_φ is the equivalent roll stiffness of the vehicle; C_φ is equivalent roll damping of automobile and ΔM_X is additional roll moment.

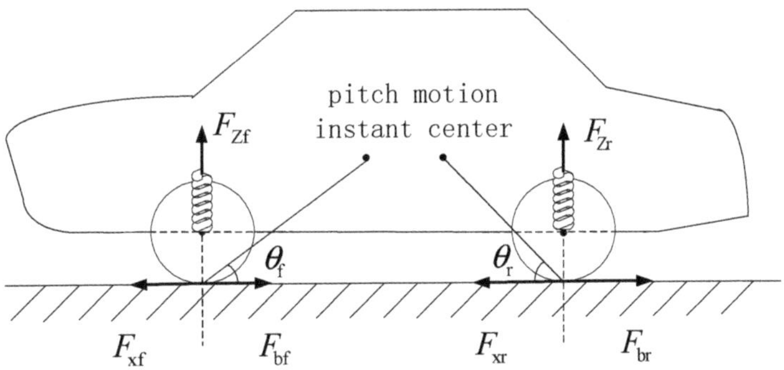

Figure 1. The diagram of additional vertical force.

2.2. *Analysis of the Rolling Effect for the Additional Vertical Force of DDEV*

Taking the vehicle turning left as an example, the transmission process of the additional vertical force in the longitudinal plane and the lateral plane of the vehicle is analyzed separately. The following assumptions are made:

(1) The left and right sides of the vehicle are symmetrical, and the front wheel angle δ is not large, that is, $\cos\delta \approx 1$, $\sin\delta \approx 1$;

(2) McPherson suspension is used for front and rear suspension;

(3) The loss is ignored during force or torque transmission.

In the longitudinal plane of the vehicle, the left suspension is used as a force body, the force is shown in Figure 2.

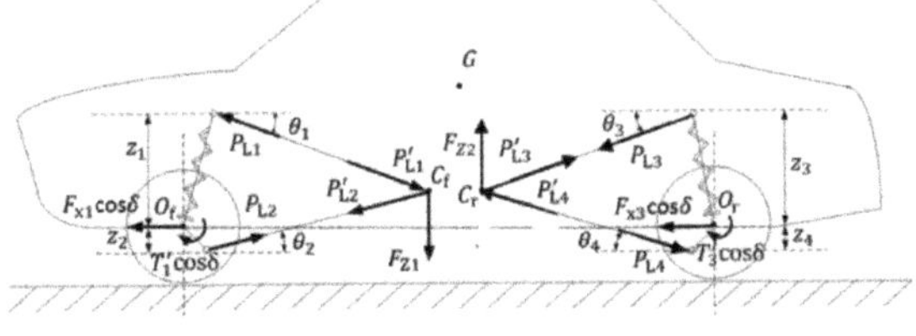

Figure 2. The force analysis of vehicle longitudinal plane.

At the center of the front left wheel O_f, the force balance equation and the moment balance equation can be obtained as

$$\begin{cases} F_{x1}\cos\delta + P_{L1}\cos\theta_1 - P_{L2}\cos\theta_2 = 0 \\ T_1\cos\delta - P_{L1}z_1/\cos\theta_1 - P_{L2}z_2/\cos\theta_2 = 0 \end{cases} \quad (2)$$

where, F_{x1} and T_1 represent the ground driving force and motor torque transmitted from the front wheel to the vehicle body via the suspension; P_{Li} (i = 1, 2) is the force exerted by the left body on the front suspensions; z_i (i = 1, 2) and θ_i (i = 1, 2) are the corresponding distance and angle in the Figure 2.

According to Equation (1), P_{L1}, $P_{Li}{}'$ and P_{L2}, $P_{Li}{}'$ can be expressed as

$$\begin{cases} P_{L1}{}' = P_{L1} = \frac{T_1\cos\delta\cos\theta_1\cos^2\theta_2 - F_{x1}\cos\delta\cos\theta_1 z_2}{z_1\cos^2\theta_2 + z_2\cos^2\theta_1} \\ P_{L2}{}' = P_{L2} = \frac{T_1\cos\delta\cos^2\theta_1\cos\theta_2 + F_{x1}\cos\delta\cos\theta_2 z_1}{z_1\cos^2\theta_2 + z_2\cos^2\theta_1} \end{cases} \quad (3)$$

where, P_{Li} (i = 1, 2) and $P_{Li}{}'$ (i = 1, 2) are acting force and reaction force; $P_{Li}{}'$ (i = 1, 2) is the force exerted by the front suspensions on the left body.

Then the vertical force of the left front wheel transmitted to the body through the suspension in the longitudinal plane can be described as

$$F_{Z1} = P_{L1}{}'\sin\theta_1 + P_{L2}{}'\sin\theta_2 \quad (4)$$

The roll moment generated by the left front wheel via the suspension in the longitudinal plane can be described as

$$M_{X1} = \frac{1}{2}BF_{Z1} = F_{x1}\frac{B\cos\delta[\sin\theta_1\cos\theta_1(r\cos^2\theta_2 - z_2) + \sin\theta_2\cos\theta_2(r\cos^2\theta_1 + z_1)]}{2(z_2\cos^2\theta_1 + z_1\cos^2\theta_2)} \quad (5)$$

where r is the wheel rolling radius; B is track width.

The Equation (5) can be abbreviated as

$$M_{X1} = K_1 F_{x1} \quad (6)$$

where $K_1 = \frac{B\cos\delta[\sin\theta_1\cos\theta_1(r\cos^2\theta_2 - z_2) + \sin\theta_2\cos\theta_2(r\cos^2\theta_1 + z_1)]}{2(z_2\cos^2\theta_1 + z_1\cos^2\theta_2)}$ Similarly, the roll moment generated by wheel via the suspension in the longitudinal plane can be described as

$$\begin{cases} M_{X1} = K_1 F_{x1} \\ M_{X2} = K_2 F_{x2} \\ M_{X3} = K_3 F_{x3} \\ M_{X4} = K_4 F_{x4} \end{cases} \quad (7)$$

where M_{X1} is the roll moment the left front wheel; M_{X2} is the roll moment of the left rear wheel; M_{X2} is the roll moment of the right front wheel; M_{X4} is the roll moment of the right rear wheel; K_i (i = 1, 2, 3, 4) is the corresponding coefficient.

The force of lateral plane is shown in Figure 3.

At the connection point of the inner suspension kingpin, there is

$$\begin{cases} F_{x1}\sin\delta + P_{i1}\cos\theta_{i1} - P_{i2}\cos\theta_{i2} = 0 \\ T_1\sin\delta - P_{i1}z_{i1}/\cos\theta_{i1} - P_{i2}z_{i2}/\cos\theta_{i2} = 0 \end{cases} \quad (8)$$

where P_{i1} and P_{i2} represent the force exerted by the vehicle body on the inner side of the front suspension; z_{i1} and z_{i2} are the corresponding distance; θ_{i1} and θ_{i2} are the corresponding angle.

P_{i1}, $P_{i1}{}'$ and, P_{i2}, $P_{i2}{}'$ can be expressed as

$$\begin{cases} P_{i1}{}' = P_{i1} = \frac{T_1 \sin\delta \cos\theta_{i1} \cos^2\theta_{i2} - F_{x1} \sin\delta \cos\theta_{i1} z_{i2}}{z_{i1} \cos^2\theta_{i2} + z_{i2} \cos^2\theta_{i1}} \\ P_{i2}{}' = P_{i2} = \frac{T_1 \sin\delta \cos^2\theta_{i1} \cos\theta_{i2} + F_{x1} \sin\delta \cos\theta_{i2} z_{i1}}{z_{i1} \cos^2\theta_{i2} + z_{i2} \cos^2\theta_{i1}} \end{cases} \tag{9}$$

where, P_{ij} (j = 1, 2) and P_{ij} (i = 1, 2) are a pair of acting force and reaction force, and P_{ij} (i = 1, 2) are the force exerted by the front-inner suspension on the vehicle body.

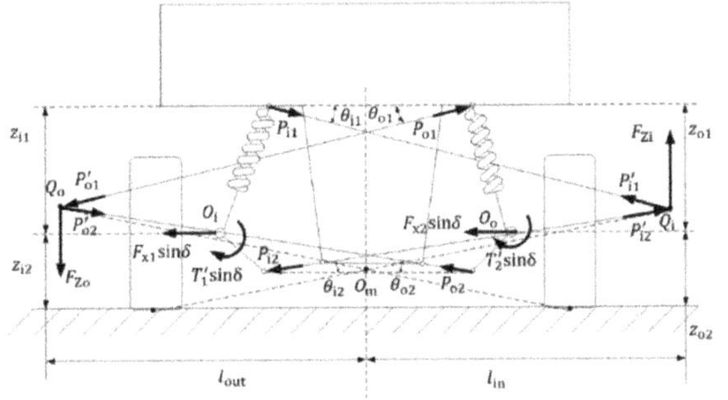

Figure 3. The force analysis of roll mechanism of lateral plane.

Then the vertical force transmitted by the front inner wheel to the body through the suspension in the transverse plane can be expressed as

$$F_{Zi} = P_{i1}{}' \sin\theta_{i1} + P_{i2}{}' \sin\theta_{i2} \tag{10}$$

And the roll moment generated by the front-inner wheel via the suspension in the transverse plane to the vehicle body is stated as

$$M_{Xi} = -(P_{i1}{}' \sin\theta_{i1} + P_{i2}{}' \sin\theta_{i2}) \cdot l_{in} \tag{11}$$

where, l_{in} is the distance from the roll center to the instantaneous center of roll motion on the inside of the front suspension.

Similarly, the roll moment generated by the front-outer wheel via the suspension is stated as

$$M_{Xo} = -(P_{o1}{}' \sin\theta_{o1} + P_{o2}{}' \sin\theta_{o2}) \cdot l_{out} \tag{12}$$

where $P_{o1}{}'$ and $P_{o2}{}'$ represent the force exerted by the front-outer suspension on the vehicle body, l_{out} is the distance from the roll center to the instantaneous center of roll motion on the outside of the front suspension, θ_{o1} and θ_{o2} are the corresponding angle.

According to the method of solving Equation (6), we can get

$$\begin{cases} M_{Xi} = K_5 F_{x1} \\ M_{Xo} = K_6 F_{x2} \end{cases} \tag{13}$$

where K_5 and K_6 are the coefficient term of the roll moment generated by the front-inner wheel and front-outer wheel.

Conclusively, the roll moment acting on the body of the DDEV can be changed by controlling the magnitude of the driving torque of the motor, so as to achieve the function of adjusting the body posture.

Based on the above analysis, it is plain that the roll posture control of the DDEV during cornering can be achieved by actively adjusting the size of the wheel driving torque.

3. Roll Stability Control Algorithm

3.1. Yaw and Roll Decoupling Control Algorithm

When the lateral acceleration is not large, the rolling effect of the vehicle body can be controlled by the wheel torque. However, the size of the wheel driving torque also directly affects the yaw moment of the vehicle, which affects the yaw stability. The distribution of the driving torque of each DDEV wheel must be comprehensive consideration of the vehicle's yaw stability and roll stability.

3.1.1. Yaw Stability Control

The linear two-degree-of-freedom vehicle can explain the vehicle's handling characteristics. The ideal side angle of mass center and ideal yaw rate of the vehicle can be expressed as

$$\beta_{\mathrm{d}} = \frac{b + mav_x^2/k_{\mathrm{r}} \cdot L}{L \cdot (1 + Kv_x^2)} \cdot \delta \tag{14}$$

$$\omega_{\mathrm{rd}} = \frac{v_x/L}{1 + K \cdot v_x^2} \cdot \delta \tag{15}$$

where, b is distance from center of mass to rear axle, a is distance from center of mass to front axle, m is the vehicle quality, v_x is the vehicle's longitudinal speed, k_r is the cornering stiffness of the rear wheel, k_f is the cornering stiffness of the front wheel, L is the vehicle wheelbase and K is the stability factor, $K = \frac{m}{L^2}(\frac{a}{k_r} - \frac{b}{k_f})$.

However, the vehicle will not be always in the small-angle operation. When the tire model is in the nonlinear region, the steady-state response value of the two-degree-of-freedom vehicle model is not suitable for the ideal value. It should be replaced by the limit value. The size of the limit is constrained by the road surface adhesion coefficient μ. Considering these factors, the ideal yaw angular velocity and the ideal centroid lateral declination angle can be expressed as

$$|\omega| = \min\{|\omega_{rd}|, |\omega_{r\max}|\} \cdot sign(\delta) \tag{16}$$

$$|\beta| = \min\{|\beta_{rd}|, |\beta_{r\max}|\} \cdot sign(\delta) \tag{17}$$

where $\omega_{r\max} = \frac{\mu g}{v_x}$ and $\beta_{r\max} = \tan^{-1}(0.02\mu g)$.

The state space equation of the linear two-degree-of-freedom vehicle model take the ideal yaw rate ω_{rd} and ideal side slip angle β_d as state variables described as

$$\begin{bmatrix} \dot{\beta}_{\mathrm{d}} \\ \dot{\omega}_{\mathrm{rd}} \end{bmatrix} = \mathbf{A} \cdot \begin{bmatrix} \beta_{\mathrm{d}} \\ \omega_{\mathrm{rd}} \end{bmatrix} + \mathbf{B} \cdot (\delta) \tag{18}$$

where $\mathbf{A} = \begin{bmatrix} \frac{k_f + k_r}{mv_x} & \frac{ak_f - bk_r}{mv_x^2} - 1 \\ \frac{ak_f - bk_r}{I_z} & \frac{a^2k_f + b^2k_r}{I_z v_x} \end{bmatrix}$, $\mathbf{B} = \begin{bmatrix} -\frac{k_f}{mv_x} & -\frac{ak_f}{I_z} \end{bmatrix}^T$, and I_z is the moment of inertia of the vehicle around the Z axis.

The yaw instability of the vehicle mostly occurs in the non-linear region of the tire. At this time, the lateral force of the tire is gradually saturated, and the vehicle begins to appear side slip phenomenon, which deviates from the driver's desired trajectory. At this time, the vehicle can be actively compensated for an additional direct yaw moment ΔMz to make the vehicles' yaw rate ω_r and side slip angle β re-track the change of the ideal value. The relationship between the vehicle's steering characteristics and the compensated additional yaw moment is shown in Table 1.

Table 1. Relationship between vehicle steering and additional yaw moment.

Steering Condition	Yaw Velocity	Additional Yaw Moment
left understeer	$\omega_r > 0$	$\Delta M_z > 0$
left oversteer	$\omega_r > 0$	$\Delta M_z < 0$
light understeer	$\omega_r < 0$	$\Delta M_z < 0$
light oversteer	$\omega_r < 0$	$\Delta M_z > 0$

With the actual yaw velocity ω_r and the actual side slip angle β as state variables, the equation of state of automobile motion described as

$$\begin{bmatrix} \dot{\beta} \\ \dot{\omega}_r \end{bmatrix} = \mathbf{A} \cdot \begin{bmatrix} \beta \\ \omega_\mathrm{r} \end{bmatrix} + \mathbf{B} \cdot (\delta) + \mathbf{B}_1 \cdot \Delta M_Z \tag{19}$$

where, ΔM_z is additional direct yaw moment, and $\mathbf{B}_1 = [0\ 1/I_z]^\mathrm{T}$.

Subtract (19) from (18), we can get the following Equation

$$\begin{bmatrix} \Delta\dot{\beta} \\ \Delta\dot{\omega}_\mathrm{r} \end{bmatrix} = \mathbf{A} \cdot \begin{bmatrix} \Delta\beta \\ \Delta\omega_\mathrm{r} \end{bmatrix} + \mathbf{B}_1 \cdot \Delta M_z \tag{20}$$

where $\Delta\beta$ is the difference between the actual side slip angle and the ideal side slip angle; $\Delta\omega_r$ is the difference between the actual yaw velocity and the ideal yaw velocity.

Equation (20) describes the dynamic relationship between the direct yaw moment and the yaw velocity deviation and the side slip angle deviation. So the optimal direct yaw moment [24–26] can be determined by LQR control theory as

$$\Delta M_Z = -\mathbf{K}\mathbf{x}(t) = -k_1\Delta\beta(t) - k_2\Delta\omega_r(t) \tag{21}$$

where $\mathbf{K}$ is the feedback matrix, and $\mathbf{K} = [k_1\ k_2]^\mathrm{T}$.

3.1.2. Roll Stability Control

Aiming at the roll phenomenon of the vehicle body, an active control strategy is applied to improve the roll attitude of the vehicle when cornering. Based on the sliding mode variable structure control theory, this paper implements the design of the roll stability controller. It can be seen from the three-degree-of-freedom vehicle model of DDEV that its roll motion equation is as follows

$$I_\mathrm{x}\ddot{\phi} = m_\mathrm{s}h_\mathrm{s}(\ \dot{\mathrm{v}}_y + v_x\omega_\mathrm{r}) + m_\mathrm{s}h_\mathrm{s}g\sin\phi - (K_\phi\phi + C_\phi\dot{\phi}) - \Delta M_\mathrm{X} \tag{22}$$

where v_y is the vehicle lateral speed.

When the body has a serious roll instability phenomenon, according to the Equation (22), the vehicle can be compensated with an anti-roll moment ΔM_X to recover the roll stability.

In order to reduce the roll angle and roll velocity, the sliding mode surface can be defined as

$$s = \dot{e} + \xi e \tag{23}$$

where, ξ is the weight coefficient between the roll angle and roll angular velocity; e is the error of roll angle.

Derivation of Equation (23) can be obtained

$$\dot{s} = \ddot{\phi} + \xi\dot{\phi} \tag{24}$$

The additional rolling moment is expressed as

$$\Delta M_\mathrm{X} = m_\mathrm{s}h_\mathrm{s}(\dot{v}_y + v_x\omega_\mathrm{r}) + (\xi I_x - C_\phi)\dot{\phi} + (m_\mathrm{s}h_\mathrm{s}g\sin\phi - K_\phi)\phi + \eta sat(s) \tag{25}$$

where *sat(s)* is the saturation function; η is the switching gain.

3.1.3. Torque Distribution Strategy for Decoupling Control

Assuming that the vehicle is in an unstable state, the yaw moment to be compensated is ΔM_z, and the roll moment to be compensated is ΔM_x, and the yaw moment and roll moment that can be compensated by adjusting the driving torque of each wheel are shown in Table 2 respectively.

Table 2. Compensating roll and yaw moment for 4 wheels.

Wheel	Compensated Roll Moment	Compensated Yaw Moment
front left wheel torque	ΔM_{X1}	ΔM_{Z1}
rear left wheel torque	ΔM_{X3}	ΔM_{Z3}
front right wheel torque	ΔM_{X2}	ΔM_{Z2}
rear right wheel torque	ΔM_{X4}	ΔM_{Z4}

The distribution of the roll moment is as follows

$$\begin{cases} \Delta M_{X1} + \Delta M_{X3} = \Delta M_X/2 \\ \Delta M_{X2} + \Delta M_{X4} = \Delta M_X/2 \end{cases} \tag{26}$$

The distribution of the yaw moment is as follows

$$\begin{cases} \Delta M_{Z1} + \Delta M_{Z3} = \Delta M_Z/2 \\ \Delta M_{Z2} + \Delta M_{Z4} = \Delta M_Z/2 \end{cases} \tag{27}$$

Suppose that the increment of driving force applied to each wheel is as follows

$$\begin{cases} \Delta M_{X1} = K_1 \Delta F_{x1} + K_5 \Delta F_{x1} \\ \Delta M_{X2} = K_2 \Delta F_{x2} + K_6 \Delta F_{x6} \\ \Delta M_{X3} = K_3 \Delta F_{x3} \\ \Delta M_{X4} = K_4 \Delta F_{x4} \end{cases} \tag{28}$$

$$\begin{cases} \Delta M_{Z1} = -\frac{1}{2} B \Delta F_{x1} \cos\delta + \Delta F_{x1} \sin\delta \cdot a \\ \Delta M_{Z2} = \frac{1}{2} B \Delta F_{x2} \cos\delta + \Delta F_{x2} \sin\delta \cdot a \\ \Delta M_{Z3} = -\frac{1}{2} B \Delta F_{x3} \\ \Delta M_{Z4} = \frac{1}{2} B \Delta F_{x4} \end{cases} \tag{29}$$

According to Equations (26)–(29), the driving force distribution strategy of the inner and outer wheels can be solved as in the following.

Inside wheels:

$$\begin{cases} K_1 \Delta F_{x1} + K_5 \Delta F_{x1} + K_3 \Delta F_{x3} = \Delta M_X/2 \\ -\frac{1}{2} B \Delta F_{x1} \cos\delta + \Delta F_{x1} \sin\delta \cdot a - \frac{1}{2} B \Delta F_{x3} = \Delta M_Z/2 \end{cases} \tag{30}$$

Outer wheels:

$$\begin{cases} K_2 \Delta F_{x2} + K_6 \Delta F_{x2} + K_4 \Delta F_{x4} = \Delta M_X/2 \\ \frac{1}{2} B \Delta F_{x2} \cos\delta + \Delta F_{x2} \sin\delta \cdot a + \frac{1}{2} B \Delta F_{x4} = \Delta M_Z/2 \end{cases} \tag{31}$$

Rewrite equation (30) and equation (31) into the following matrix form:

$$\mathbf{Ax} = \mathbf{B} \tag{32}$$

where, $\mathbf{K} = [\Delta F_{x1} \quad \Delta F_{x2} \quad \Delta F_{x3} \quad \Delta F_{x4}]^{\mathrm{T}}$

$$\mathbf{A} = \begin{bmatrix} K_1 + K_5 & 0 & K_3 & 0 \\ -\frac{1}{2}B\cos\delta + \sin\delta \cdot a & 0 & -\frac{1}{2}B & 0 \\ 0 & K_2 + K_6 & 0 & K_4 \\ 0 & \frac{1}{2}B\cos\delta + \sin\delta \cdot a & 0 & \frac{1}{2}B \end{bmatrix},$$

$\mathbf{B} = \left[\Delta M_X/2 \quad \Delta M_Z/2 \quad \Delta M_X/2 \quad \Delta M_Z/2\right]^{\mathrm{T}}$.

Considering the limitation of motor power and pavement condition, ΔF_{xi} should meet the limits as following:

$$\begin{cases} |\Delta F_{xi} \cdot r| \leq T_{\max} \\ |\Delta F_{xi} \cdot r| \leq \mu \mathrm{mg} \end{cases} \quad (33)$$

where ΔF_{xi} is the increment of each driving force. r is the radius of the wheel; $T_{\max}$ is the maximum driving moment of the motor; μ is the road adhesion coefficient.

And the increment of each wheel drive torque can be expressed as following:

$$\begin{cases} \Delta T_1 = \Delta F_{x1} \cdot r \\ \Delta T_2 = \Delta F_{x2} \cdot r \\ \Delta T_3 = \Delta F_{x3} \cdot r \\ \Delta T_4 = \Delta F_{x4} \cdot r \end{cases} \quad (34)$$

Finally, the decoupling control of the roll stability and yaw stability of the DDEV can be achieved by distributing the increment of the driving torque of each wheel.

3.2. Anti-Rollover Control Algorithm Based on Differential Brake

On a good level road with high adhesion coefficient, the lateral acceleration of the vehicle can reach more than 0.8 g when turning. At the same time, the centrifugal force of the vehicle is large enough, which is likely to cause rollover. Although DDEV can control the vehicle's roll attitude by controlling the driving force, it is difficult to avoid the vehicle rollover phenomenon only by controlling the driving force in an emergency situation of high-speed sharp turns. Applying brake control and reducing the speed are often the safest control strategy.

As shown in Figure 4, the brake control is applied to the target wheels by differential braking. The fuzzy controller outputs the compensated yaw moment, and the torque distribution controller outputs the braking pressure applied on the front outer wheel.

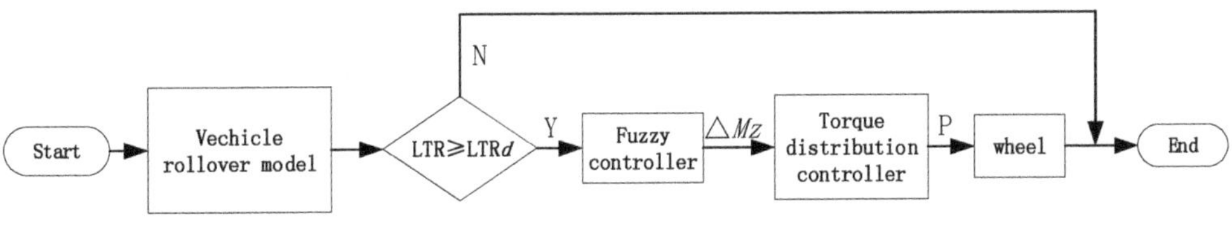

Figure 4. Anti-rollover control flow.

The vehicle's longitudinal speed will obviously change after performing differential brake control. At the time, the vehicle model should be extended from three-degree-of-freedom vehicle model to four-degree-of-freedom vehicle model, and the equation of motion is

$$\begin{cases} m\dot{v}_x = F_b \\ ma_y = F_{yf} + F_{yr} \\ I_z\dot{\omega}_r = aF_{yf} - bF_{yr} + \Delta M_Z \\ I_x\ddot{\phi} = m_s h_s a_y + m_s h_s g \sin\phi - K_\phi \phi - C_\phi \dot{\phi} \end{cases} \quad (35)$$

where F_b represents the braking force exerted on the front outer wheel, a_y is the vehicle lateral acceleration, F_{yf} and F_{yr} are the side force of vehicle front and rear wheels, ΔM_Z represents the compensated additional yaw moment.

We can get from Equation (35) that the essence of differential braking control is applying a braking force to the front outer wheel to make the vehicle generate an additional yaw moment. The longitudinal speed and yaw rate will be improved. Obviously, the correctness of the additional yaw moment is directly related to the quality of the control effect. ΔM_Z will be calculated by fuzzy control algorithm.

3.2.1. Evaluation Index of Vehicle Rollover

LTR is used as the evaluation index of vehicle rollover [15]. The definition of LTR is

$$\text{LTR} = \frac{(F_{z1} + F_{z3}) - (F_{z2} + F_{z4})}{F_{z1} + F_{z2} + F_{z3} + F_{z4}} \tag{36}$$

where F_{z1}, F_{z2}, F_{z3} and F_{z4} are the vertical load of each driving wheel.

When LTR = 0, it means that the vertical load on the left and right sides of the vehicle is equal, and there is no roll phenomenon;

When LTR = 1, it means that the vertical load of the right wheel is just 0, and the vehicle has a tendency to turn to the left;

When LTR = −1, it means that the vertical load of the left wheel is just 0, and the vehicle has a tendency to turn to the right.

The value of the lateral load transfer rate LTR should be as close to 0 as possible. In order to prevent the vehicle from entering a rollover state, LTR should be satisfied that $|\text{LTR}| \leq 1$. In general, $|\text{LTR}| = 0.8$ is taken as the critical state of automobile rollover to ensure the safety of the vehicle and prevent the negative impact of excessive lateral load transfer.

3.2.2. Fuzzy Control Algorithm

Fuzzy control is a control method based on fuzzy mathematics. Its great advantage is that it does not require accurate mathematical models. A series of variables describing the driving state of the vehicle such as the yaw rate, the side angle, the roll angle and the lateral acceleration are difficult to express with a precise mathematical equation. At this time, the concept of fuzzy mathematics can be used to deal with similar control problems. The fuzzy controller outputs the compensated yaw moment ΔM_Z who is entered into torque distribution controller. The torque distribution controller outputs the braking pressure applied on the front outer wheel.

And the fuzzy rules in this paper are listed in Table 3.

Table 3. Fuzzy control rule table.

ΔM_Z		e						
		NB	**NM**	**NS**	**ZO**	**PS**	**PM**	**PB**
ec	PB	ZO	ZO	NW	NS	NM	NB	NB
	PS	PM	PS	ZO	NW	NS	NB	NB
	ZO	PB	PM	PW	ZO	NW	NM	NB
	NS	PB	PB	PS	PW	ZO	NS	NM
	NB	PB	PB	PM	PS	PW	ZO	ZO

3.2.3. Distribution Strategy of Yaw Moment

Taking the vehicle turning left as an example to illustrate the distribution strategy of compensated yaw moment ΔM_Z. When the vehicle is in danger of rollover, a braking torque will be applied to the right front wheel of the vehicle separately. The relationship between compensated yaw moment and braking force is

$$F_b\left(\frac{B}{2}\cos\delta + a\sin\delta\right) = \Delta M_Z \tag{37}$$

The kinematic equation of the right front wheel during braking is as follows

$$I_w\dot{\omega} = T_b - F_b r \tag{38}$$

where I_w is the rotating inertia of front outer wheel; ω is the angular velocity of the wheel; T_b is the vehicle braking torque.

The mathematical expressions of braking torque and wheel cylinder pressure of the braking system are as following:

$$T_b = S \cdot P \tag{39}$$

where P is the braking pressure; S is the braking efficiency coefficient.

The relationship between the yaw moment and brake pressure can be obtained from Equations (37)–(39) as following:

$$P = \frac{1}{S}\left(\frac{2 \cdot \mathrm{r} \cdot \Delta M_Z}{B\cos\delta + 2a\sin\delta} + I_w\dot{\omega}\right) \tag{40}$$

As long as the braking pressure of the size P is applied to the front outer wheels, the vehicle can generate an additional yaw moment according to the Equation (40), so that the DDEV can achieve the effect of anti-rollover control.

4. Simulation and Verification

4.1. Vehicle Model

The 18 DOF vehicle model established in literature [27] is used in this paper. 6 freedoms of vehicle body, 4 vertical freedoms for suspension, 4 rotary motion freedoms and 4 vertical freedoms of wheels are included. The main simulation parameters of the vehicle model are shown in Table 4.

Table 4. Main parameters of vehicle model.

Parameters	Value
Vehicle mass	1380
sprung mass	900
un-sprung mass	480
Distance from center of mass to front axle	1.05
Distance from center of mass to rear axle	1.57
front wheel tread	1.4
rear wheel tread	1.4
height of centroid	0.6
Tire diameter load radius	0.33
tire type	255/75 R16

The Magic-Formula tire model is used as the tire model. The brushless DC motor is selected as the driving motor of the vehicle. Since the research is focused on the roll stability of the vehicle, the motor torque control can be simplified into the transfer function model of the actual electromagnetic torque T_m to the target electromagnetic torque $T_m{}^*$. The transfer function is

$$G(s) = \frac{T_m(s)}{T_m^*(s)} = \frac{1}{2\zeta^2 s^2 + 2\zeta s + 1} \tag{41}$$

where ξ is determined by motor characteristics, it can be obtained by fitting test results.

4.2. Simulation Verification of Yaw and Roll Decoupling Control Algorithm

4.2.1. Angular Step Input Condition

At speed of 60 km/h input the steering wheel angle which set to 50°, and the road surface adhesion coefficient is set to 0.25. The simulation results are shown in Figure 5.

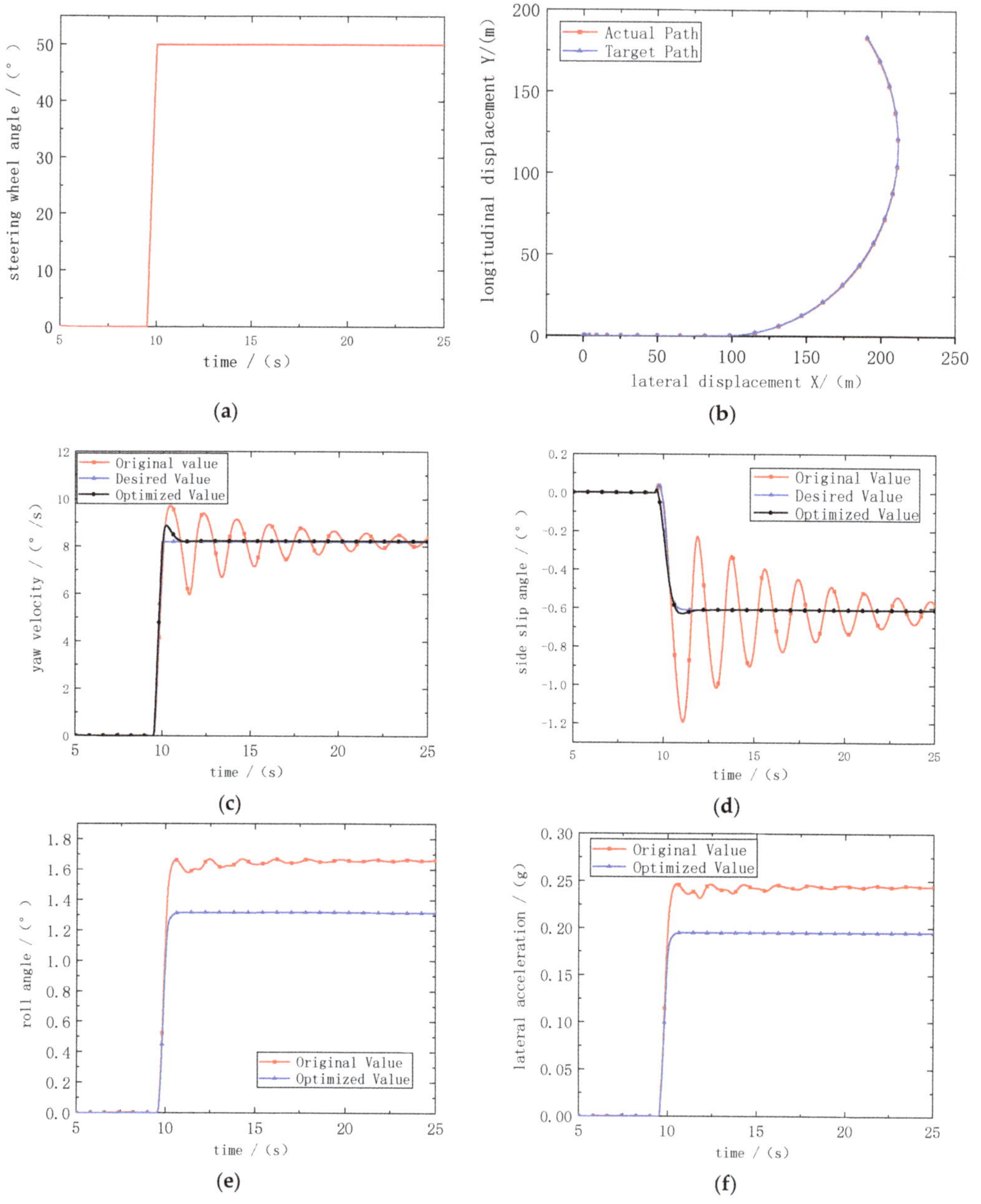

Figure 5. Simulation results of angular step input condition. (**a**) Steering wheel angle. (**b**) Vehicle track. (**c**) Yaw velocity. (**d**) Side slip angle. (**e**) Lateral acceleration. (**f**) Roll angle.

Figure 5b shows the change of the trajectory of the vehicle before and after the control. There is no obvious difference between the two curves, indicating that the vehicle is not completely unstable; Figure 5c,d show the yaw rate and the side slip angle of the vehicle respectively. After the control strategy is applied, the vehicle's yaw rate and side slip angle can ideally track the change of the ideal value. When the control strategy is not applied, both the yaw rate and side slip angle of show large fluctuations. It indicates that the vehicle has not completely destabilized. The controller can control the vehicle's lateral well under this condition showed by Figure 5e,f. It can be seen that the roll angle of the vehicle is reduced by about 80% after the control is applied, and the lateral acceleration is also suppressed to a certain extent. In summary, the roll and yaw decoupling controller designed in this paper can gradually stabilize the vehicle that is not completely unstable under the angular step condition.

4.2.2. Sine Input Condition

Sine input condition at the speed of 60 km/h. The road surface adhesion coefficient is 0.20. The simulation results are shown in Figure 6.

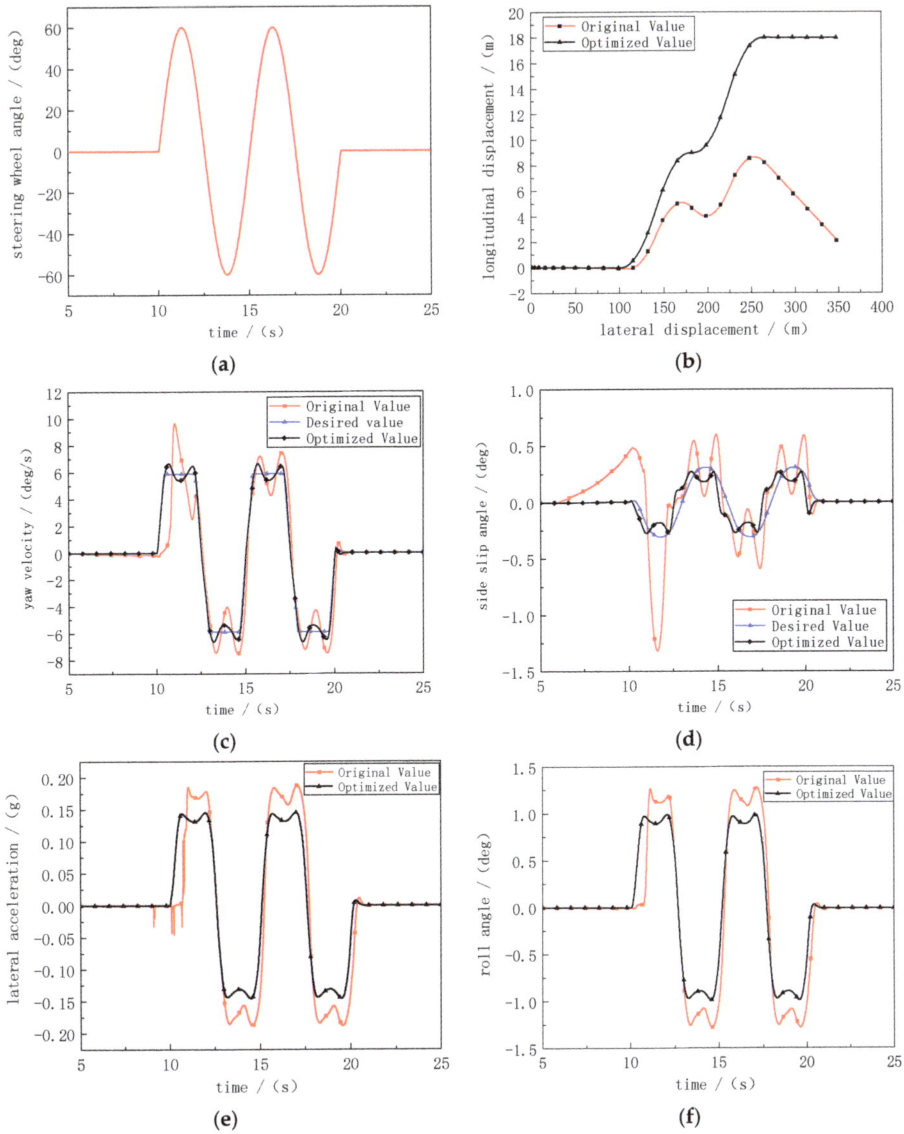

Figure 6. Simulation results of sine input condition. (**a**) Steering wheel angle. (**b**) Vehicle track. (**c**) Yaw velocity. (**d**) Side slip angle. (**e**) Lateral acceleration. (**f**) Roll angle.

Figure 6b shows the driving trajectory curve of the vehicle. Without the control strategy the vehicle has been off tracking; Figure 6c,d show the yaw rate and side slip angle of the vehicle. Both of them have been better corrected with control. It indicates that the vehicle's yaw stability has been improved. Figure 6e,f show the changes of the vehicle's lateral acceleration and roll angle. It can be seen both of them have been significantly reduced.

4.2.3. Fish Hook Test Condition

Fish hook test condition at the speed of 60 km/h. Let the vehicle turn left sharply for 160° at 10 s, and then quickly turn right for 320°. The road surface adhesion coefficient is 0.20. The simulation results are shown in Figure 7.

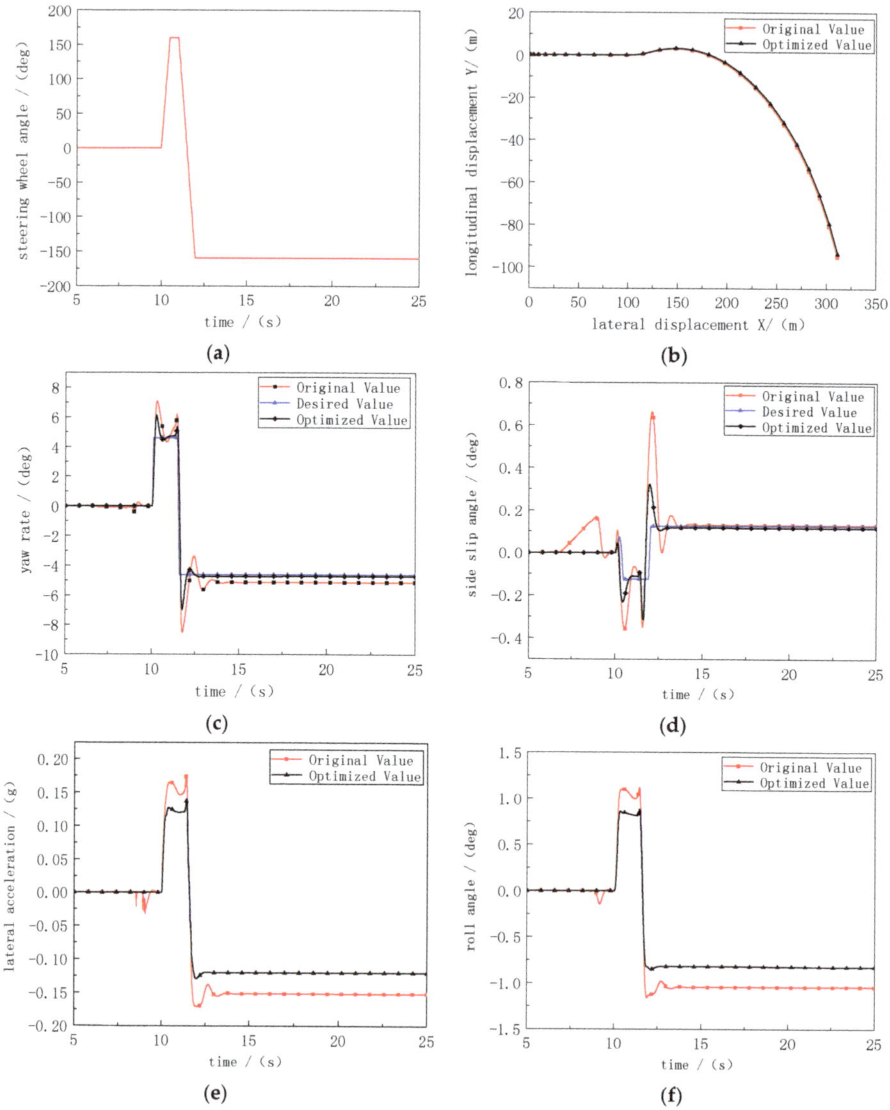

Figure 7. Simulation results of fish hook test condition. (**a**) Steering wheel angle. (**b**) Vehicle track. (**c**) Yaw rate. (**d**) Side slip angle. (**e**) Lateral acceleration. (**f**) Roll angle.

Figure 7c,d show the curves of the vehicle's yaw rate and side slip angle. It can be seen that the vehicle is not seriously stability at this time, and the control strategy of the vehicle's yaw rate is ideal to track the expected value changes while there is a large deviation in the yaw rate and side slip angle without control. Figure 7e,f show the curves of the lateral acceleration and roll angle of the vehicle. After the control strategy is applied, both of them have been better corrected, indicating that the roll yaw control strategy designed in this paper is feasible.

4.3. Simulation Verification of Anti-Rollover Control Algorithm

High-speed sharp turning will cause the vehicle to roll over. Therefore, J-Turn condition and fish hook condition are used as test conditions for simulation verification.

4.3.1. J-Turn Condition

Let the vehicle turn left sharply for 120° at the speed of 80 km/h. It make the vehicle enter the J-turn condition. The road surface adhesion coefficient is 0.85. The simulation results are shown in Figure 8.

It can be seen from Figure 8a,b the LTR value of the vehicle that will reach 0.8 at about 9.5 s and about 1 at 13 s without control. It indicates that the vehicle has already experienced serious roll. At about 17.5 s, the vehicle rolls over. But the LTR value does not fluctuate significantly with control. It keep stable at around 0.8 at 13 s and then remains stable. The vehicle does not roll over. The lateral acceleration reaches 0.78 g at 10 s without control in Figure 8c. Such a large lateral acceleration will inevitably cause the vehicle to generate a greater centrifugal force. At 17.5 s, the curve disappears. It indicates that the vehicle has roll over. The lateral acceleration is significantly reduced and remains stable with control. Figure 8d is the roll angle curve of the vehicle. It can be intuitively judged from the change of the vehicle's roll angle that the body's roll attitude has been significantly suppressed. The controlled roll angle will stabilize at 4.5° after entering the J-turn, while the uncontrolled vehicle roll angle will continue to increase until vehicle roll over. Figure 8e,f are the changes of the vertical load of the four wheels before and after the control. The vertical load of uncontrolled vehicle wheels fluctuates greatly, and the vertical load of the left wheel decreases rapidly after entering a turn. The vertical load of the left rear wheel even drops to 0 which indicates the left rear wheel left the ground. After about 17 s, the vehicle rolls over and all four curves disappear, and the vertical load of the right wheel of the vehicle is significantly reduced with control, especially the peak load of the right front wheel has dropped to about 9000 N. In summary, the anti-rollover control strategy applied to DDEV in this paper is basically effective and feasible under the J-turn operating condition.

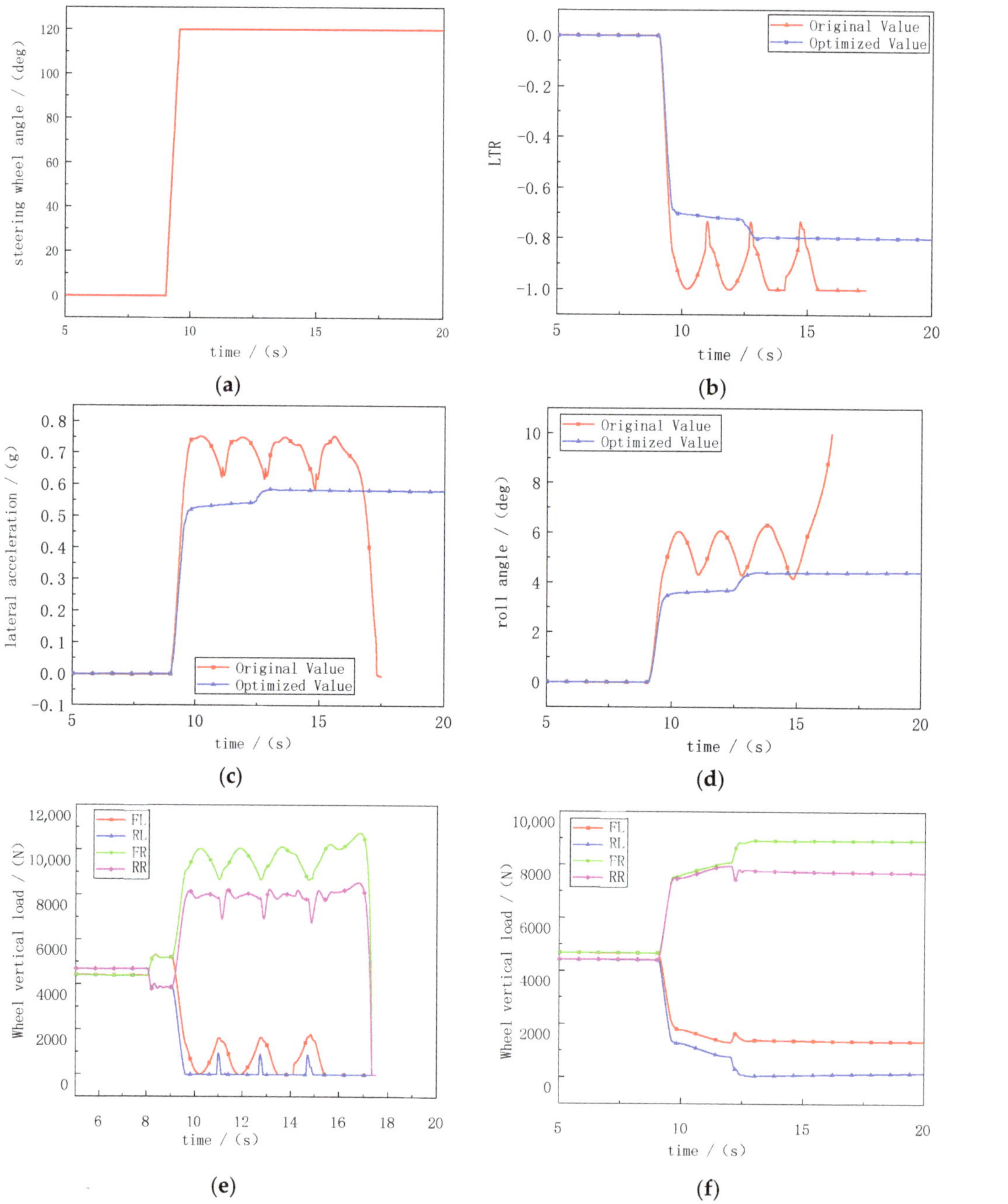

Figure 8. Simulation results of J-turn condition. (**a**) Steering wheel angle. (**b**) LTR. (**c**) Lateral acceleration. (**d**) Roll angle. (**e**) Wheel vertical load before control. (**f**) Wheel vertical load after control.

4.3.2. Fish Hook Test Condition

Fish hook input at the speed of 80 km/h. The road surface adhesion coefficient is 0.85. The simulation results are shown in Figure 9.

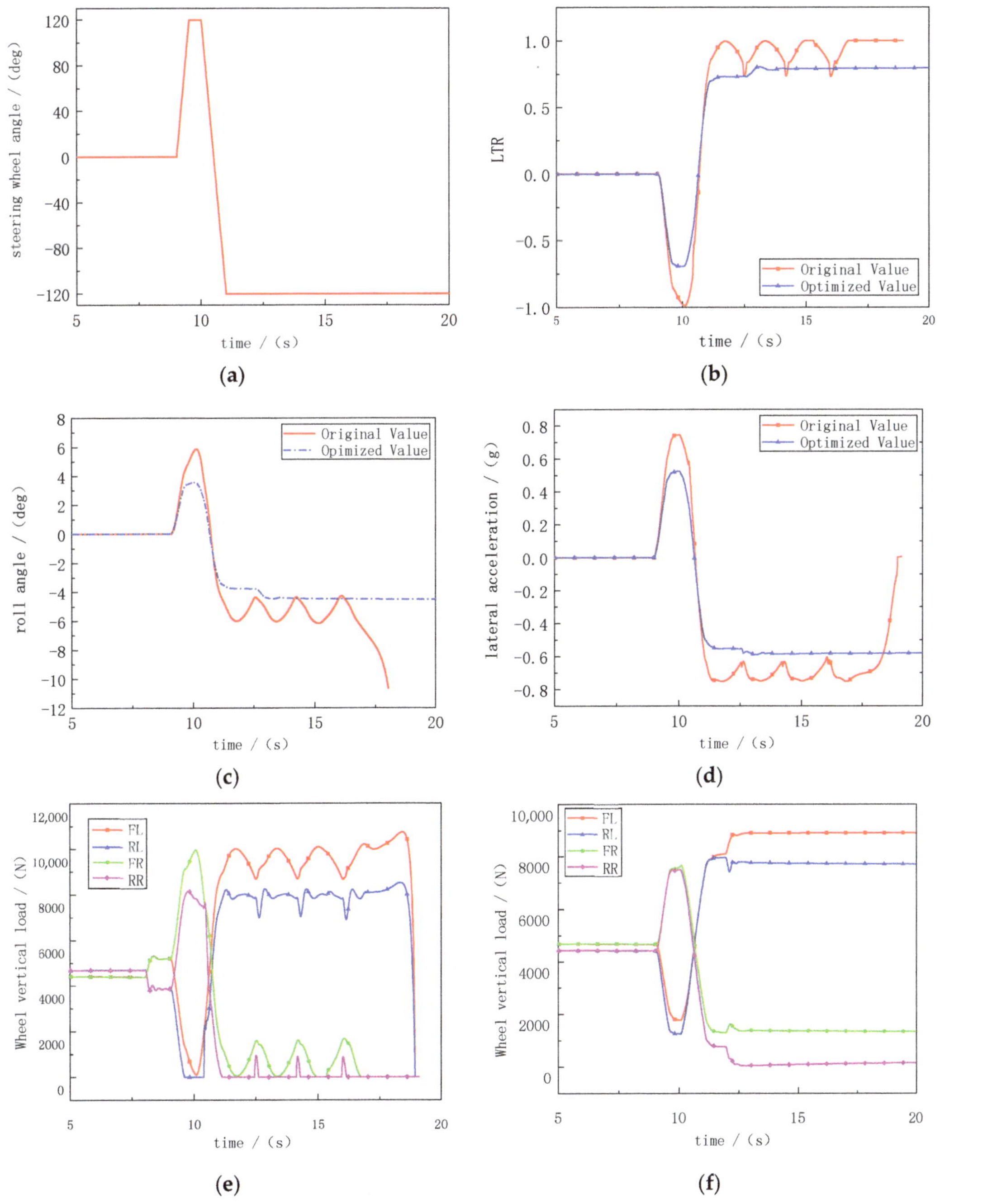

Figure 9. Simulation results of fish hook test condition. (**a**) Steering wheel angle. (**b**) LTR. (**c**) Lateral acceleration. (**d**) Roll angle. (**e**) Wheel vertical load before control. (**f**) Wheel vertical load after control.

Figure 9b shows that the LTR value fluctuates greatly without control. With control the LTR decreases by about 20%, and the curve changes smoothly and stabilizes at 0.8 s. It can be seen from Figure 9c the peak value of lateral acceleration drops from 0.8 g to 0.5 g with control, and the vehicle no longer rolls over. The lateral acceleration finally stabilizes at 0.55 g. Figure 9d shows the curve of the roll angle of the vehicle. The roll angle continues to increase until the vehicle rolls over without control. The peak value of the roll angle decreases significantly with control. The roll angle at the 20 s still stable at 4.5°. It indicates that the vehicle did not roll over at this time. Figure 9e,f show the changes of the vertical load of the four wheels before and after the control. The vertical load no longer fluctuates greatly with control Moreover, after the vehicle enters a right turn, the peak value of the inner wheel increases, and there is no 0 value. The vertical load of the outer wheel also decreases from 11,000 N at the maximum peak to 9000 N. The vehicle can maintain stable driving without rollover phenomenon.

Therefore the simulation results show that the differential braking anti-rollover control strategy proposed for DDEV in this paper can effectively prevent the vehicle from rolling over under high-speed sharp turns.

5. Conclusions

(1) Active distribution of wheel drive torque will affect both the roll and yaw movements of the vehicle, a decoupling control strategy for roll and yaw is proposed. The yaw stability controller and the roll stability controller are designed based on the LQR control theory and the sliding mode control theory. The control strategy of the compensated yaw and roll moment is evenly distributed in the left and right wheels.

(2) For maneuvering rollover caused by excessive lateral acceleration, an anti-rollover control strategy based on differential braking is designed. The vehicle generates a reverse yaw moment to achieve the control effect of reducing vehicle speed and changing steering characteristics by separately applying a braking torque to the front outer wheel. The lateral load transfer rate is used as the main evaluation index to simulate the typical working conditions. It shows that the differential braking anti-rollover control strategy proposed for DDEV is effective.

Author Contributions: Methodology: S.Y.; formal analysis: S.H. and Y.M.; Preparation of the original draft: X.C.; Writing review and editing: H.Z. All authors have read and agreed to the published version of the manuscript.

Funding: This research was funded by National Natural Science Foundation of China grant number 51705306.

Data Availability Statement: Not applicable.

Conflicts of Interest: The authors declare no conflict of interest.

Nomenclature

Symbol	
I_x (kg·m^2)	rotational inertia of the vehicle around the x axis
ϕ (deg)	vehicle roll angle
m_s (kg)	vehicle sprung mass
h_s (m)	distance from the center of the sprung mass to the roll center of the car
b (m)	distance from center of mass to rear axle
a (m)	distance from center of mass to front axle
m (kg)	vehicle mass
v_y (km/h)	vehicle lateral speed
v_x (km/h)	vehicle longitudinal speed
a_y (m/s^2)	vehicle lateral acceleration
ω (rad/s)	angular velocity of the wheel
k_r, k_f (N/rad)	cornering stiffness of the rear and front wheel
L (m)	vehicle wheelbase
r (m)	wheel rolling radius
B (m)	track width

δ (deg)	front wheel angle
K	stability factor
g_y (m/s^2)	gravitational acceleration
F_{yf}, F_{yr} (N)	side force of vehicle front and rear wheels
K^{Φ} (N/rad)	equivalent roll stiffness of the car
C^{Φ} (N/(km/h))	equivalent roll damping of automobile
ΔM_X (N·m)	additional roll moment.
F_{x1} (N)	ground driving force
T_1 (N·m)	motor torque transmitted from the front wheel to the vehicle body via the suspension
P_{Li} (i = 1, 2, 3, 4) (N)	force exerted by the left body on the front suspensions
P_{Li} (i = 1, 2, 3, 4) (N)	force exerted by the front suspensions on the left body
P_{ij} (j = 1, 2) (N)	force exerted by the car body on the side of the front-inner suspension
P_{ij} (j = 1, 2) (N)	force exerted by the front-inner suspension on the car body
$P_{oi}{}'$ (j = 1, 2) (N)	force exerted by the front-outer suspension on the car body
K_i (i = 1, 2, 3, 4, 5, 6)	the corresponding coefficient of the roll moment
z_i (i = 1, 2, 3, 4) (m)	
z_{ij} (j = 1, 2) (m)	the corresponding distance
l_{out}, l_{in} (m)	
θ_i (i = 1, 2, 3, 4) (rad)	
θ_{ij} (j = 1, 2) (rad)	the corresponding angle
θ_{oi} (i = 1, 2) (rad)	
M_{Xj} (j = 1, 2, 3, 4) (N·m)	roll moment generated by the suspension to the vehicle body
M_{Xi}, M_{Xo} (N·m)	roll moment generated by the front wheel via the suspension
ω_r (dge/s)	vehicle yaw rate
ω_{rd} (dge/s)	ideal yaw rate
ω_{rmax} (dge/s)	the maximum values of yaw rate
β_d (dge)	ideal side slip angle
β_{rmax} (dge)	the maximum values of side slip angle
ξ	weight coefficient between the roll angle and roll angular velocity
e (dge)	the error of roll angle
η	switching gain
μ	road adhesion coefficient
sat (s)	the saturation function
ΔF_{xi} (i = 1, 2, 3, 4) (N)	increment of each driving force
ΔT_i (i = 1, 2, 3, 4)	wheels drive torque
T_{max} (N·m)	maximum driving moment of the motor
F_b (N)	braking force exerted on the front outer wheel
I_w (kg·m^2)	rotating inertia of front outer wheel
T_b (N·m)	vehicle braking torque.
F_{zi} (i = 1, 2, 3, 4) (N)	wheels vertical load
P	braking pressure
S	braking efficiency coefficient.

References

1. Chakraborty, S.; Kumar, N.M.; Jayakumar, A.; Dash, S.K.; Elangovan, D. Selected Aspects of Sustainable Mobility Reveals Implementable Approaches and Conceivable Actions. *Sustainability* **2021**, *13*, 12918. [CrossRef]
2. Boada, B.L.; Boada, M.J.L.; Vargas-Melendez, L. A robust observer based on H ∞ filtering with parameter uncertainties combined with neural networks for estimation of vehicle roll angle. *Mech. Syst. Signal Process.* **2018**, *99*, 611–623. [CrossRef]
3. Zhou, B.; Lü, X.N.; Fan, L. Integrated control of active suspension system and active roll stabilizer. *China Mech. Eng.* **2014**, *14*, 1978–1983.
4. Chen, S.; Xia, C.G.; Pan, D.Y. A study on hybrid rollover control of vehicle with active anti-roll bar. *Automot. Eng.* **2017**, *39*, 667–674.
5. Parsons, K.G.R.; Pask, M.; Burdock, W. The development of ACE for discovery II. In Proceedings of the SAE 2000 World Congress, Detroit, MI, USA, 6–9 March 2000.
6. Karnopp, D. Active and semi-active vibration isolation. *Curr. Adv. Mech. Des. Prod. VI* **1995**, *117*, 409–423.

7. Xiao, L.J.; Wang, M.; Zhang, B.J. Vehicle roll stability control with active roll-resistant electro-hydraulic suspension. *Front. Mech. Eng.* **2020**, *15*, 43–54. [CrossRef]
8. Yakub, F.; Abu, A.; Mori, Y. Enhancing the yaw stability and the manoeuvrability of a heavy vehicle in difficult scenarios by an emergency threat avoidance manoeuvre. *Proc. Inst. Mech. Eng. Part D J. Automob. Eng.* **2017**, *231*, 615–637. [CrossRef]
9. Lua, C.A.; Castillo-Toledo, B.; Cespi, R.; Gennaro, S.D. Nonlinear observer-based active control of ground vehicles with non negligible roll dynamics. *Int. J. Control Autom. Syst.* **2016**, *14*, 743–752. [CrossRef]
10. Dal Poggetto, V.F.; Serpa, A.L. Vehicle rollover avoidance by application of gain-scheduled LQR controllers using state observers. *Veh. Syst. Dyn.* **2016**, *54*, 191–209. [CrossRef]
11. Imine, H.; Fridman, L.M.; Madani, T. Steering control for rollover avoidance of heavy vehicles. *IEEE Trans. Veh. Technol.* **2013**, *61*, 3499–3509. [CrossRef]
12. Chen, B.C.; Peng, H. Differential-braking-based rollover prevention for sport utility vehicles with human-in-the-loop evaluations. *Veh. Syst. Dyn.* **2001**, *36*, 359–389. [CrossRef]
13. Solmaz, S.; Akar, M.; Shorten, R. Adaptive rollover prevention for automotive vehicles with differential braking. *IFAC Proc. Vol.* **2008**, *41*, 4695–4700. [CrossRef]
14. Feng, F. Study on Roll Stability of Electric Wheel Drive Vehicle on Uneven Pavement. Master's Thesis, Yanshan University, Qinhuangdao, China, 2017.
15. Li, G.; Zong, C.F.; Chen, G.Y.; Wei, H.; Lei, H. Integrated control for X-by-wire electric vehicle with 4 independently driven in-wheel motors. *J. Jilin Univ. (Eng. Technol. Ed.)* **2012**, *42*, 796–802.
16. Wang, Z. The Integrated Control of Active Suspension and Drive Force Distribution Based on 4 In-Wheel Motor Driven Electric Vehicle. Master's Thesis, Southeast University, Nanjing, China, 2016.
17. Kiyotaka, K.; Toshiyuki, U.; Yoichi, H. Normal force stabilizing control using small EV powered only by Electric double layer capacitor. *World Electr. Veh. J.* **2008**, *1*, 62–67.
18. Fujimoto, H.; Sato, S. Pitching control method based on quick torque response for electric vehicle. In Proceedings of the 2010 International Power Electronics Conference, Sapporo, Japan, 21–24 June 2010.
19. Kawashima, K.; Uchida, T.; Hori, Y. Rolling stability control of in-wheel electric vehicle based on two-degree-of-freedom control. In Proceedings of the IEEE International Workshop on Advanced Motion Control, Trento, Italy, 26–28 March 2008.
20. Kawashima, K.; Uchida, T.; Hori, Y. Rolling stability control based on electronic stability program for in-wheel-motor electric vehicle. *World Electr. Veh. J.* **2009**, *3*, 34–41. [CrossRef]
21. Katsuyama, E. Decoupled 3D moment control using in-wheel motors. *Veh. Syst. Dyn.* **2013**, *51*, 18–31. [CrossRef]
22. Murata, S. Innovation by in-wheel-motor drive unit. *Veh. Syst. Dyn.* **2012**, *50*, 807–830. [CrossRef]
23. Yu, Z.S. *Automobile Theory*, 5th ed.; China Machine Press: Beijing, China, 2009.
24. Liu, B.; Tang, W.S. *Modern Control Theory*; China Machine Press: Beijing, China, 2006.
25. Zhang, X.Z.; Zheng, L. Vehicle stability control of distributed-driven electric vehicles based on optimal torque allocation. *China Mech. Eng.* **2018**, *29*, 1780–1787.
26. Ren, Z.Y.; Zhao, W.Q.; Zong, C.F. Research on vehicle yaw stability control based on improved LQR method. *Sci. Technol. Eng.* **2017**, *17*, 97–102.
27. Yan, S. Research on Roll Stability Control of Distributed Drive Electric Vehicle. Master's Thesis, Shanghai University of Engineering Science, Shanghai, China, 2020.

World Electric Vehicle Journal

MDPI

Article

Regenerative Braking Strategy of Dual-Motor EV Considering Energy Recovery and Brake Stability

Tonglie Wu, Feng Wang * and Peng Ye

Automotive Engineering Research Institute, Jiangsu University, Zhenjiang 212013, China
* Correspondence: bewater@ujs.edu.cn

Abstract: The dual-motor EV (Electric Vehicle) is increasingly favored by manufacturers for its excellent performance in terms of power and economy. How to further reduce its energy consumption and make full use of the dual-motor energy recovery is an important support to improve the overall vehicle economy and realize the "dual carbon" strategy. For the dual-motor EV architecture, the motor model, power battery loss model and vehicle longitudinal braking force model are established and the energy recovery-dominated regenerative braking torque distribution (RBD) rule of the dual motors is designed. Based on genetic algorithm (GA) theory and taking into account SOC, vehicle speed and braking intensity, a regenerative-braking torque optimization method is proposed that integrates energy recovery and braking stability. The braking intensity of 0.3 and the initial vehicle speed of 90 km/h are selected for verification. Compared with the rule method, the energy recovery and stability are improved by 22.8% and 4.8%, respectively, under the genetic algorithm-based and energy recovery-dominated regenerative-braking torque distribution (GA-RBD) strategy. A variety of conditions are selected for further strategy validation and the result shows that compared with the rule-based method, both energy recovery and braking stability are improved as braking speed and braking intensity increase under the GA-RBD strategy.

Keywords: electric vehicles; regenerative braking; energy recovery; genetic algorithm; braking stability

Citation: Wu, T.; Wang, F.; Ye, P. Regenerative Braking Strategy of Dual-Motor EV Considering Energy Recovery and Brake Stability. *World Electr. Veh. J.* **2023**, *14*, 19. https://doi.org/10.3390/wevj14010019

Academic Editors: Yong Li, Xing Xu, Lin Zhang, Yechen Qin and Yang Lu

Received: 29 November 2022
Revised: 16 December 2022
Accepted: 4 January 2023
Published: 9 January 2023

1. Introduction

The Energy Saving and New Energy Vehicle Technology Roadmap (Version 2.0) indicates the development direction of the automotive industry during the next 15 years. By 2035, the annual sales of energy-efficient vehicles and new energy vehicles will each account for 50% and the transformation of the automotive industry towards electrification. Dual-motor EVs are favored by an increasing number of manufacturers for their outstanding performance in terms of power and economy. According to statistics, the vehicle braking energy loss accounts for more than 43% in typical urban conditions, and making full use of the dual-motor EV regenerative braking for energy recovery is an important supportive role in improving the economy of the whole vehicle and realizing the "double carbon" strategy [1]. As a result, regenerative braking systems [2–5] are being studied in greater depth by national and international scholars.

To maximize energy recovery, Pennycott et al. [6] designed a constant proportional regenerative braking strategy based on the control distribution, which considered the influence of motor operating characteristics on regenerative braking. Considering the different braking conditions, Pei [7] proposes a coordinated control strategy for the electro-hydraulic braking of distributed electric vehicles, aiming to improve the comprehensive performance of the system in terms of energy regeneration and braking stability. Maia et al. [8] proposed a fuzzy controller-based distribution strategy for regenerative braking torque, taking into account vehicle acceleration, bumpiness and road inclination and verifying the effectiveness of this distribution strategy under the real road experiments. Xu [9] proposed a new braking torque distribution strategy based on model predictive control which aimed to achieve both braking stability and optimal energy recovery under the constraints of regenerative braking. Chen [10] proposed a hierarchical cooperative control for the electromechanical brake-by-wire system (EBW) to solve the coordination of mechanical and regenerative braking and ensure vehicle stability and maximum energy regeneration.

Existing studies on regenerative braking energy recovery have generally focused on the effects of braking intensity, vehicle speed and residual power on energy recovery and braking safety during regenerative braking. The relevant research on composite braking has mainly focused on models with the single-motor configuration where the energy recovery power flow path is relatively simple, and less research has been conducted on braking energy recovery in the dual-motor configuration. The research object of this paper has characteristics that dual motors can participate in energy recovery. Compared to the single-motor configuration, the dual-motor driven vehicles have the multiple power flow paths for energy recovery, and by distributing the motor braking torque, the motors can work to a greater extent in the high efficiency zone. Thus, fully considering the structural characteristics of dual-motor EVs and establishing a regenerative braking strategy for dual-motor EVs that integrates energy recovery and braking stability to further improving the regenerative braking energy recovery rate has important theoretical significance and engineering value. The contributions of the proposed regenerative braking method lie in the following three aspects:

(i) A energy recovery-dominated regenerative-braking torque distribution rule of the dual motors is designed, which takes into account the characteristics that both motors can participate in the energy recovery characteristics.

(ii) Considering the variation of SOC, vehicle speed and braking intensity, a dual-motor EV regenerative-braking optimization method that integrates energy recovery and braking stability is proposed.

(iii) The energy recovery rate and braking stability are integrated into one control objective by weighting coefficients, and the optimal value of torque distribution is solved by genetic algorithm under the corresponding weighting coefficients.

Therefore, this paper proposes a dual-motor EV regenerative braking strategy that integrates energy recovery and braking stability. Both the energy recovery rate and braking stability are improved compared to the rule-based method. The rest of the paper is organized as follows. In Section 2, the powertrain configuration and main component models of the dual-motor EV are presented. In Section 3, the energy-recovery rate-dominated regenerative-braking torque-distribution rule is designed and a regenerative-braking torque-optimization strategy that incorporates braking energy recovery and braking stability is proposed. In Section 4, the results of the two distribution strategies are compared and the effectiveness of the strategies is demonstrated. At last, conclusions are given in Section 5.

2. Model of a Dual-Motor EV Regenerative Braking System

2.1. Dual-Motor EV System Configuration

The configuration of one dual-motor EV in this paper is shown in Figure 1, which mainly include a coupled structure of two motors and two gear pairs, hydraulic lines, control units, battery packs and other components. Motor 1 and motor 2 can individually or jointly provide the braking torque in regenerative mode.

When the driver applies the brake pedal to apply the brakes, the Vehicle Control Unit (VCU) determines the braking torque to be assumed by the Motor Brake System (MBS) and the Hydraulic Brake System (HBS) based on information such as current vehicle speed, braking intensity and battery SOC. The motor braking torque and hydraulic braking torque are controlled by the Motor Control Unit (MCU) and the Hydraulic Control Unit (HCU), respectively.

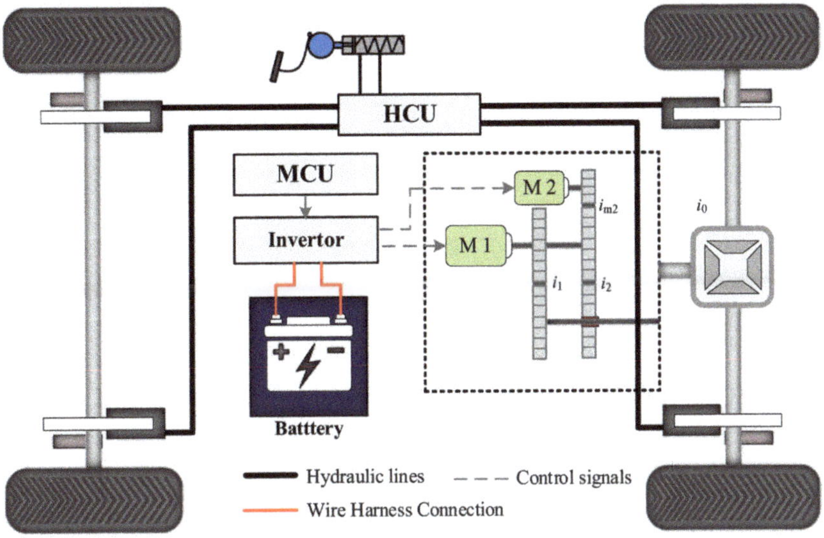

Figure 1. Schematic diagram of an EV system structure.

2.2. Motor Model

Permanent magnet synchronous motors are used in this configuration of EV. It offers the advantages of small size, high-speed, high-power density and flexibility in shape and size. Compared to other types of motors, permanent magnet synchronous motors are more efficient and have a longer range. Moreover, China is rich in rare earth resources and the cost of using permanent magnet-synchronous stand-alone machines is lower.

Since the focus of this paper is on the regenerative-braking torque-distribution strategy, the transient characteristics of the motors are simulated by first-order delays. The equations are shown as follows:

$$T_{m1} = \frac{1}{\tau_{m1}s + 1} \tag{1}$$

$$T_{m2} = \frac{1}{\tau_{m2}s + 1} \tag{2}$$

where τ_{m1} and τ_{m2} denote the time constants of the first-order system; T_{m1}, T_{m2} denote the actual output torque

The generation efficiency maps of motor 1 and motor 2 are measured experimentally, as shown in Figures 2 and 3.

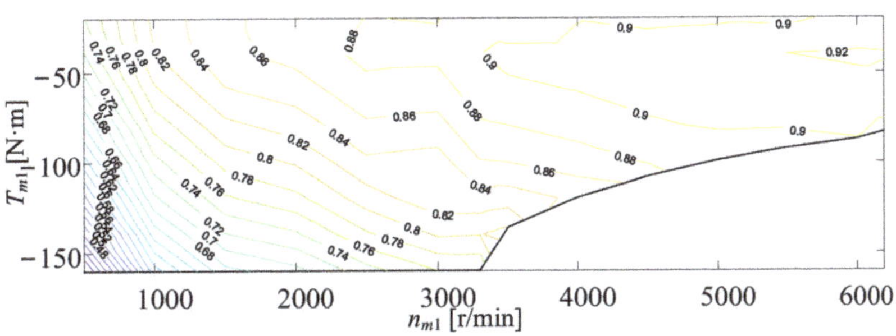

Figure 2. Efficiency map of motor 1.

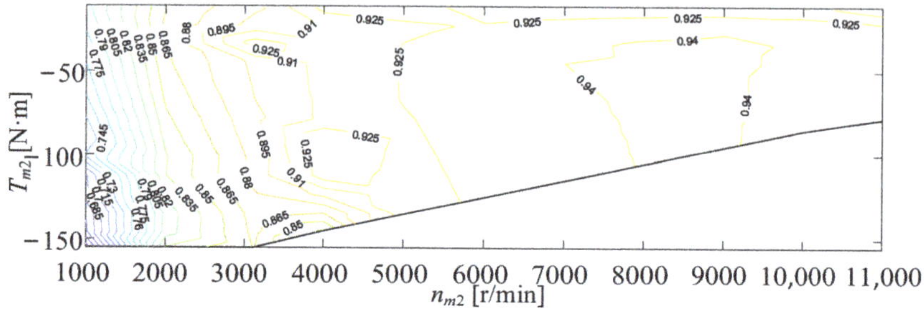

Figure 3. Efficiency map of motor 2.

2.3. *Power Battery Loss Model*

The charging efficiency of the power battery directly influences the regenerative-braking energy recovery, and its power loss can be expressed as follows:

$$P_{los} = \frac{\left(V_o - \sqrt{V_o^2 - 4R_{c\_dis} \cdot P_m}\right)^2}{4R_c} \tag{3}$$

where V_o denotes the battery terminal voltage; R_c and R_{c_dis} denote the battery equivalent resistance and the battery discharge resistance, respectively; and P_m denotes the total generated power of the motor.

2.4. *Vehicle Braking Model*

The force analysis of the braking process is shown in Figure 4.

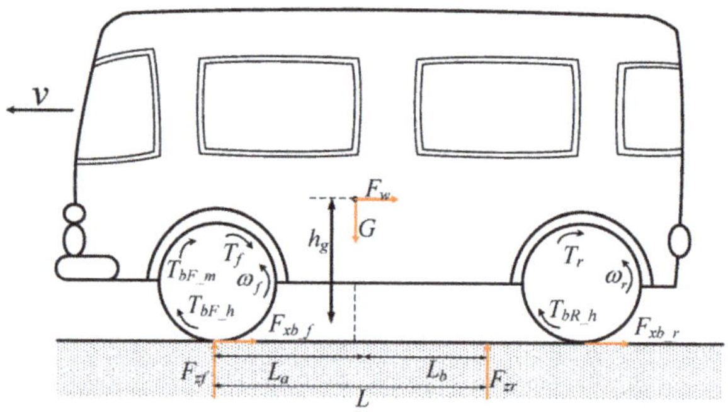

Figure 4. Vehicle force analysis diagram of the braking process.

The vehicle longitudinal dynamics equations are as follows:

$$m\dot{v} = -F_{xb\_f} - F_{xb\_r} - F_w - F_f \tag{4}$$

$$F_w = \frac{C_d \cdot A_v}{21.15} v^2 \tag{5}$$

$$F_f = f_v \cdot m \cdot g \tag{6}$$

where v denotes the longitudinal speed of the vehicle; m denotes the overall vehicle mass; F_{xb_f} and F_{xb_r} denote the ground braking force of the front wheel and the rear wheel, respectively; F_w and F_f denote the wind resistance and the rolling resistance of the vehicle; C_d and f_v indicate the wind resistance coefficient and the rolling resistance coefficient; and A_v is expressed as the equivalent wind resistance area.

The torsional dynamics equations of the front wheel and the rear wheel are expressed as follows:

$$\begin{cases} J_f\dot{\omega}_f = \frac{1}{2}\left(R_v \cdot F_{xb\_f} - T_{bF\_m} - T_{bF\_h}\right) \\ J_r\dot{\omega}_r = \frac{1}{2}(R_v \cdot F_{xb\_r} - T_{bR\_h}) \end{cases} \tag{7}$$

where J_f and J_r denote the equivalent rotational inertia of the front and rear wheels, respectively; R_v indicates the wheel radius; ω_f and ω_r represent the front and rear wheel rotational speeds; T_{bF_h} and T_{bR_h} represent the hydraulic braking force of the front and rear wheels. T_{bF_m} is the regenerative braking torque applied to the wheel end by the dual motor, expressed as follows:

$$T_{bF\_m} = \left(T_{m1}i_{g1} + T_{m2}i_{m2}i_{g2}\right) \cdot i_0 \tag{8}$$

where T_{m1} and T_{m2} denote the output torque of motor 1 and motor 2; i_{m2} and i_0 denote the output reduction ratio of motor 2 and the final drive ratio, respectively; and i_{g1} and i_{g2} represent the corresponding reduction ratios of motor 1 and motor 2 in the current operating mode of the coupling mechanism.

The detailed parameters of the EV and related components studied in this paper are shown in Table 1.

Table 1. Key parameters of vehicle powertrain.

Components	Description
Transmission	Reduction ratios (i_1, i_2): 2.11/1.31 Final drive ratio (i_0): 3.91 Reduction ratio of the motor 2 end (i_{m2}): 1.72
Motor	Type: Permanent magnet synchronous motor (PMSM) Maximum power: 55 kW (M1); 75 kW (M2)
Battery	Type: NiMH Voltage: 387 V Capacity: 25 kW·h Internal resistance: 0.015 Ω
Vehicle	Vehicle mass: 1570 kg Frontal area of vehicle: 1.26 m^2; Aerodynamic drag: 0.35 Tire rolling resistance coefficient: 0.018 Drive wheel radius: 0.3 m

3. Regenerative Braking Strategy for Dual-Motor EV

3.1. Energy Recovery-Dominated Regenerative Braking Torque Distribution (RBD) Rule

Combined with the dual-motor EV configuration in this paper, an energy recovery rate-dominated regenerative braking torque distribution rule is proposed, and its specific distribution strategy is shown in Figure 5. Where F_{bF} and F_{bR} denote the front wheel braking torque and the rear wheel braking torque, respectively.

From the Figure 5, the braking force operating points under the RBD rule are between the I curve and the ECE regulation curve. With the braking intensity increases, the front and rear wheel braking forces are distributed along the OABCDE curve. The term z denotes braking severity.

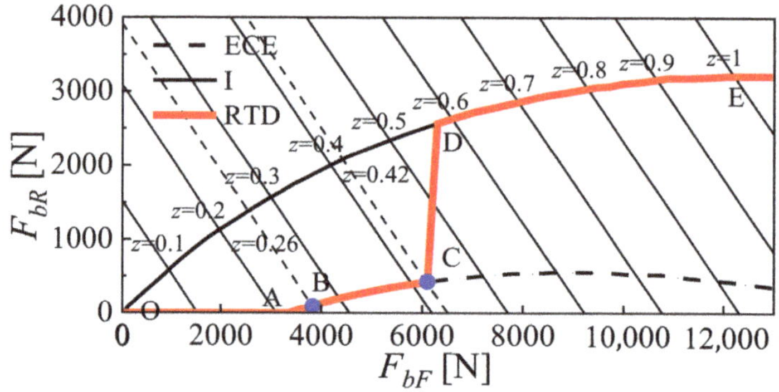

Figure 5. Regenerative braking torque distribution rules dominated by energy recovery.

(1) OA segment: When $z < 0.21$, the braking intensity is so light that braking stability does not need to be considered. In this case, the braking force is supplied exclusively by motor 2 to the front wheels and the braking force of the rear wheels is 0.

(2) AB segment: When $0.21 \leq z < 0.26$, with the braking intensity increasing, the rear wheels start to engage the brakes and the braking torque is supplied exclusively by the hydraulic system. In addition, the braking force of the front wheels is still provided only by motor 2, and the braking torque of motor 2 reaches a maximum at point B.

(3) BC segment: When $0.26 \leq z < 0.42$, the braking force provided by motor 2 is no longer sufficient for the braking of the front wheels, so that motor 1 and motor 2 together provide braking force to the front wheels. At the point C, the braking torque of motor 1 and motor 2 reaches its maximum value at the same time.

(4) CD segment: When $0.42 \leq z < 0.58$, motor 1 and motor 2 are all involved in the braking process and have reached their peak state. At this point, the energy recovery rate of the system has been ensured, and in order to take into account the braking stability at the same time, the braking force is distributed according to the f-curve with $\varphi = 0.58$. At this stage, the hydraulics start to participate in the front wheel braking, and as the braking intensity increases, the curve gradually approaches the I curve.

(5) DE segment: When $0.58 \leq z \leq 1$, in this phase, the braking stability is predominant, the braking force distribution curve exactly follows the I-curve and the braking torque is provided by the hydraulic system and the motor.

3.2. Regenerative Braking Torque Optimisation Incorporating Energy Recovery and Braking Stability

GA-RBD distribution strategy architecture diagram is shown in Figure 6.

Through the brake intention module, the driver's required braking torque and braking intensity are calculated. Based on the current vehicle speed, battery SOC and other information, the weighting coefficients under the current state are obtained according to the weighting allocation table. The optimal values of the assigned braking torque under this weighting factor are calculated by the genetic algorithm, and the corresponding reference torque information of each actuator is sent to the motor system and the hydraulic braking system. Finally, the actuator outputs the braking torque to complete the braking process.

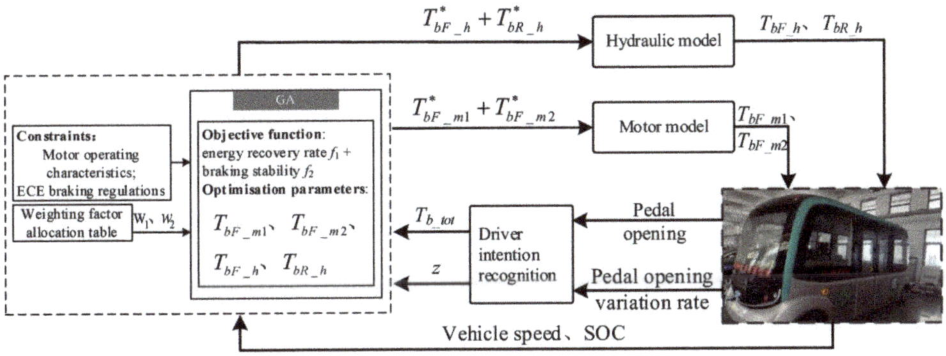

Figure 6. GA-RBD distribution strategy architecture diagram.

In order to take into account both the braking stability and the energy recovery rate of the vehicle, both the energy recovery rate and the braking stability were selected as optimization objectives in this study.

The energy recovery rate is defined as the ratio of the energy recovered by the battery to the kinetic energy lost during braking, expressed as follows:

$$f_1 = \frac{\int \left(\frac{n_w i_0 i_{g1} T_{m1}}{9550}\eta_1 + \frac{n_w i_0 i_{g2} i_{m2} T_{m2}}{9550}\eta_2 - P_{los}\right)dt}{\frac{1}{2}m\left(v_t{}^2 - v_0^2\right) - \int \left(mgf_v + \frac{C_d \cdot A_v}{21.15}v^2\right)vdt} \tag{9}$$

where n_w denotes wheel end rotational speed; η_1 and η_2 denote the generation efficiency of motor 1 and motor 2, respectively; v_t and v_0 represent the current vehicle speed and the initial braking speed; and P_{los} denotes the loss of power for battery charging.

Braking stability is defined as the degree of deviation between the front and rear axle braking torque distribution curves and the ideal braking torque distribution curve, which is expressed as follows:

$$f_2 = \frac{\sqrt{\left(T_{bF\_h} + T_{bF\_m} - T_{If}\right)^2 + \left(T_{bR\_h} - T_{Ir}\right)^2}}{zmgR_v} \tag{10}$$

where T_{If} and T_{Ir} represent the front and rear wheel braking torques corresponding to when the braking force distribution curve is on the ideal distribution curve, respectively.

The comprehensive optimization objectives of the final design are as follows:

$$f(T) = w_1 \cdot f_1 + w_2 \cdot f_2 \tag{11}$$

where w_1 and w_2 denote the corresponding weighting factors.

The regenerative braking system will stop working when the vehicle speed is below 20 km/h or when the battery SOC value is high, the former to take into account that the motor does not recover enough braking energy at low speeds, and the latter to prevent the battery from being overcharged. Focusing on energy recovery at low speeds or at low braking intensities, the weighting factor w_1 will be increased. Conversely, at high speeds or high braking intensities the focus is on braking stability and the weighting factor w_1 will be reduced, w_2 increased. When braking under normal operating conditions, the weighting coefficients are taken to be 0.5, respectively, in order to take into account the effects of both braking energy recovery and braking stability.

Taking into account SOC, vehicle speed and braking intensity, the weighting factors are set as shown in Figure 7.

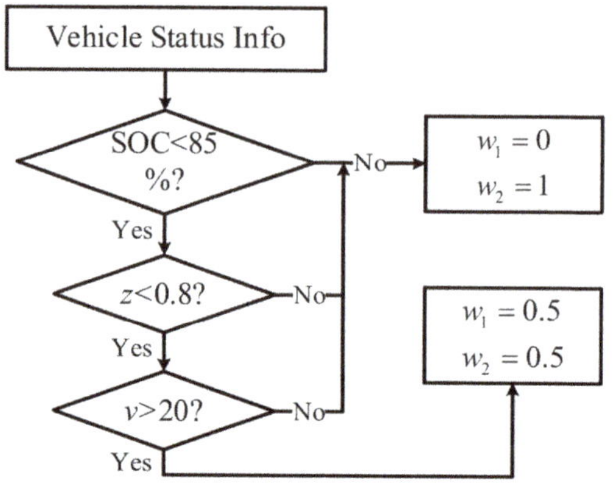

Figure 7. Rules for setting weighting factors.

The optimization objective selected by the genetic algorithm is a comprehensive index of braking energy recovery and braking stability after considering weighting factors (as shown in Equation (11)). The motor 1 braking torque, motor 2 braking torque, front wheel hydraulic braking force and rear wheel hydraulic braking force are used as genes for the individuals in the genetic algorithm, and the objective function is solved using the genetic algorithm to obtain the torque distribution corresponding to the optimal value.

The Genetic Algorithm (GA) [11] process is shown in Figure 8 and mainly consists of three parts as follows:

(1) Initialization

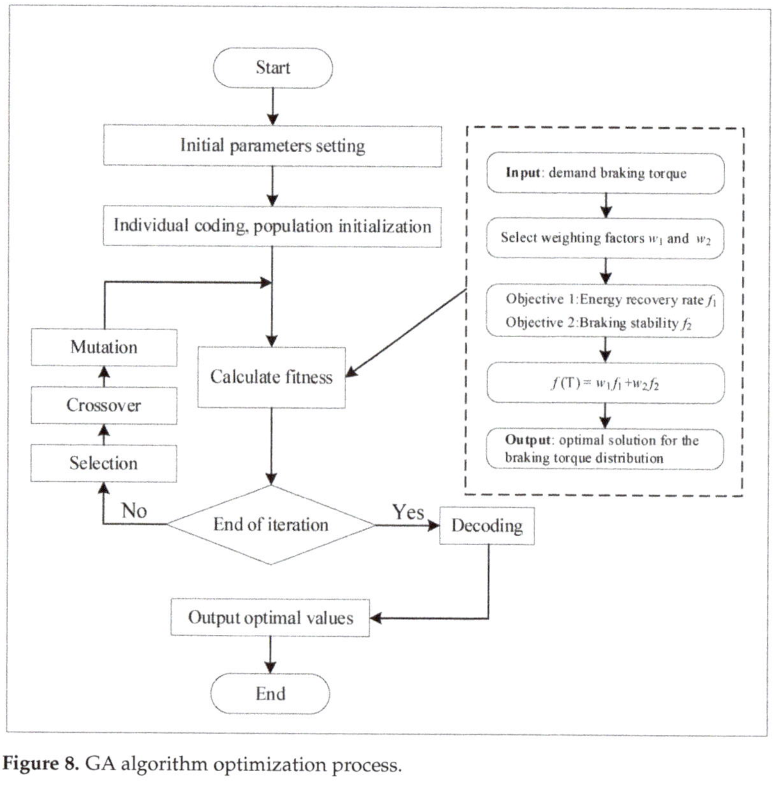

Figure 8. GA algorithm optimization process.

Individual coding and population and initialization. The coding method adopted in this paper is real number coding and the optimisation parameters of the genetic algorithm are set as follows: the population size is set to 30, the crossover probability and variation probability are set to 0.95 and 0.1, respectively, and the number of stopping iterations is 50.

Where the size of the population is related to the degree of dispersion for the optimised problem, and the larger the dispersion, the larger the population size to improve the speed of convergence [12]. In this paper the linear relationship between the torque and the objective function is obvious, so the population can be relatively small and is set to 30.

(2) Calculation of the fitness

Calculate the value of the fitness function corresponding to each individual $f(T_I)$, and determine whether it satisfies the termination condition of the genetic algorithm; if it does, then output the optimal solution, otherwise continue to evolutionary operations.

(3) Evolutionary operations

Evolutionary operations are the heart of genetic algorithms and include selection, crossover and mutation. In nature, the further adapted individuals are, the more likely they are to reproduce offspring. Based on fitness, the system selects a certain number of individuals to cross and mutate in order to produce offspring and form new populations, and repeats the operation in (2).

4. Results Verification

In order to verify the effectiveness of the GA-RBD allocation strategy algorithm, the optimization was first carried out at different braking intensities and speeds to obtain the vehicle speed-braking intensity-braking torque maps, as shown in Figure 9.

Based on the maps of braking torque distribution obtained from the above optimization, the brake intensity of 0.3 and the initial vehicle speed of 90 km/h were selected for verification. As shown in Figure 10, compared to the rule-based regenerative braking strategy with motor 2 working first and then motor 1, the GA-RBD strategy enables both motors to participate in the braking process more evenly. As can be seen from Figure 11, the motor efficiency under the GA-RBD strategy moves towards the high efficiency zone benefiting from the genetic algorithm's optimization for the dual motor operating point.

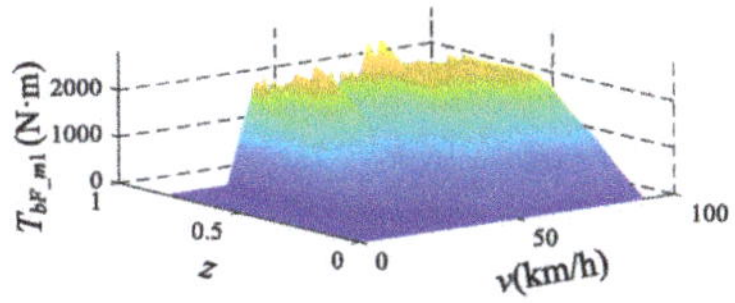

(a)

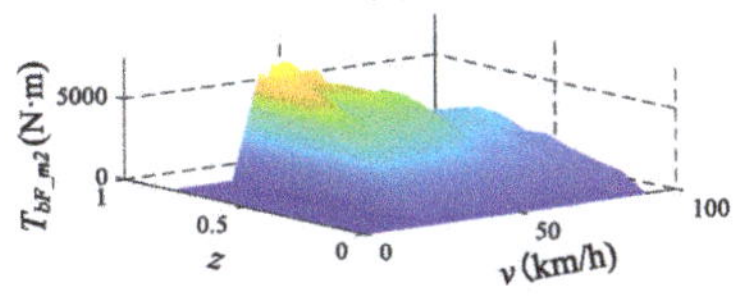

(b)

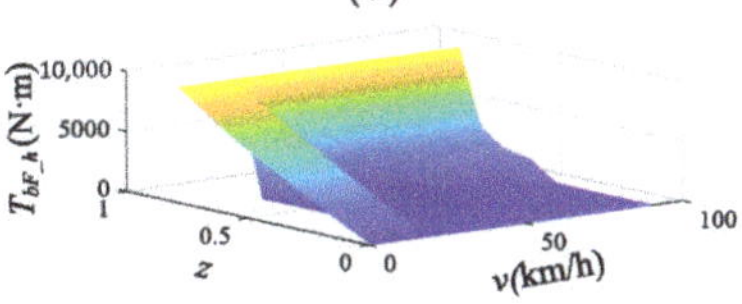

(c)

Figure 9. *Cont.*

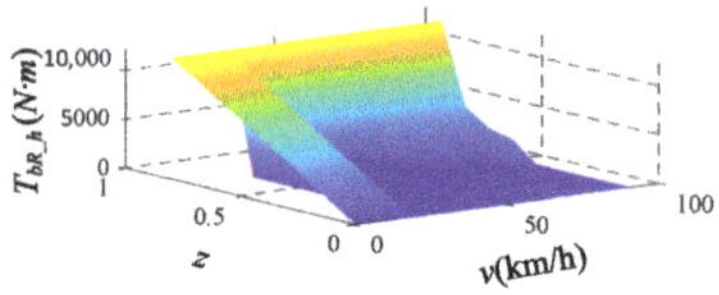

(d)

Figure 9. Vehicle speed-braking intensity-braking torque distribution maps. (**a**) Motor 1 braking torque. (**b**) Motor 2 braking torque. (**c**) Front wheel hydraulic braking torque. (**d**) Rear wheel hydraulic braking torque.

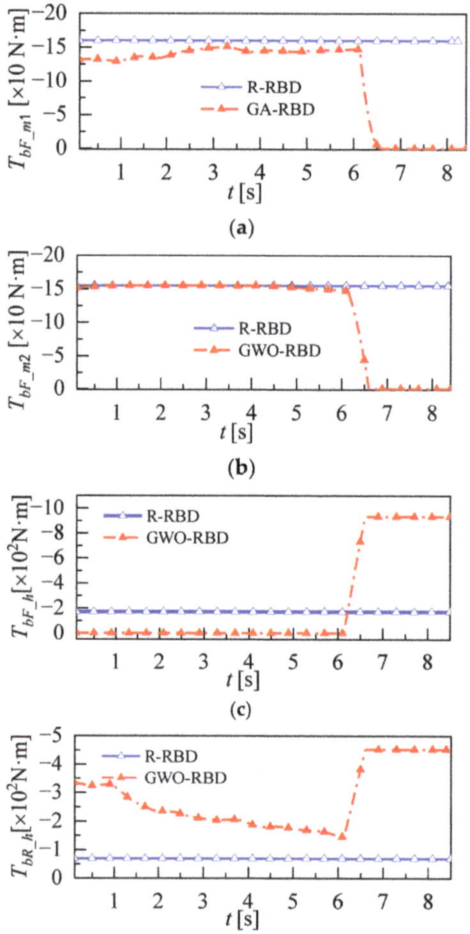

(d)

Figure 10. Comparative graph of the different braking torque distribution strategies. (**a**) Motor 1 braking torque. (**b**) Motor 2 braking torque. (**c**) Front wheel hydraulic braking torque. (**d**) Rear wheel hydraulic braking torque.

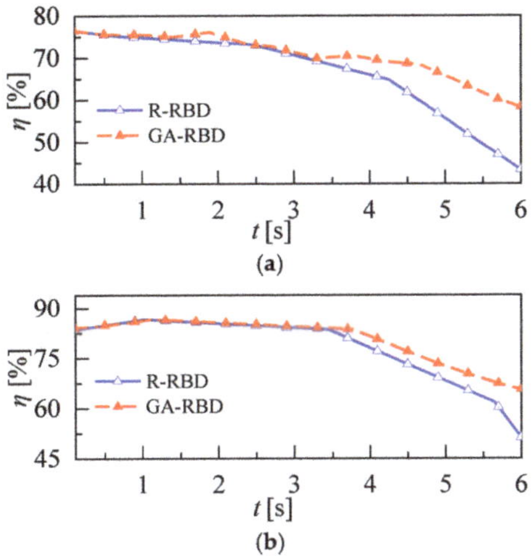

Figure 11. Comparative graphs of motor efficiency under the different strategies. (**a**) Motor 1 efficiency. (**b**) Motor 2 efficiency.

The comparison of energy recovery and stability of the vehicle under different strategies is depicted in (a) and (b) of Figure 12, respectively. Compared to the rule-based method, energy recovery under the GA-RBD strategy (shown in Figure 12a) is improved by 16.3% and the root-mean-square value of the stability coefficient ε (shown in Figure 12b) is reduced by 4.5% (stability is improved).

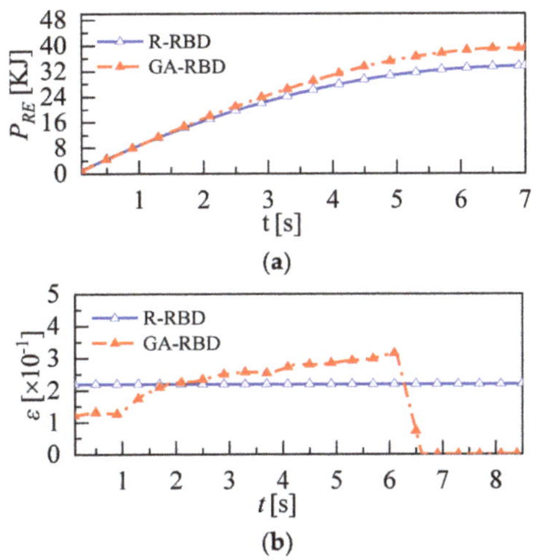

Figure 12. Comparative graphs of energy recovery and stability under the different strategies. (**a**) Cumulative energy recovery. (**b**) Stability coefficient.

Where braking stability ε [13]: During braking, the situation with full use of the ground adhesion is defined as the ideal situation, therefore the deviation of the front- and rear-axle braking torque from the ideal braking torque is used to express braking stability. The symbol is denoted as ε.

In order to further verify the effectiveness of the two distribution strategies under different vehicle speeds and braking intensity operating conditions, the two strategies are compared and verified in Table 2. Under the braking intensity of 0.2 to 0.6 and vehicle speeds between 60 km/h and 90 km/h, the GA-RBD strategy improved the braking stability by a maximum of 5.4% compared to the rule method and the braking energy recovery improvement rate varied from 8.3% to 20.2% as the braking intensity and initial vehicle speed increased. The effectiveness of the GA-RBD strategy is further verified.

Table 2. Results comparison under the different braking conditions.

Braking Intensity	Vehicle Speed [km/h]	Distribution Strategy	ε-RMS	Change Rate	Recovered Energy [KJ]	Change Rate
0.2	$v = 60$	R-RBD GA-RBD	0.482 0.469	−2.7%	30.43 32.96	8.3%
	$v = 75$	R-RBD GA-RBD	0.473 0.455	−3.8%	34.63 38.16	10.2%
	$v = 90$	R-RBD GA-RBD	0.452 0.433	−4.3%	42.09 47.60	13.1%
0.4	$v = 60$	R-RBD GA-RBD	0.331 0.321	−2.9%	21.32 23.77	11.5%
	$v = 75$	R-RBD GA-RBD	0.321 0.311	−3.1%	21.12 24.12	14.2%
	$v = 90$	R-RBD GA-RBD	0.311 0.300	−3.4%	19.71 23.28	18.1%
0.6	$v = 60$	R-RBD GA-RBD	0.150 0.143	−4.6%	8.29 9.08	13.5%
	$v = 75$	R-RBD GA-RBD	0.140 0.133	−5.0%	11.01 12.79	16.2%
	$v = 90$	R-RBD GA-RBD	0.128 0.121	−5.4%	13.71 16.48	20.2%

5. Conclusions

This paper fully considered the structural characteristics of dual-motor EVs and established a regenerative braking strategy for dual-motor EVs that integrates energy recovery and braking stability, which further improves the regenerative braking energy recovery rate and braking stability. The main conclusions show the following:

(1) Based on the dual-motor EV architecture, an electric motor model, a power battery loss model and a vehicle longitudinal braking force model are established, and an energy recovery rate-dominated regenerative braking torque distribution rule considering the dual motors is designed.

(2) Based on the theory of genetic algorithm, a regenerative braking torque optimization method integrating energy recovery and braking stability is proposed, which considering SOC, vehicle speed and braking intensity. The braking intensity of 0.3 and the initial vehicle speed of 90 km/h are selected for validation. Compared with the rule-based method, the energy recovery and the stability under the GA-RBD strategy are improved by 22.8% and 4.8%.

(3) Various conditions were further selected for strategy validation and the results show that as vehicle speed increases and braking intensity increases, both the energy recovery rate and braking stability under the GA-RBD strategy are improved compared to the rule-based method.

In short, the proposed dual-motor EV regenerative braking that combines energy recovery and braking stability strategy can significantly improve the energy recovery rate and braking stability during braking, and can provide a theoretical reference for the EV in engineering practice. Considering the many uncertainties in real vehicle testing, the effectiveness of the GA-RED strategy could be tested on different types of real vehicles in future research.

Author Contributions: Conceptualization, T.W. and F.W.; methodology, P.Y.; software, T.W.; validation, F.W. All authors have read and agreed to the published version of the manuscript.

Funding: This research was funded by the National Natural Science Foundation of China [No. 52172358], Postdoctoral Science Foundation of China (No. 2022T150339), Project of Faculty of Agricultural Equipment of Jiangsu University [No. NZXB20210103].

Data Availability Statement: Not applicable.

Conflicts of Interest: The authors declare no conflict of interest.

References

1. Fang, G.; Tian, L.; Fu, M.; Sun, M.; Du, R.; Lu, L. The effect of energy construction adjustment on the dynamical evolution of energy-saving and emission-reduction system in China. *Appl. Energ.* **2017**, *196*, 180–189. [CrossRef]
2. Guo, J.; Dong, H.; Sheng, W.; Tu, C. Optimum control strategy of regenerative braking energy for electric vehicle. *J. Jiangsu Univ. Nat. Sci. Ed.* **2018**, *39*, 132–138.
3. Ahmed, T.; Mehmet, F. An overview of regenerative braking systems. *J. Energy Storage* **2022**, *52*, 105033.
4. Chen, L.; Zeng, L.; Pan, C.; Li, Z. Control of regenerative braking force for EV equipped with ultra-capacitor. *J. Jiangsu Univ. Nat. Sci. Ed.* **2014**, *35*, 508–512.
5. He, R.; Li, M. Integrated control strategy of combined braking system and ABS based on road identification. *J. Jiangsu Univ. Nat. Sci. Ed.* **2020**, *41*, 20–26.
6. Pennycott, A.; Novellis, L.D.; Gruber, P.; Sorniotti, A. Optimal braking force allocation for a four-wheel drive fully electric vehicle. *J. Syst. Control Eng.* **2014**, *228*, 621–628. [CrossRef]
7. Pei, X.; Pan, H.; Chen, Z. Coordinated control strategy of electro-hydraulic braking for energy regeneration. *Control Eng. Pract.* **2020**, *96*, 104324. [CrossRef]
8. Maia, R.; Silva, M.M.; Araújo, R.; Nunes, U. Electrical vehicle modeling: A fuzzy logic model for regenerative braking, Expert Systems with Applications. *Expert. Syst. Appl.* **2015**, *42*, 8504–8519. [CrossRef]
9. Xu, W.; Chen, H.; Zhao, H.; Ren, B. Torque optimization control for electric vehicles with four in-wheel motors equipped with regenerative braking system. *Mechatronics* **2019**, *57*, 95–108. [CrossRef]
10. Chen, X.; Wei, L.; Wang, X.; Li, L.; Wu, Q.; Xiao, L. Hierarchical cooperative control of anti-lock braking and energy regeneration for electromechanical brake-by-wire system. *Mech. Syst. Signal Process.* **2021**, *159*, 107796. [CrossRef]
11. Banaei, A.; Alamatian, J.; Tohidi, R.Z. Active control of structures using genetic algorithm with dynamic weighting factors using in the constrained objective function. *Structures* **2022**, *47*, 189–200. [CrossRef]
12. Wang, C. Research on Energy Management Strategy of Hybrid Electric Bus Based on Genetic Algorithm. Master's Thesis, Shandong University, Jinan, China, 2020.
13. Guo, H.; He, H.; Sun, X. Hierarchical optimization method for regenerative braking stability of hybrid electric vehicles. *J. Beijing Inst. Technol. Engl. Ed.* **2014**, *23*, 1–7.

World Electric Vehicle Journal

MDPI

Article

Torque Distribution Based on Dynamic Programming Algorithm for Four In-Wheel Motor Drive Electric Vehicle Considering Energy Efficiency Optimization

Oluwatobi Pelumi Adeleke [1], Yong Li [1,2,*], Qiang Chen [1], Wentao Zhou [1], Xing Xu [1] and Xiaoli Cui [3]

1 Automotive Engineering Research Institute, Jiangsu University, 301 Xuefu Road, Zhenjiang 212013, China
2 Suzhou Automotive Research Institute, Tsinghua University, Suzhou 215200, China
3 Research Institute of Automotive Parts Technology, Hunan Institute of Technology, Hengyang 421002, China
* Correspondence: liyongthinkpad@outlook.com

Abstract: The improvement of both the stability and economy of the four in-wheel motor drive (4IWMD) electric vehicle under complex drive cycles is currently a difficult problem in this field. A torque distribution method with the comprehensive goals of optimal torque distribution and energy efficiency, considering economy through energy efficiency for the 4IWMD electric vehicle, is proposed in this paper. Each component of the 4IWMD electric vehicle is modelled. The dynamic programming (DP) control algorithm is utilized for torque distribution between the front and rear in-wheel motors to obtain optimal torque distribution and energy efficiency in the 4IWMD electric vehicle. The simulation is performed on a co-simulation platform with the software of AVL Cruise and MATLAB/Simulink, considering a straight road. Compared to the fuzzy logic control algorithm, the simulation results are very promising, as the energy consumption of the electric vehicle was reduced by 22.68%, 20.73% and 21.84% under the WLTC, NEDC and customized IM240 driving cycle conditions, respectively, with the proposed DP control algorithm. The hardware-in-the loop (HIL) experimental results also indicate that the effectiveness of the proposed DP algorithm is verified under the NEDC, WLTC and IM240 driving cycles, when a straight road is considered. The proposed DP control algorithm not only reduces the vehicle energy consumption and guarantees the optimization of torque distribution, but also increases the driving range of the vehicle.

Keywords: energy consumption optimization; torque distribution; energy efficiency; motor efficiency; four in-wheel motor drive electric vehicle

Citation: Adeleke, O.P.; Li, Y.; Chen, Q.; Zhou, W.; Xu, X.; Cui, X. Torque Distribution Based on Dynamic Programming Algorithm for Four In-Wheel Motor Drive Electric Vehicle Considering Energy Efficiency Optimization. *World Electr. Veh. J.* **2022**, *13*, 181. https://doi.org/10.3390/wevj13100181

Academic Editor: Joeri Van Mierlo

Received: 8 September 2022
Accepted: 26 September 2022
Published: 30 September 2022

1. Introduction

Over the years, owing to the increasingly severe energy crisis and environmental pollution, there has been an increase in the demand and manufacture of electric vehicles. Researchers have carried out extensive research on electric vehicles, especially on in-wheel motor drive (IWMD) electric vehicles. Amongst electric vehicles, IWMD electric vehicles possess distinct advantages, some of which are a simple and compact structure, flexible maneuverability and steering, high transmission efficiency and easy control including the independent torque control of each wheel.

The possibility for individual torque control has led to ample research in this area, including research focus on torque distribution, with a large focus on safety. Safety-based torque distribution utilizes torque vector/torque distribution to improve traction, handling and stability performances in vehicles. Li et al. [1] proposed an optimal torque distribution approach for the improvement of vehicle handling and stability in spite of slippery road conditions. Joa et al. [2] presented an integrated chassis control method for front/rear torque distribution and four-wheel independent braking based on tire slip which improves handling performance. A novel torque vectoring algorithm was proposed by Park et al. [3] to improve cornering performance in electronic-four-wheel drive vehicles, meanwhile

Deng et al. [4] and Chatzikomis et al. [5] studied a torque vectoring algorithm, with a consideration of stability and economy, as well as safety and energy efficiency improvement, respectively. The total longitudinal slip of an electric vehicle can also be reduced through novel torque distribution strategies to improve the vehicle safety [6]. When considering safety, results from a torque distribution study indicate that the improved stability of the vehicle can be derived when a greater weight coefficient is applied to the rear wheels, as this causes the rear axle to bear a larger weight [7]. Four IWMD electric vehicles which have four in-wheel motors positioned inside each of the vehicle wheel possess several advantages as stated earlier, including the delivery of the desired torque directly to each wheel [8], providing increased possibilities for economy management and improvement. Based on the vehicle requirement for propulsion, the torque distributed to each motor has to be well controlled and distributed to ensure the in-wheel motor functions efficiently and prevents energy loss. Furthermore, part of the efficiently distributed energy can be saved through regenerative braking, ensuring maximal energy saving [9–11].

Considering vehicle economy, methods for acquiring optimal torque distribution are diverse, and this includes utilizing motor loss models for gaining optimal torque distribution [12,13]. The motor loss model has the possibility of increasing system efficiency by some margin. However, the boundary conditions, such as motor parameters and control algorithms, could affect the possibility of getting the desired positive results [14,15]. The steering controlled by the driver and longitudinal forces restricted from yaw result in the increase in the vehicle's maximum acceleration ability [16]. Furthermore, mathematical models and quadratic programming methods are used to provide driving force [17–19]. Additionally, energy management strategies, such as strategies based on optimal driving torque distribution, can be considered to reduce electric energy consumption [20]. Braking torque distribution between the front and the rear wheels in an electric vehicle can also significantly improve energy regeneration efficiency [21].

It is necessary to investigate the torque distribution approach of the 4IWMD electric vehicle considering the coordinated control of stability and economy [22]. However, there are few scholars working from this perspective. Deng et al. proposed a novel torque vectoring algorithm based on a novel type of mechanical elastic electric wheel, which ensures the stability of the vehicle and reduces the energy consumption of the powertrain.

An optimized torque distribution method considering energy efficiency optimization based on DP strategy for a 4IWMD EV is proposed in this paper. This was considered under the constraint of straight-line driving. The main contributions made by this research are stated as follows:

(1) A torque distribution method with the comprehensive goals of optimal torque distribution and energy efficiency considering economy through energy efficiency is proposed in this paper.
(2) The DP control algorithm is utilized for toque distribution between the front and rear in-wheel motors to obtain optimal torque distribution and energy efficiency in the 4IWMD EV.
(3) The proposed torque distribution based on the DP algorithm for the 4IWMD electric vehicle considering energy efficiency optimization is effectively verified through simulation and experiment under the NEDC, WLTC and IM240 driving cycles.

The rest of the paper is organized as follows. First, in Section 2, the four in-wheel electric vehicle model is discussed, meanwhile Section 3 exposes the torque distribution strategies applied in the study, followed by Sections 4 and 5 which detail the simulation results and the experimental validation results, respectively, and in Section 6, the conclusion of the study is drawn and considerations for future studies are presented.

2. IWMD Electric Vehicle Model

A four-in-wheel motor drive (4IWMD) electric vehicle model is built to verify the effectiveness of the proposed strategies under various driving cycles. The AVL Cruise and MATLAB/Simulink platforms are utilized in building the co-simulation model. The complete vehicle model consists of the vehicle, in-wheel motor and battery model established in the AVL Cruise software, meanwhile the vehicle dynamics model and the torque distribution control models are built in the MATLAB/Simulink software.

2.1. Vehicle Dynamics Model

The vehicle model consists of four electric motors located inside each wheel of the electric vehicle. The total vehicle mass can be expressed using the following equation:

$$M_v = m_c + m_{em} + m_{bat} \tag{1}$$

where $m_{em} = m_{fl} + m_{fr} + m_{rl} + m_{rr}$, m_c is vehicle curb weight, m_{em} is total mass of the in-wheel motors, m_{bat} is battery mass, m_{fl} is the mass of the front left in-wheel motor, m_{fr} is the mass of the front right in-wheel motor, m_{rl} is the mass of the rear left in-wheel motor and m_{rr} is the mass of the rear right in-wheel motor [23].

Table 1 shows the parameters of the four IWMD electric vehicles modelled in this study.

Table 1. Vehicle parameters for four IWMD electric vehicles.

Vehicle Parameter	Symbol	Value (Unit)
Curb weight	M	1270 kg
Coefficient of rolling friction	C_r	0.017
Cross-sectional area	A	1.97 m^2
Aerodynamic drag coefficient	C_D	0.35
Rolling radius	R	0.31 m

The vehicle torque calculation is carried out in MATLAB/Simulink, in which the input variables are vehicle speed u and acceleration a_x. Figure 1 shows the free body diagram of the 4IWMD electric vehicle on a slope, with the resistance forces that act on the vehicle [24,25].

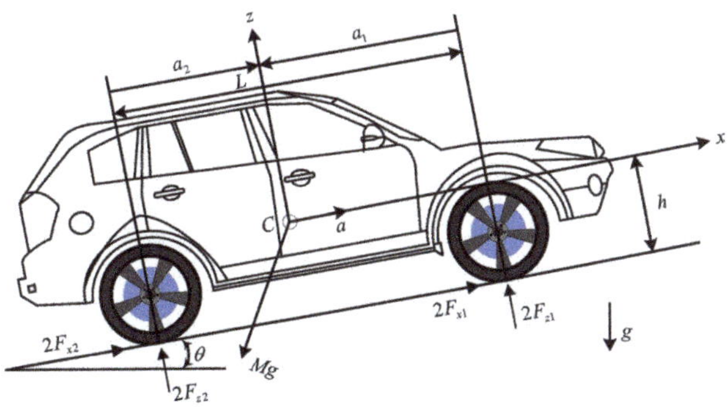

Figure 1. Longitudinal model diagram of vehicle accelerating up a slope.

The total driving torque (F_t) is derived through the following equations:

$$F_t = F_a + F_G + F_r + F_w \tag{2}$$

The forces that make up the necessary overall total driving force can be expressed as follows:

$$F_a = M\dot{v} \tag{3}$$

$$F_G = Mg\sin\theta \tag{4}$$

$$F_r = C_r Mg\cos\theta \tag{5}$$

$$F_w = \frac{1}{2}\rho A C_a v^2 \tag{6}$$

where F_a is the acceleration resistance, F_G is the grade resistance, F_r is the vehicle rolling resistance, F_w is the aerodynamic resistance, C_r represents the rolling resistance coefficient, C_a represents the aerodynamic resistance coefficient, ρ is the density of air and A is the frontal area of the 4IWMD electric vehicle.

The power consumption of the 4IWMD electric vehicle can be calculated through the following equation:

$$P_{EV} = F_t v = \sum_{j=1}^{2} 2\eta(j) P_{motor}(j) \tag{7}$$

where $P_{motor} = T\omega$ and it represents the power of a single in-wheel motor, meanwhile η represents the in-wheel motor efficiency, v is the vehicle speed and j represents the polynomial order number. T represents torque, and ω represents the angular velocity of the in-wheel motor.

The torque required on the drive wheel can be calculated as follows:

$$T = F_t \times r_{wheel} \tag{8}$$

where r_{wheel} is radius of the drive wheel.

2.2. In-Wheel Motor Model

The efficiency map of the in-wheel motor is shown in Figure 2. The output power of the in-wheel motor is bounded by the output torque and motor speed conditions, which can be defined as:

$$T_m(t) \in [T_{m,\min}(\omega_m(t)), T_{m,\max}(\omega_m(t))] \tag{9}$$

$$\omega_m(t) \in [0, \omega_{m,\max}(t)] \tag{10}$$

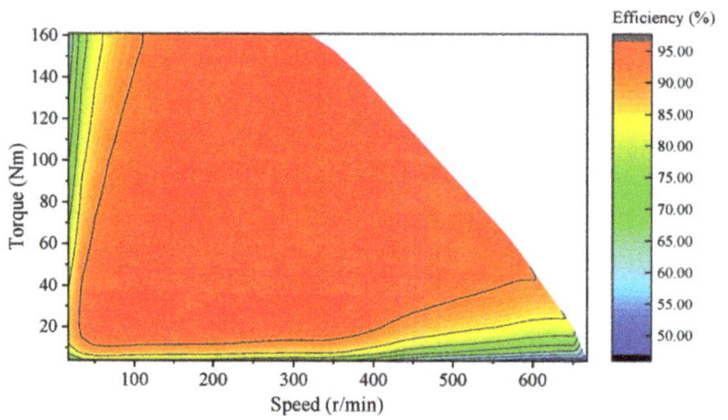

Figure 2. Efficiency map of in-wheel motor.

Where P_m is the output power of the in-wheel motor, T_m is the output torque of the in-wheel motor and ω_m is the speed of the in-wheel motor.

The in-wheel motor power demanded by the 4IWMD electric vehicle need to be supplied by the battery through the following equation:

$$P_{dem} = P_m + P_{loss.m} \tag{11}$$

where P_{dem} represents the power demanded by the power system of the 4IWMD electric vehicle, and $P_{loss.m}$ is the power loss of the in-wheel motor, especially as a result of motor heat losses and mechanical losses.

The efficiency η of the in-wheel motor can be calculated by the formula:

$$\eta = P_{out}/P_{in} \tag{12}$$

where P_{out} and P_{in} are the output and input power of the in-wheel motor, respectively [26].

Table 2 shows the parameters of the in-wheel motor used for the test bench experimental study.

Table 2. Parameters of the in-wheel motor utilized for the experimental test bench studies.

In-Wheel Motor Parameter	Value (Unit)
Rated voltage	72 V
Rated current	110 A
Maximum speed	1200 rpm
Rated power	8 kW
Rated frequency	50 Hz
Maximum torque	250 Nm

2.3. Battery Model

The battery pack is made up of rows and columns of battery cells modelled as voltage sources with resistance. The total power of the battery can be described as follows:

$$P_{em,tot} = P_{dem} + P_{aux} \tag{13}$$

where P_{aux} is the auxiliary power demanded by the vehicle.

The simplified battery equivalent circuit model utilized by theoretically deriving the state of charge (SOC) of the EV is shown in Figure 3 below.

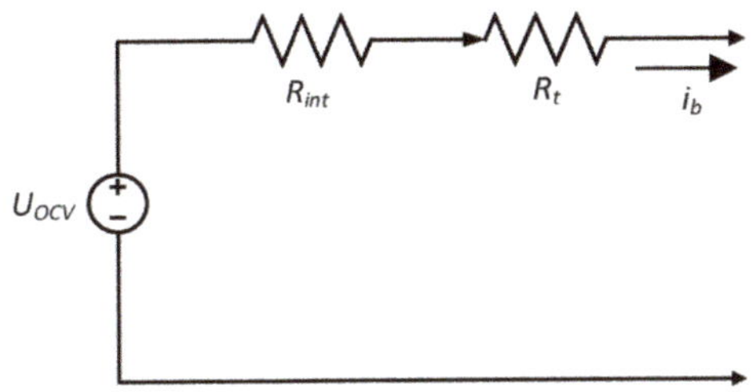

Figure 3. Simplified battery equivalent circuit model.

The simplified equation of the battery SOC is shown as follows:

$$\dot{SOC} = -\frac{I(k)}{Q_{bat}} \tag{14}$$

$$\dot{SOC} = -\eta_{SOC}\frac{U_{OCV}(SOC) - \sqrt{U_{OCV}^2(SOC) - 4(R_{int}(SOC) + R_t)P_m(k)}}{2(R_{int}(SOC) + R_t)Q_{bat}} \tag{15}$$

where I represents the battery current, Q_{bat} represents the battery capacity, η_{SOC} represents the coulomb efficiency, U_{OCV} represents the battery's open circuit voltage, R_{int} represents the battery's polarization internal resistance, R_t represents the battery's ohmic internal resistance and P_m represents the required power of the in-wheel motor. R_{int} and U_{OCV} are the function of the battery's SOC as a variable [27]. Equation (15) is the broadened expression of Equation (14).

3. Torque Distribution Strategies

3.1. Torque Optimization Approach

Torque distribution and energy saving can be achieved by the proper control allocation method [28]. The overall framework for torque distribution control is illustrated in Figure 4. Torque can be allocated utilizing different control allocation methods. In this study, the DP algorithm is proposed for optimizing the torque distributed to the in-wheel motors. Meanwhile, the fuzzy logic control (FLC) algorithm, based on the fuzzy set theory which operates very precisely and responds rapidly, ascertained by other studies, is used in comparison. The triangular membership function with the input variables of vehicle speed, vehicle acceleration and output variable being the coefficient of torque distribution k, were utilized for the fuzzy logic controller. Additionally, the simulated FLC algorithm is designed considering equal torque distribution between front and rear axles under a straight-line driving scenario. The optimal operation of the in-wheel motor is improved by the optimal distribution of the required drive torque, so as to ensure that the in-wheel motor operates in high efficiency areas during operation at specified motor working speeds. As a result, optimal torque distribution control can be expressed as a problem of determining the torque distribution coefficient k of the front wheels and rear wheels. The coefficient k is described as the torque distribution characteristic between the front wheels and rear wheels, which can be expressed by the following Equation:

$$k = \frac{T_f}{T_f + T_r} \tag{16}$$

The boundary conditions guiding the above equation are shown below:

$$\begin{cases} T_f + T_r = T_{req} \\ 0 \leq T_f \leq T_{req} \\ 0 \leq T_r \leq T_{req} \\ 0 \leq k \leq 1 \end{cases} \tag{17}$$

where T_f represents the torque of the front axle motor, T_r represents the torque of the rear axle motor and T_{req} represents the torque demand of the 4IWMD electric vehicle.

Note that k can express different driving modes. When k = 1, it expresses a separate front wheel drive. When k = 0, it expresses a separate rear wheel drive. When k = 0.5, it expresses a four-wheel average torque distribution mode [18,29].

The driving energy utilization efficiency under the in-wheel motor driving condition can be defined by Equation (18).

$$\max\eta = \left[\frac{k}{\eta_f(kT_{dem}, n)} + \frac{(1-k)}{\eta_r((1-k)T_{dem}, n)}\right] \tag{18}$$

where η_f represents the efficiency of the front axle motor, η_r represents the efficiency of the rear axle motor and n represents the equivalent speed of the axle motor.

$$\begin{cases} n < n_{\max} \\ 0 < T_f < T_{dem} \\ 0 < T_r < T_{dem} \\ T_{dem} < T_{\max} \end{cases} \tag{19}$$

The boundary condition that satisfies the above equation is shown above.

Considering the interference of other factors, the overall efficiency of the system η can be acquired. Then, the energy consumption of the of the in-wheel motor drive system under driving cycles can be simplified as below:

$$E = \int_0^t (P_{dem} \times \eta)dt \tag{20}$$

The torque distribution coefficient k is the output value derived by the control allocation method and used for optimal torque distribution to ensure high working efficiency of the front and rear in-wheel motors, in straight line driving conditions as considered in this study [30].

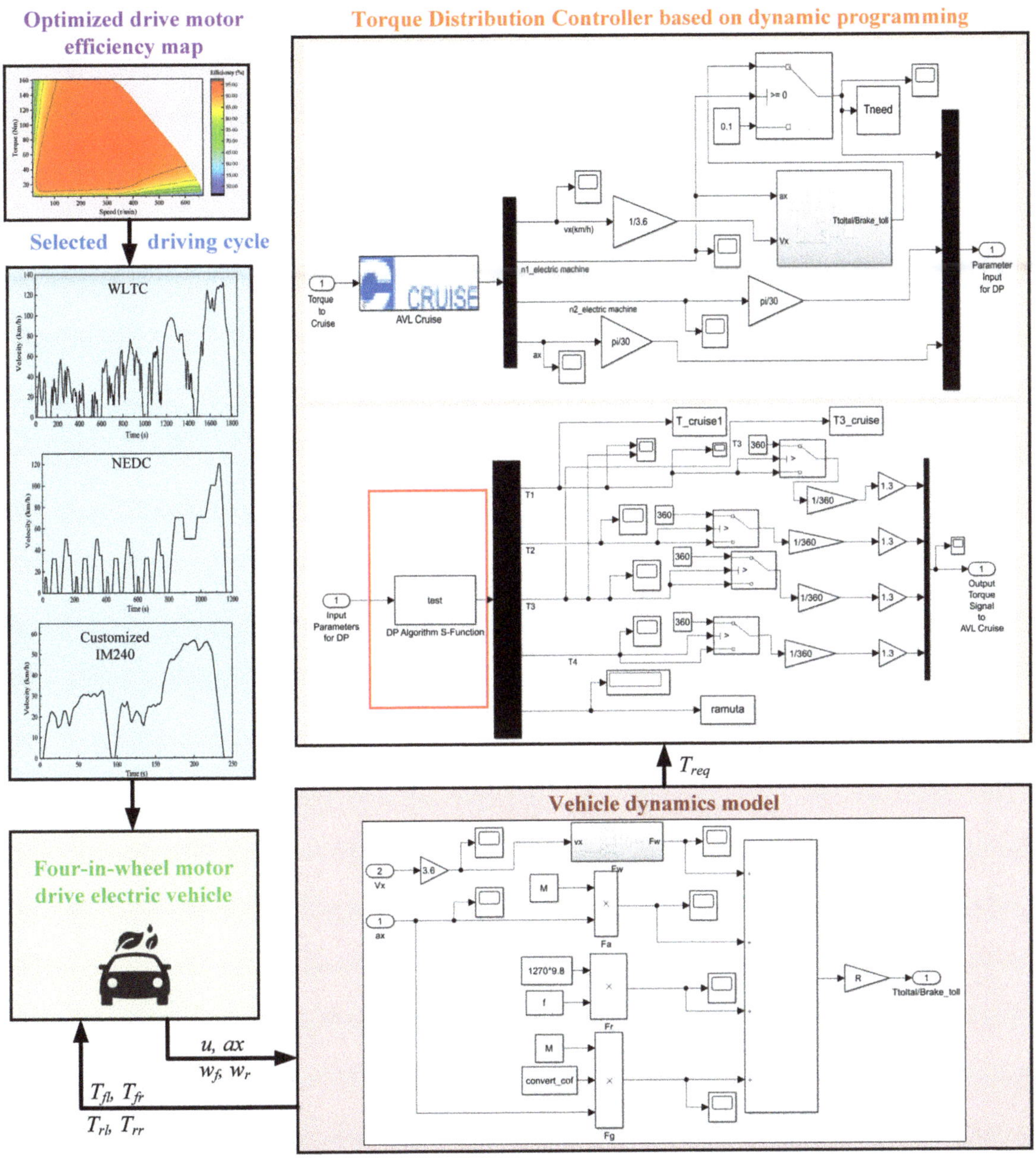

Figure 4. Framework of torque distribution control scheme.

3.2. Torque Distribution Based on DP

To derive a better efficiency optimizing effect, the torque distribution by DP is utilized to solve the torque optimization problem.

Torque distribution through DP optimization involves the establishment of the constrained optimization problem and solving the numerical solution [31]. The utilization of the mathematical optimization and computer programming method developed by Richard Bellman in the 1950s for torque distribution in an electric vehicle requires utilizing the DP algorithm to solve and derive an optimal shorter path for the working points of the in-wheel motor, in such a way that it brings about increased in-wheel motor and vehicle efficiency.

The DP computational technique extends the decision-making concept to sequences of decisions, which as a whole define an optimal policy and trajectory. To determine the optimal trajectory and enable the in-wheel motors to work in high efficiency regions, the DP algorithm is defined by these mathematical equations:

$$C^*_{ax_ih} = J_{ax_i} + J^*_{x_ih} \tag{21}$$

$$J^*_{\alpha h} = \min\{C^*_{ax_1h}, C^*_{ax_2h}, \ldots, C^*_{ax_ih}, \ldots\} \tag{22}$$

where α is the current state, u_i is an allowable decision elected at the state α, x_i is the state adjacent to α that is replaced by the application of u_i at α, h is the final state, J_{ax_i} is the cost to move from α to x_i, $J^*_{x_ih}$ is the minimum cost to reach the final state h from x_i, $C^*_{ax_ih}$ is the minimum cost to go from α to h via x_i, $J^*_{\alpha h}$ is the minimum cost to go from α to h (by any allowable path), $u^*(\alpha)$ is the optimal decision (control) at α.

The principle of optimality utilized to find an optimal policy, represented by Equation (23), can be represented by the state equation of the 4IWMD electric vehicle presented in Equation (24).

$$u^*(t) = f(x(t), t) \tag{23}$$

where f is a functional relationship referred to as the optimal control law or the optimal policy, and t is the time.

The vehicle speed u and torque T_{dem} of the 4IWMD electric vehicle are the state variables in the range of actual domain $[t_0, t_f]$ of power system of the 4IWMD. The speed of the 4IWMD electric vehicle can be determined according to the driving cycle utilized for the optimization. Therefore, the state variable is noted as $x(t) = [T_{dem}(t), n]'$, meanwhile the vehicle demand power is utilized as the control variable, which is noted as $u(t) = [P_{dem}(t)]$ discrete state. The powertrain of the 4IWMD electric vehicle can then be described by the following Equation (24).

$$\frac{dx}{dt} = f(x(k), u(k)) \tag{24}$$

where f is the equation of the power system of the 4IWMD electric vehicle, consisting of the vehicle dynamics Equations (2)–(7) and (11),

$$\begin{aligned} &0 \le P_{dem} \le P_{dem,max} \\ &T_{m,min} \le T_{dem}(t) \le T_{m,max} \\ &n_{m,min} \le n(t) \le n_{m,max} \end{aligned} \tag{25}$$

where T_{dem} represents the in-wheel motor torque, $T_{m,min}$ represents the minimum torque, $T_{m,max}$ represents the maximum torque of the in-wheel motor, $n_{m,min}$ and $n_{m,max}$ represent the minimum and maximum speed of the in-wheel motor, respectively, $P_{dem,max}$ represents the maximum output power of the in-wheel motor.

The energy efficiency of the in-wheel motor drive system that is necessary for the optimal energy consumption is taken as the objective function in this study. Additionally, the objective function is shown as follows:

$$J = \min\sum_{i=0}^{N}\left[\frac{nT_f}{\eta_f\left(T_f, n\right)} + \frac{nT_r}{\eta_r(T_r, n)}\right] = \min\sum_{i=0}^{N}\left[\frac{nkT_{dem}}{\eta_f(kT_{dem}, n)} + \frac{n(1-k)T_{dem}}{\eta_r((1-k)T_{dem}, n)}\right] \tag{26}$$

where T_f represents the output torque value of the front axle in-wheel motor, T_r represents the output torque value of the rear axle in-wheel motor, $\eta_f\,(T_f,n)$ represents the efficiency of the front axle in-wheel motor and $\eta_r\,(T_r,n)$ represents the efficiency of the rear axle in-wheel motor [32].

The dynamics model of the 4IWMD electric vehicle needs to be represented in terms of change of state for the investigation of DP. The procedure needs to be performed in a discrete format, rather than in a continuous format. Therefore, the discretization needs to be carried out due to the numerical solution of DP. The time and system state are firstly discretized, and then the calculation grid of the torque distribution ratio state is divided along the time direction of the vehicle driving cycle.

According to the drive cycle, the vehicle model is utilized to calculate the power demand and speed of the drive cycle along the time direction. Considering the constraints of the in-wheel motor, the system's reachable boundary of the entire driving cycle is acquired from the initial state and the termination state. The system constraints are met within the reachable boundary range. Additionally, the forward function is calculated according to the designed objective function. Through the recursive call method, the final state is reversed to the initial state when the torque distribution ratio state matrix of the in-wheel motor for the entire drive cycle is obtained.

The transversal optimization process is completed when the optimal torque distribution trajectory of the in-wheel motor is obtained. The calculation approach of the DP method is shown in Figure 5, and it can be seen that the backward recursion process of the dynamic program is used to calculate the minimal cost for torque distribution along the entire drive cycle by proceeding backward. This is done in order to output the dynamic and optimal path for the calculation process, in such a way that the optimal torque distribution coefficient can be derived for every time step of the state space of the dynamic programming algorithm [33].

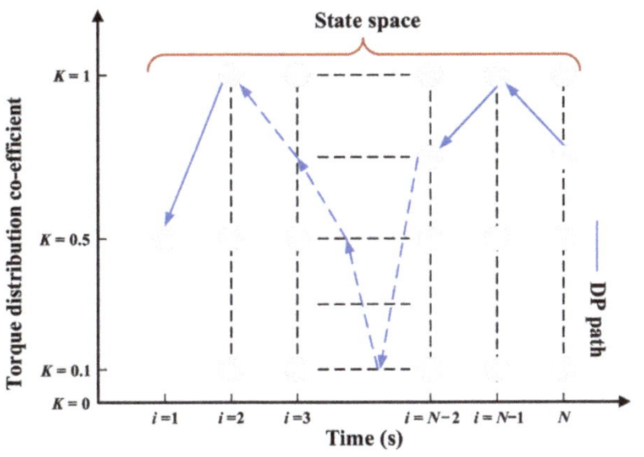

Figure 5. DP control method calculation approach.

The process of utilizing the DP method to conduct the torque distribution of the in-wheel motor is shown in Figure 6. For the simulation's implementation, the inputs for the DP optimization are the calculated total torque and wheel speeds for the front and rear in-wheel motors, while the output is the coefficient of torque distribution. Considering the in-wheel motor, stator resistance, rated motor power, wheel speed and loss model, the in-wheel motor efficiency map is linearly interpolated and the distribution coefficient are greatly utilized to find the optimal outputs. The output is the optimal torque for each in-wheel motor, with the utilized distribution coefficient [34].

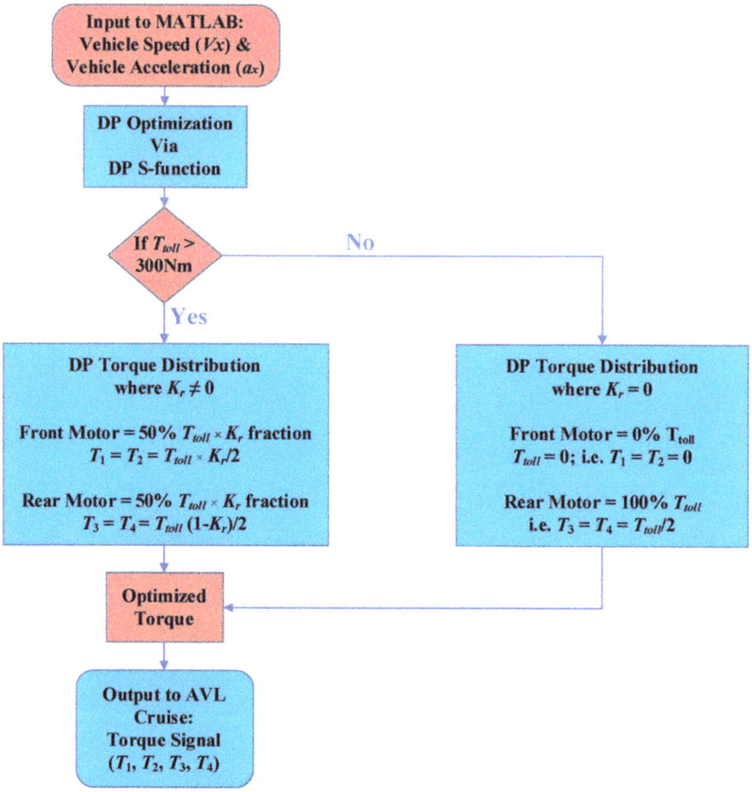

Figure 6. DP optimization diagram.

The equations for the calculation of the output are expressed as follows:

$$\begin{aligned} MT_1 &= \frac{T_{toll} \times K_r}{2}; \\ MT_2 &= MT_1; \\ MT_3 &= \frac{T_{toll} \times (1 - K_r)}{2}; \\ MT_4 &= MT_3 \end{aligned} \tag{27}$$

where MT_i represents the in-wheel motor torque and i = 1, 2, 3, 4 represent each of the four motors (front left, front right, rear left, rear right), respectively. T_{toll} represents the required total torque for propulsion and braking and K_r is the torque distribution coefficient.

The discretized dynamic model produces the output indicated by the above equations to provide the front left, front right, rear left and rear right total motor torque utilized by the in-wheel motor when the in-wheel motor torque is greater than T_1 (at $T_{toll} > T_1$). Meanwhile, at other instances of $T_{toll} < T_1$, the torque is distributed to the in-wheel motors as indicated below:

$$\begin{aligned} MT_1 &= 0; \\ MT_2 &= 0; \\ MT_3 &= \frac{T_{toll}}{2}; \\ MT_4 &= \frac{T_{toll}}{2} \end{aligned} \tag{28}$$

The vehicle dynamic model may optimally distribute no torque to the front electric motors when the vehicle torque is lower than T_1. However, it only distributes the total required torque to the vehicle by the rear wheels, as it enhances the optimal performance of the in-wheel motor and electric vehicle overall, as the in-wheel motor operates with a higher torque.

For the vehicle torque to be optimally controlled, the dynamic programming model utilizes the above mode for the derivation of the vehicle torque. This is further indicated as below,

$$MT_i(k+1) = MT_3 = \frac{T_{toll}(k)}{2} = MT_4 = \frac{T_{toll}(k)}{2} \quad (29)$$

where k represents the index of the current node, MT_i $(k + 1)$ represents the motor torque for the next node and T_{toll} (k) is the required motor torque.

4. Simulation Results and Analysis

To compare the effectiveness of the proposed DP-based torque distribution strategies for the 4IWMD electric vehicle, simulations were carried out using the co-simulation platform of MATLAB/Simulink and AVL Cruise. The simulations ran under the WLTC driving cycle, the NEDC driving cycle and the IM240 driving cycle to simulate the driving. The simulation results of the torque distribution based on DP control algorithm carried out under these different driving cycle conditions are compared with the torque distribution based on the FLC strategy. This comparison documents a comprehensive energy saving analysis of both torque distribution strategies.

The FLC distribution strategy is developed considering the equal distribution of torque, as well as the effective torque distribution to both sets of in-wheel motors of the front and rear axle [35]. When the required torque is low, the rear in-wheel motors supply most of the torque. When the required torque is supposed to increase and enlarge, the front in-wheel motors will supply more torque and compensate for the remaining required torque. The DP controller is developed considering maximum in-wheel motor efficiency, in which both of the rear in-wheel motors handle the vehicle propulsion request when the calculated total required torque is less than 300 Nm. When the required torque is over 300 Nm, both of the front in-wheel motors assist the rear in-wheel motors and supply the left-over share of the required torque that is needed to propel the vehicle. The simulations are carried out under the assumption that the 4IWMD electric vehicle drives on a straight line without cornering. The low-speed–low-torque characteristic of the in-wheel motor keeps the electric vehicle within a maximum speed under 60 km/h. The main parameters of the 4IWMD electric vehicle utilized in the co-simulation studies are shown in Table 1.

4.1. WLTC Driving Cycle

The worldwide harmonized light vehicle test cycle (WLTC) with a distance of 23,266 m, a duration of 1800s and a maximum speed around 130 km/h is utilized, as it is the classified test cycle for a broader category of vehicles and diverse electric powertrain vehicles. The speed profile of the WLTC drive cycle is shown in Figure 7.

The required total torque under the WLTC drive cycle is shown in Figure 8, as the control distribution method controls the amount of torque that is needed to navigate through the entire driving cycle. The torque that is individually distributed to the pair of front and rear motors when FLC is utilized for torque distribution is shown in Figure 9a. The torque that is individually distributed to the pair of front and rear motors when the DP algorithm is utilized for torque distribution is depicted in Figure 9b. It can be noted that the calculated required torque by vehicle dynamics (T_{toll}) is different from the required total torque to navigate through the driving cycles, as the latter is the total torque which the control algorithm utilizes to navigate through the driving cycle, considering the control parameters, constraints and objectives. Therefore, it is the result calculated by the torque distribution control algorithm, using the total desired torque by vehicle dynamics and in-wheel motor speed for the front and rear in-wheel motors.

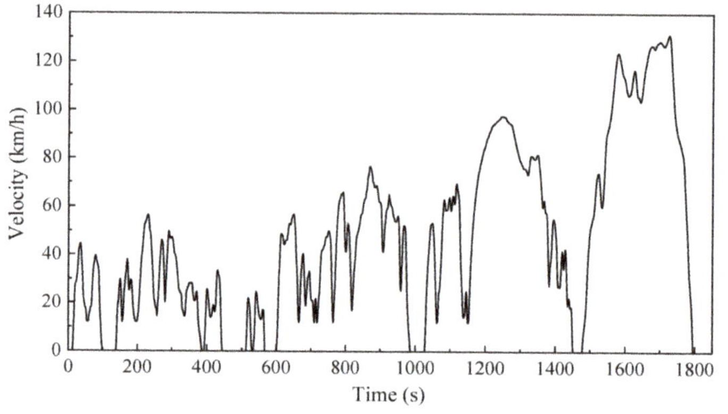

Figure 7. WLTC driving cycle.

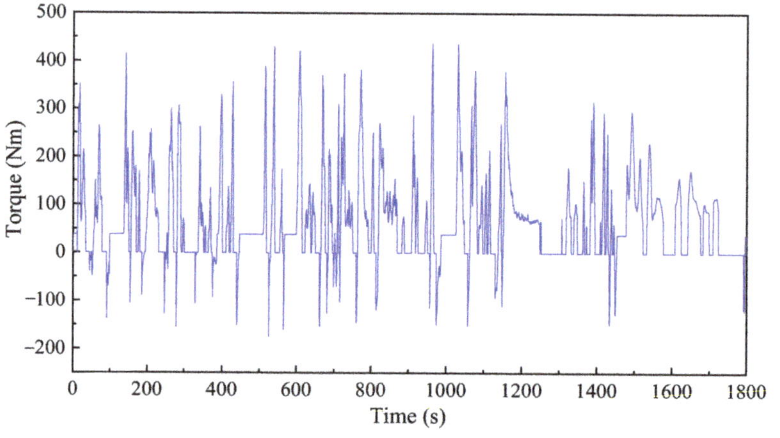

Figure 8. Total torque desired under WLTC drive cycle.

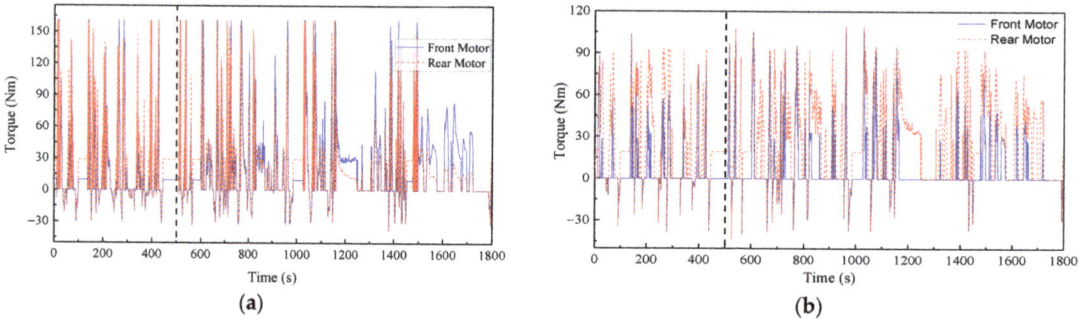

Figure 9. Torque distribution to front and rear in–wheel motors in WLTC: (**a**) FLC algorithm; (**b**) DP algorithm.

Figure 10 shows a section of the torque distribution figure in Figure 9b, from 0 s to 500 s. Figure 10 shows that the rear in-wheel motors work all through the drive cycle, while the front in-wheel motors assist the rear motors as intended, when the calculated required total torque is over 300 Nm. At other time periods of lower torque, the front in-wheel motors remain inactive and do not engage in torque distribution. It is noted that both the front and rear in-wheel motors engage in energy recuperation, as the negative torque recorded shows. It is also visible in Figure 9 that the torque distribution using the proposed DP algorithm is well coordinated, efficient in distribution and better in comparison to the FLC method. The results indicate that a lesser amount of torque with a maximum of 109.3 Nm is utilized to achieve the torque distribution task in comparison with FLC with the maximum of 160.927 Nm. Good regenerative braking/energy recuperation is also visible, as the above figures indicate.

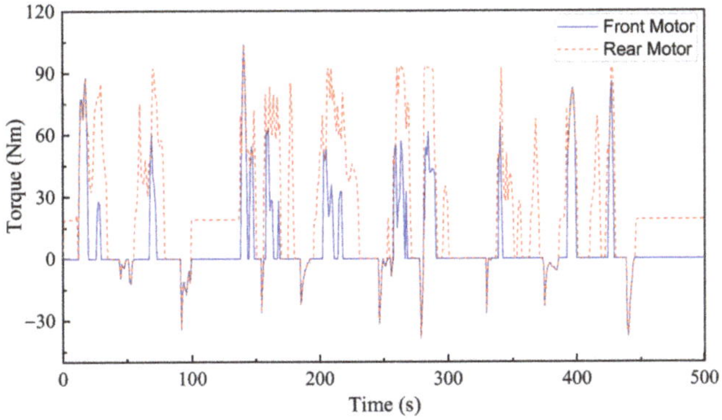

Figure 10. Torque distribution to front and rear electric motors using DP in WLTC (cropped).

The DP controller works with the in-wheel motor to ensure it works within its designated operation capacity, while working with the end goal of improving torque distribution to the front and rear axle through maximum in-wheel motor efficiency.

The in-wheel motor operation points for both front in-wheel motors and both rear in-wheel motors in the WLTC driving cycle are shown in Figure 11a,b, respectively. The figures show a comparison of operating points between the FLC and the proposed DP torque distribution strategy.

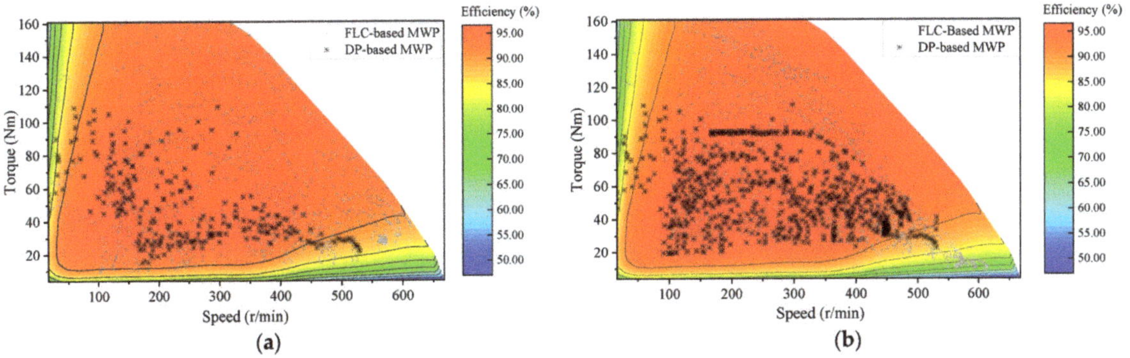

Figure 11. Distribution of motor working points of the front and rear motors under WLTC driving cycle: (**a**) Front motor; (**b**) Rear motor.

As the operating points have significantly shown in Figure 11, important efforts are being undertaken by the front and rear in-wheel electric motors. The in-wheel motors work effectively as required in very high efficiency working regions. The proposed DP-based torque distribution strategy ensures the in-wheel motor works in higher efficiency regions than that with the FLC-based torque distribution strategy. Therefore, compared to the FLC-based torque distribution method, the proposed DP-based torque distribution method achieves a better efficiency optimizing effect, as both of the front and rear in-wheel motors work mostly in highly efficient regions. The proposed DP-based method always works in high efficiency regions effectively, even when only the front motor is in operation. The WLTC driving cycle motor operating points with the proposed DP-based torque distribution strategy presents better results in comparison to the motor operating points with the FLC-based torque distribution strategy, as depicted especially by the WLTC driving cycle motor operating point map in Figure 11a,b.

4.2. NEDC Driving Cycle

The new European driving cycle (NEDC) with a total distance of about 11,017 m, a duration of 1180 s and a maximum speed of 120 km/h, is the one used for the determination of a vehicle's consumption and emission values. The speed profile of the NEDC driving cycle is shown in Figure 12.

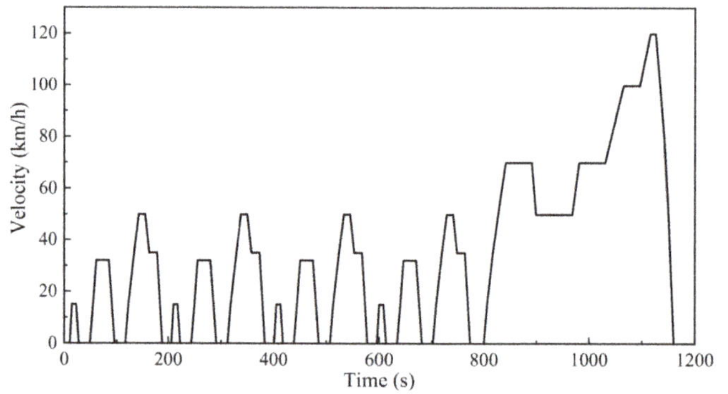

Figure 12. NEDC driving cycle.

The required total torque utilized by the 4IWMD electric vehicle to navigate through the entire NEDC cycle is shown in Figure 13. Meanwhile, shown in Figure 14a,b is the torque individually distributed to the pair of front and rear motors when FLC and the proposed DP strategy are utilized for torque distribution, respectively. Figure 15 shows one section, from 0 s to 430 s of the torque distribution result presented in Figure 14.

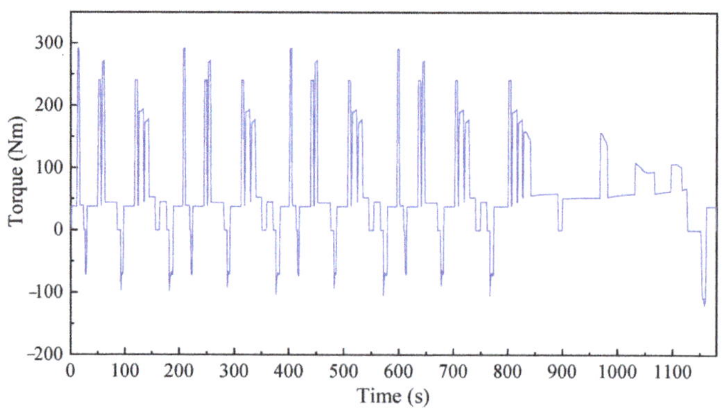

Figure 13. Total torque desired under NEDC driving cycle.

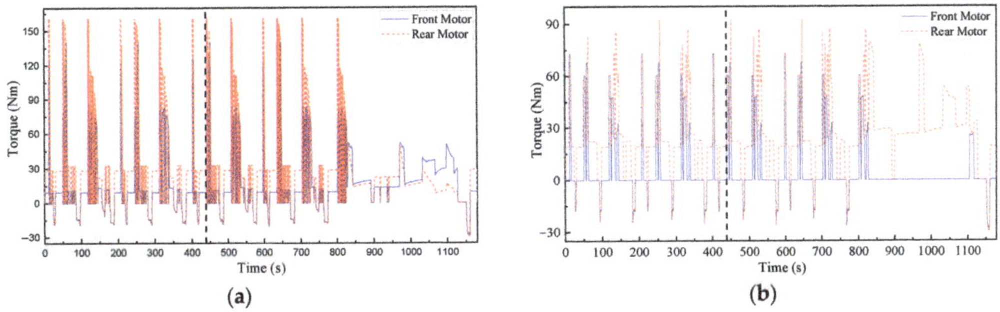

Figure 14. Torque distribution to front and rear electric motors in NEDC: (**a**) FLC algorithm; (**b**) DP algorithm.

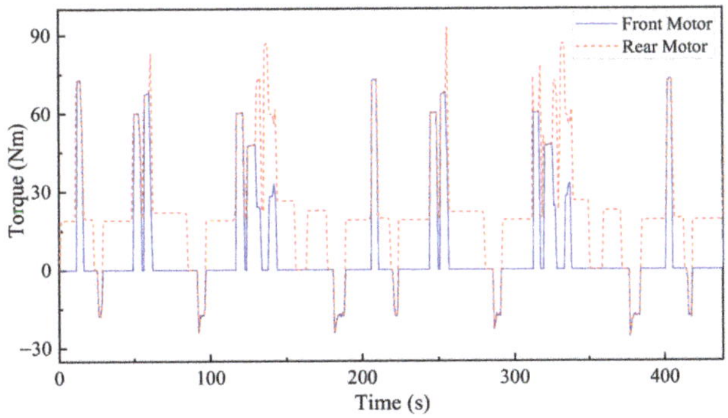

Figure 15. Torque distribution to front and rear electric motors using DP in NEDC.

The rear in-wheel motor works all through the NEDC driving cycle, while the front in-wheel motor assists the rear motor as intended. It can be visibly seen in Figure 14 from 830 s to 1100 s that the front in-wheel motors remain inactive and do not engage in torque distribution at other time periods of lower torque, where the front in-wheel motors' torque becomes idle after assisting the vehicle in reaching the higher acceleration from low torque. The rear in-wheel motor is able to provide sufficient torque to sustain the vehicle at the attained desired vehicle speed. The in-wheel motor works as a generator at deceleration periods, enabling energy recuperation through regenerative braking.

The in-wheel motor operating points for both of the front motors and rear motors in the NEDC driving cycle are shown in Figure 16. The figures show a comparison between the operating points, when torque distribution is based on FLC and the proposed DP strategy.

The in-wheel motors effectively work as required in higher efficiency working regions of the electric motors when the DP-based torque distribution strategy is applied. Therefore, the proposed DP-based torque distribution method achieves a better efficiency optimization effect than that with the FLC-based torque distribution method, as both the front and rear in-wheel motors work mostly in highly efficient regions. The motor operating points in the NEDC driving cycle, under the proposed DP-based torque distribution strategy, clearly shows that the rear in-wheel motor provides more torque for vehicle propulsion than the front in-wheel motor.

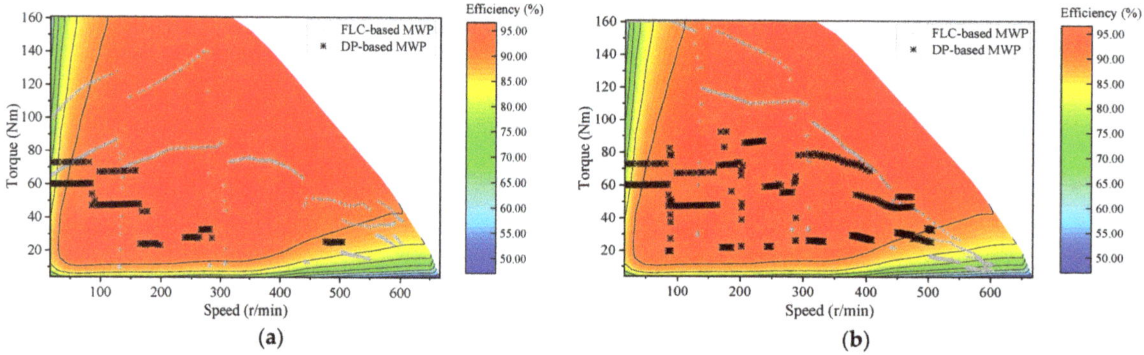

Figure 16. Distribution of motor working points of the front and rear motors under NEDC driving cycle: (**a**) Front motor; (**b**) Rear motor.

4.3. Customized IM240 Driving Cycle

A customized driving cycle called the custom IM240 drive cycle based on the inspection and maintenance driving cycle with a total distance of about 3100 km, a duration of 240 s and maximum speed of 56.7 km/h is also utilized in this study as it is a representation of a low-speed driving cycle, which represents the driving pattern in urban areas with traffic and low speed limitations. The speed profile of the custom IM240 driving cycle is shown in Figure 17. The required total torque with the entire custom IM240 drive cycle is presented in Figure 18. The torque individually distributed to the front and rear motors based on FLC and the proposed DP algorithm strategy are shown in Figure 19a,b. The torque requirement with the IM240 drive cycle is low and the rear in-wheel motor supplies most of the required torque throughout the driving cycle. This has the benefit of reducing the battery consumption, as only the rear in-wheel motor supplies the needed torque, therefore increasing the final *SOC* of the battery, which can be used to cover more travel range.

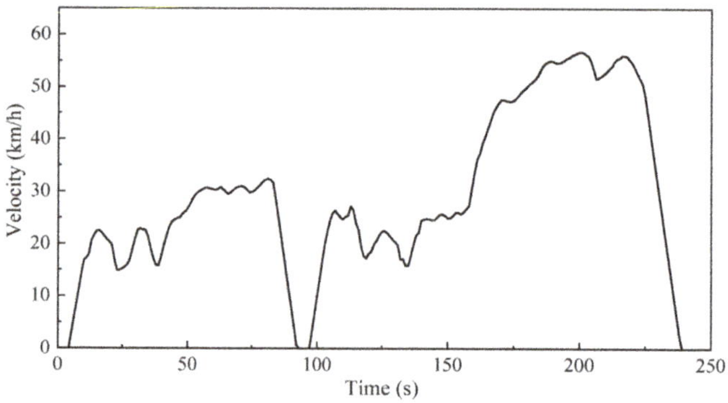

Figure 17. Custom IM240 driving cycle.

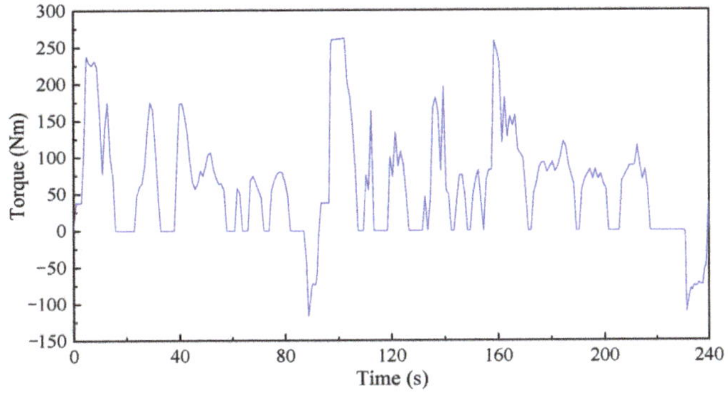

Figure 18. Total torque required to navigate through custom IM240 driving cycle.

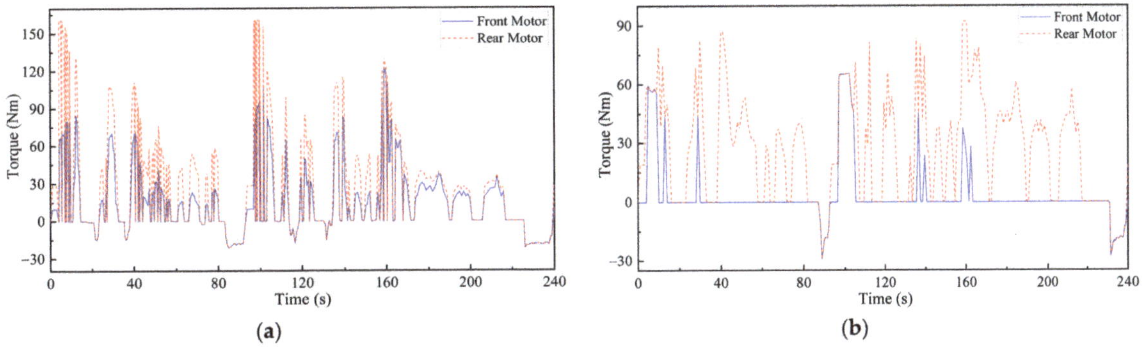

Figure 19. Torque distribution to front and rear electric motors using FLC and DP in custom IM240 driving cycle: (**a**) FLC algorithm; (**b**) DP algorithm.

The motor operating points for both the front and rear in-wheel motors in the custom IM240 driving cycle are shown in Figure 20. The figures show a comparison between the motor operating points, when torque distribution based on FLC and on the proposed DP are utilized for torque distribution.

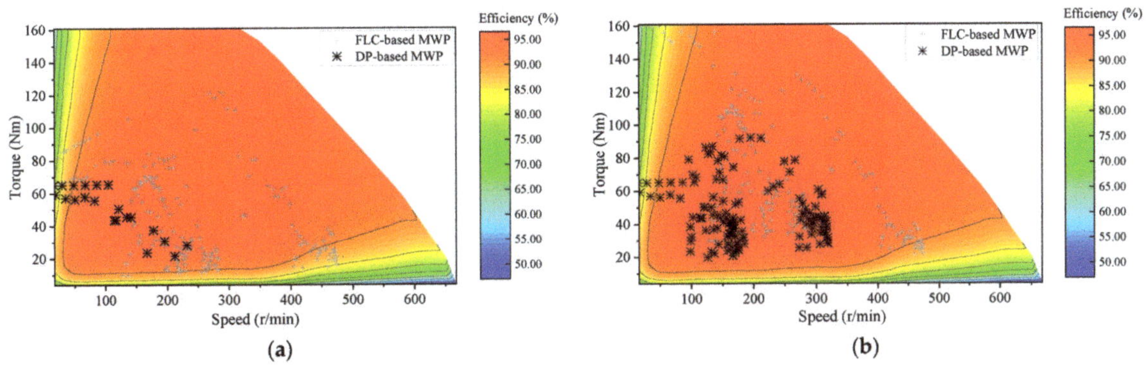

Figure 20. Distribution of motor working points of the front motor and rear motor under custom IM240 driving cycle: (**a**) Front motor; (**b**) Rear motor.

The motor operating points under the custom IM240 driving cycles shown in Figure 20 indicate that the in-wheel motors work in high efficiency working areas when the proposed

DP-based torque distribution is utilized. The custom IM240 driving cycle operating points confirm that an increase in vehicle speed will lead to an increase in torque, which allows the in-wheel motor to work at higher efficiency regions. The in-wheel motor contributes to the high motor efficiency, with a maximum efficiency of approximately 96.48% by the front motors and 96.52% by the rear motors, which is obtained by the proposed DP-based torque distribution method in the 4IWMD electric vehicle.

4.4. Energy Saving Analysis

The energy consumption of the 4IWMD electric vehicle with FLC and the proposed DP strategy for torque distribution under different drive cycles are shown in Table 3. From Table 3, it can be seen that the energy consumption of the 4IWMD electric vehicle with FLC and the proposed DP strategy for torque distribution under the WLTC drive cycle are 10.01 kWh/100 km and 7.74 kWh/100 km, respectively. It can be seen that the energy consumption with the proposed DP strategy is less than that with the FLC strategy, in which the energy consumption is reduced by 22.68%. The energy consumption of the vehicle with FLC and the proposed DP strategy for torque distribution under the NEDC drive cycle are 9.89 kWh/100 km and 7.84 kWh/100 km, respectively. It also can be seen that the energy consumption with the proposed DP strategy is less than that with the FLC strategy, in which the energy consumption is reduced by 20.73%. The energy consumption with FLC and the proposed DP strategy for torque distribution under the custom IM240 drive cycle are 9.11 kWh/100 km and 7.12 kWh/100 km, respectively. We see that the energy consumption is reduced by 21.84% with the proposed DP strategy compared to the FLC strategy.

Table 3. Energy consumption of the 4IWDEV over the studied driving cycles.

	FLC-Based Torque Distribution Energy Consumption (kWh/100 km)	DP-Based Torque Distribution Energy Consumption (kWh/100 km)	Improvement in Energy Consumption (%)
WLTC	10.01	7.74	22.68
NEDC	9.89	7.84	20.73
Custom IM240	9.11	7.12	21.84

The energy consumption with FLC and the proposed DP strategy for torque distribution under different drive cycles is clearly shown in Figure 21.

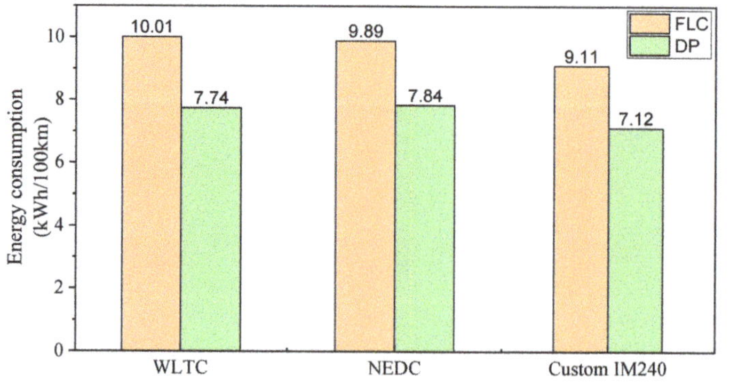

Figure 21. Energy consumption in the three drive cycles using FLC and the proposed DP strategy.

It can be seen that the proposed DP-based torque distribution method functions with high motor efficiency capability, which enables the in-wheel motor to work maximally in highly efficient working regions. The rear in-wheel motors supply most of the required torque, while the front in-wheel motors assist the rear in-wheel motors as designed.

The proposed DP-based torque distribution method brings an improvement in the torque distribution of the 4IWMD electric vehicle, resulting in reduced energy consumption, good energy recuperation and an increase in vehicle travel range.

5. Experimental Validation

Experimental studies were carried out to verify the effectiveness of the proposed DP-based torque distribution strategy, which is achieved on a NI Veristand IWMD electric vehicle test bench. The above-mentioned three drive cycles are used for experimental studies. Figure 22 shows the utilized experimental test bench setup.

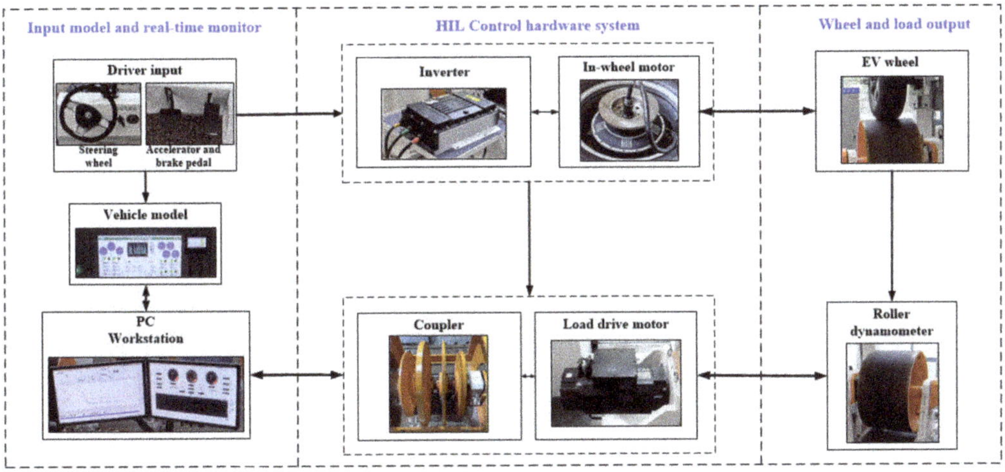

Figure 22. Experiment test bench setup.

5.1. WLTC Drive Cycle

A comparison of both the experimental and simulation results of torque distribution for the front and rear in-wheel motors with the FLC and the proposed DP strategy under the WLTC drive cycle are shown in Figures 23 and 24, respectively. It can be seen that the experimental results of torque distribution track most of the simulation results. The torque distribution results obtained by the proposed DP strategy follow those obtained by the FLC strategy, in which the track accuracy is improved and the fluctuation is reduced. It is confirmed that the proposed DP strategy is a better option for torque distribution than the FLC strategy in the 4IWMD electric vehicle.

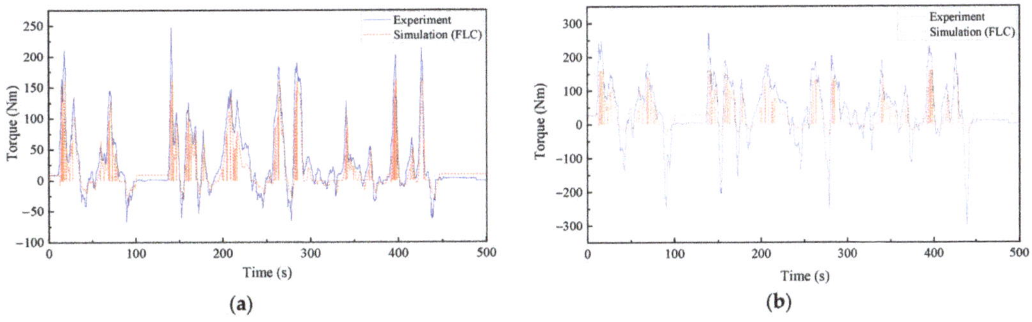

Figure 23. Torque distribution to front motor and rear motor obtained in experiment and FLC simulation of WLTC driving cycle: (**a**) Front motor; (**b**) Rear motor.

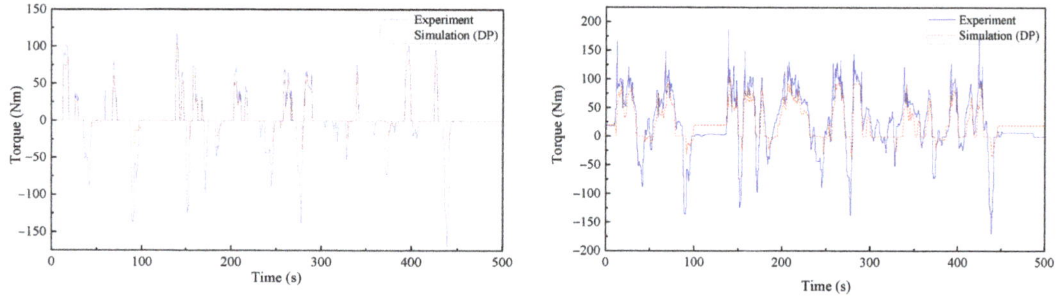

(**a**) (**b**)

Figure 24. Torque distribution to front motor and rear motor obtained in experiment and DP simulation under WLTC driving cycle: (**a**) Front motor; (**b**) Rear motor.

5.2. NEDC Drive Cycle

Figures 25 and 26 show the comparison between simulation and experimental results of torque distribution with FLC and the proposed DP strategy under the NEDC driving cycle. It is noted that the NEDC cycle requires less torque for acceleration to the required vehicle speed in comparison to the WLTC cycle, as the WLTC experiment results show in Figures 23 and 24. The experimental result validates the simulation results of torque distribution in the NEDC driving cycle.

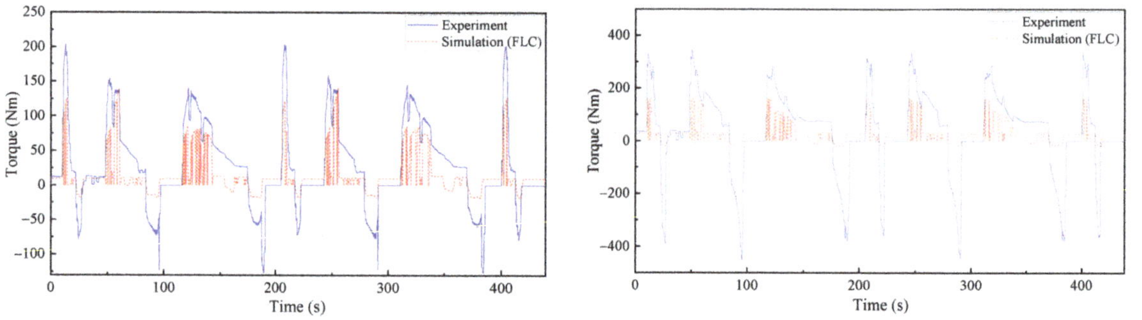

(**a**) (**b**)

Figure 25. Torque distribution to front motor and rear motor obtained in experiment and FLC simulation of NEDC driving cycle: (**a**) Front motor; (**b**) Rear motor.

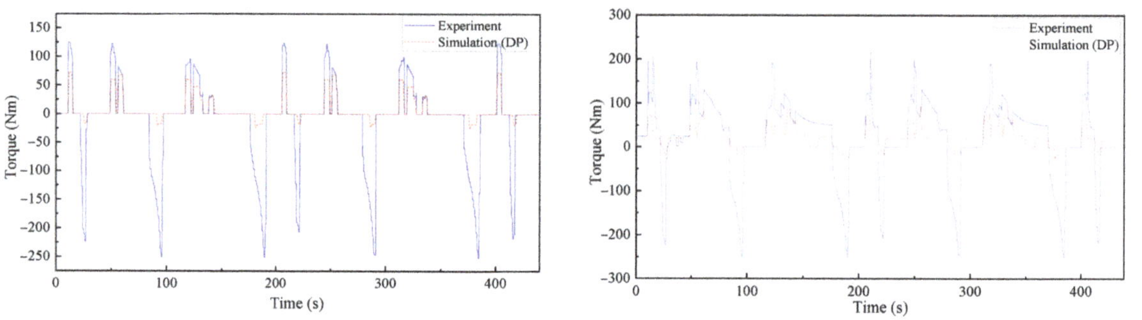

(**a**) (**b**)

Figure 26. Torque distribution to front motor rear motor obtained in experiment and simulation with DP algorithm under NEDC driving cycle: (**a**) Front motor; (**b**) Rear motor.

5.3. Customized IM240 Driving Cycle

Figures 27 and 28 highlight the torque distribution results for the experiment in the custom IM240 driving cycle in comparison with the simulation results in the custom IM240 driving cycle for torque distribution based on the FLC and DP algorithm, respectively. The experimental results of the custom IM240 driving cycle for the torque distribution, similar to the NEDC driving cycle, tracks the simulation results, with high torque exerted during deceleration.

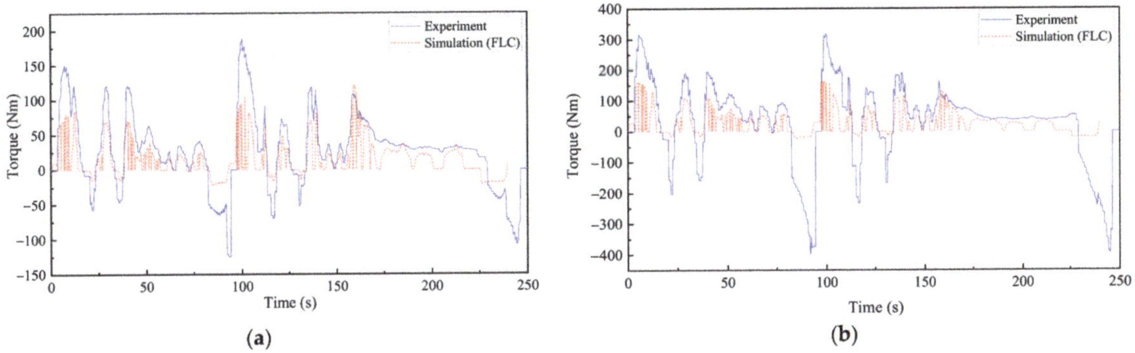

Figure 27. Torque distribution to front motor and rear motor obtained in experiment and simulation with FLC algorithm under custom IM240 driving cycle: (**a**) Front motor; (**b**) Rear motor.

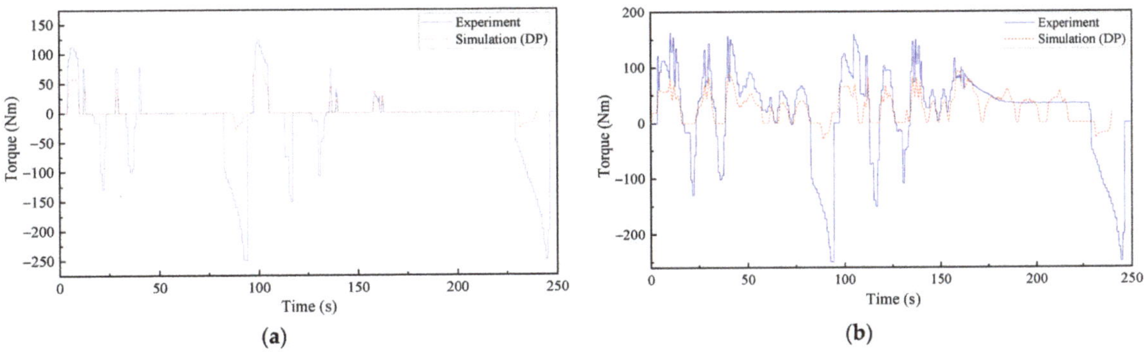

Figure 28. Torque distribution to front motor and rear motor obtained in experiment and simulation with DP algorithm under custom IM240 driving cycle: (**a**) Front motor; (**b**) Rear motor.

5.4. Energy Saving Analysis

Energy saving analysis is conducted in line with the energy consumption during the experimental studies. The quantitative analysis of the vehicle energy consumption during the experiment was derived based on the torque distribution coefficient, for torque distribution based on the FLC algorithm and the DP algorithm, respectively. The total energy consumed by the in-wheel motor during the experiment based on the FLC algorithm and the DP algorithm in the WLTC driving cycle is 359.714 kJ and 273.765 kJ, respectively. Meanwhile, in the NEDC driving cycle, the in-wheel motor consumed an energy of 528.834 kJ and 406.549 kJ during the experiment based on the FLC algorithm and the DP algorithm, respectively. For the custom IM240 driving cycle, the in-wheel motor consumed an energy of 163.051 kJ and 125.527 kJ during the experiment based on the FLC algorithm and the DP algorithm, respectively. As is shown from Figures 23–28, the WLTC urban driving experiment covered 0–500 s of the WLTC driving cycle, while that of NEDC covered 0–438 s of the NEDC driving cycle; meanwhile, the custom IM240 experiment covers the entire 240 s of the custom IM240 driving cycle.

The energy consumption in Kilowatt-hour per 100 km, calculated from the Watt-hour calculation, is assumed to showcase the consumption per km for the three driving cycles, for torque distribution methods with FLC and the proposed DP algorithm. The percentage of improvement in energy consumption during the experiments based on the FLC and DP torque distributions is shown in Table 4.

Table 4. In-wheel Motor Energy Consumption during Experimental Studies.

	Energy Consumption with FLC Algorithm (kWh/100 km)	Energy Consumption with DP Algorithm (kWh/100 km)	Improvement in Energy Consumption (%)
WLTC	9.992	7.605	23.89
NEDC	14.690	11.293	23.12
Custom IM240	4.529	3.487	23.01

The IM240 driving cycle during the experiments based on the FLC and DP torque distributions is presented in Figure 29. Compared to the FLC-based strategy, it can be seen that there is a reduction in energy consumption with the DP-based torque distribution strategy.

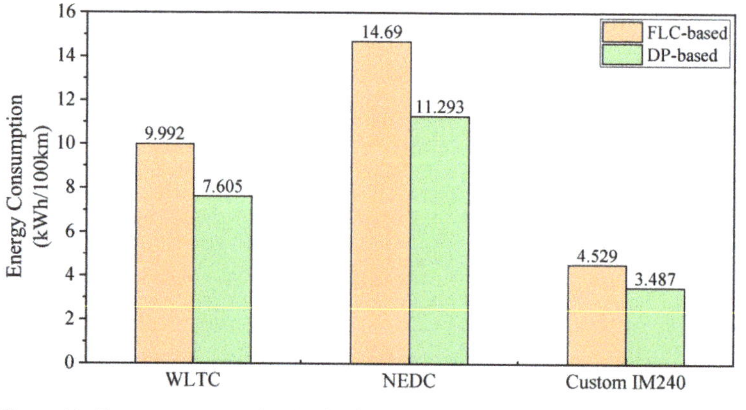

Figure 29. Energy consumption in the three drive cycles using FLC-based and DP-based torque distribution during experiment.

The energy consumed in the WLTC, NEDC and custom IM240 driving cycles during the experiment clearly verifies that DP-based torque distribution is a more optimal torque distribution method than the FLC strategy, for which less energy is consumed when the DP-based torque distribution is applied in the 4IWMD EV.

6. Conclusions

Centered on the aim of improving energy saving, the economy-based torque distribution strategy is proposed in this paper. Upon the building of a complete four-in-wheel motor drive electric vehicle model, featuring a comprehensive vehicle model, a motor model and a battery model, torque distribution methods based on the FLC algorithm and a proposed DP algorithm are investigated through co-simulation studies carried out in AVL Cruise and MATLAB/Simulink software. Additionally, further experimental studies were implemented to verify the simulation results. These were performed considering a straight-line road.

This article produces very interesting results, as shown by the simulation and experimental results. The simulation results show that the torque distribution based on the

DP algorithm is the optimal option for optimized front and rear torque distribution, as it effectively reduces the vehicle's energy consumption by 2.27 kWh, 2.05 kWh and 1.99 kWh for every 100 km of distance travelled in the WLTC, NEDC and custom IM240 driving cycle conditions, respectively, when compared to the torque distribution based on the FLC algorithm. Furthermore, compared to the FLC algorithm, the experimental results show that the energy consumption under the WLTC, NEDC and IM240 drive cycles is reduced by 23.89%, 23.12% and 23.01% with the proposed DP algorithm, respectively. Hence, the proposed DP algorithm produces an optimized front and rear torque distribution that effectively reduces vehicle energy consumption, which leads to an improved energy saving and overall vehicle efficiency in four-in-wheel motor drive electric vehicles. The online global optimization method with the proposed DP algorithm which can be monitored in real-time during simulation and the vehicle experiment studies may assist in optimization and real time control, enabling better simulation results and even experimental results to be obtained with minimal or negligible errors.

It should be noted that DP is an exhaustive search that requires more time and space for its computation. Future work will focus on the algorithm and the reduction of the computation load.

Author Contributions: Conceptualization, Y.L. and X.X.; methodology, O.P.A.; software, Q.C. and X.C.; validation, O.P.A., Q.C. and W.Z.; investigation, Y.L. and X.X.; data curation, O.P.A.; writing—original draft preparation, O.P.A.; writing—review and editing, Y.L. and X.C.; supervision, Y.L.; project administration, Y.L.; funding acquisition, Y.L. All authors have read and agreed to the published version of the manuscript.

Funding: This work is supported by National Natural Science Foundation of China (Grant No. 51705213), China Postdoctoral Science Foundation (Grant No. 2019M660105, Grant No. 2020T130360), Jiangsu Province Postdoctoral Science Foundation (Grant No. 2021K443C), Primary Research & Development Plan of Jiangsu Province (Grant No. BE2019010) and Hunan Innovation Platform Open Fund (Grant No. 20K041).

Data Availability Statement: Not applicable.

Conflicts of Interest: The authors declare no conflict of interest.

References

1. Li, B.; Goodarzi, A.; Khajepour, A.; Chen, S.K.; Litkouhi, B. An optimal torque distribution control strategy for four-independent wheel drive electric vehicles. *Veh. Syst. Dyn.* **2015**, *53*, 1172–1189. [CrossRef]
2. Joa, E.; Park, K.; Koh, Y.; Yi, K.; Kim, K. A tyre slip-based integrated chassis control of front/rear traction distribution and four-wheel independent brake from moderate driving to limit handling. *Veh. Syst. Dyn.* **2018**, *56*, 579–603. [CrossRef]
3. Park, G.; Han, K.; Nam, K.; Kim, H.; Choi, S.B. Torque Vectoring Algorithm of Electronic-Four-Wheel Drive Vehicles for Enhancement of Cornering Performance. *IEEE Trans. Veh. Technol.* **2020**, *69*, 3668–3679. [CrossRef]
4. Deng, H.; Zhao, Y.; Feng, S.; Wang, Q.; Lin, F. Torque Vectoring Algorithm Based on Mechanical Elastic Electric Wheels with Consideration of the Stability and Economy. *Energy* **2021**, *219*, 119643. [CrossRef]
5. Chatzikomis, C.; Zanchetta, M.; Gruber, P.; Sorniotti, A.; Modic, B.; Motaln, T.; Blagotinsek, L.; Gotovac, G. An energy-efficient torque-vectoring algorithm for electric vehicles with multiple motors. *Mech. Syst. Signal Process.* **2019**, *128*, 655–673. [CrossRef]
6. Debada, E.; Marcos, D.; Montero, C.; Camacho, E.F.; Bordons, C.; Ridao, M.A. Torque distribution strategy for a four In-wheel fully electric car. *Jorn. Autom.* **2015**, 517–525. Available online: https://idus.us.es/handle/11441/92138 (accessed on 6 September 2022).
7. Mokhiamar, O.; Abe, M. How the four wheels should share forces in an optimum cooperative chassis control. *Control Eng. Pract.* **2006**, *14*, 295–304. [CrossRef]
8. Yong, L.; Deng, H.; Xing, X.; Jiang, H. Review on torque distribution strategies for four in-wheel motor drive electric vehicles. *IOP Conf. Ser. Mater. Sci. Eng.* **2018**, *394*, 042041.
9. Wang, Y.; Su, Y. A research for brake strategy based on fuzzy control in pure electric vehicles. In Proceedings of the 2015 4th International Conference on Computer Science and Network Technology (ICCSNT), Harbin, China, 19–20 December 2015; pp. 689–693.
10. Wang, B.; Huang, X.; Wang, J.; Guo, X.; Zhu, X. A robust wheel slip control design for in-wheel-motor-driven electric vehicles with hydraulic and regenerative braking systems. In Proceedings of the 2014 IEEE American Control Conference, Portland, OR, USA, 4–6 June 2014; pp. 3225–3230.
11. Prajeesh, K.; Beevi, M.W. An Efficient Regenerative Braking System for BLDCM driven Electric Vehicles. In Proceedings of the 2018 4th International Conference for Convergence in Technology (I2CT), Mangalore, India, 27–28 October 2018; pp. 1–5.

12. Hannan, M.A.; Hoque, M.M.; Mohamed, A.; Ayob, A. Review of energy storage systems for electric vehicle applications: Issues and challenges. *Renew. Sustain. Energy Rev.* **2017**, *69*, 771–789. [CrossRef]
13. Tie, S.F.; Tan, C.W. A review of energy sources and energy management system in electric vehicles. *Renew. Sustain. Energy Rev.* **2013**, *20*, 82–102. [CrossRef]
14. Lu, D.; Ouyang, M.; Gu, J.; Li, J. Torque distribution algorithm for a permanent brushless DC hub motor for four-wheel drive electric vehicles. *J. Tsinghua Univ. (Sci. Technol.)* **2012**, *52*, 451–456.
15. Yang, L.; Zhang, J.W.; Guo, K.; Wu, D. Optimized Torque Distribution Algorithm to Improve the Energy Efficiency of 4WD Electric Vehicle. *SAE Tech. Pap.* **2014**, 1–10. [CrossRef]
16. Peng, H.; Hu, S.J. Traction/Braking Force Distribution for Optimal Longitudinal Motion during Curve Following. *Veh. Syst. Dyn.* **2007**, *26*, 301–320. [CrossRef]
17. Wu, D.; Tian, S. Torque distribution strategy of pure electric bus with double motors driving by front and rear axles. *J. Jiangsu Univ. (Nat. Sci. Ed.)* **2021**, *42*, 634–641.
18. Li, S.; Ding, X.; Yu, B. Optimal control strategy of efficiency for dual motor coupling drive system of pure electric vehicle. *J. Jiangsu Univ. (Nat. Sci. Ed.)* **2022**, *43*, 1–7.
19. Wang, B. *Study on Experiment Platform of Four-Wheel-Independent-Drive Ev and Its Driving Force Control System*; Tsinghua University: Beijing, China, 2009.
20. Fu, X.; Yang, F.; Huang, B.; He, Z.; Pei, B. Coordinated control of active rear wheel steering and four wheel independent driving vehicle. *J. Jiangsu Univ. (Nat. Sci. Ed.)* **2021**, *42*, 497–505.
21. Zhao, X.; Guo, G. Braking torque distribution for hybrid electric vehicles based on nonlinear disturbance observer. *Proc. Inst. Mech. Eng. Part D J. Automob. Eng.* **2019**, *233*, 3327–3341. [CrossRef]
22. Li, Y.; Adeleke, O.P.; Xu, X. Methods and applications of energy saving control of in-wheel motor drive system in electric vehicles: A comprehensive review. *J. Renew. Sustain. Energy* **2019**, *11*, 062701. [CrossRef]
23. Li, Z.; Song, X.; Chen, X.; Xue, H. Dynamic Characteristics Analysis of the Hub Direct Drive-Air Suspension System from Vertical and Longitudinal Directions. *Shock Vib.* **2021**, *2021*, 8891860. [CrossRef]
24. Kühlwein, J. Driving resistances of light-duty vehicles in Europe: Present situation, trends, and scenarios for 2025. *Communications* **2016**, *49*. Available online: https://theicct.org/publication/driving-resistances-of-light-duty-vehicles-in-europe-present-situation-trends-and-scenarios-for-2025/ (accessed on 8 September 2022).
25. Jazar, R.N. *Vehicle Dynamics: Theory and Application*; Springer: Berlin/Heidelberg, Germany, 2017; pp. 287–288.
26. Hong, J.; Yu, Z.; Hongtao, X.; Zhongxing, L. Sequential diagnosis method for bearing fault of in-wheel motor based on CDI and AHNs. *J. Jiangsu Univ. (Nat. Sci. Ed.)* **2021**, *42*, 15–21.
27. Wu, D.; Tian, S. New control strategy of motor for pure electric vehicle based on TLGI technology. *J. Jiangsu Univ. (Nat. Sci. Ed.)* **2021**, *42*, 9–14.
28. Li, Y.; Zhang, B.; Xu, X. Robust control for permanent magnet in-wheel motor in electric vehicles using adaptive fuzzy neural network with inverse system decoupling. *Trans. Can. Soc. Mech. Eng.* **2018**, *42*, 286–297. [CrossRef]
29. Zhang, J.; Wang, T.; Wang, L.; Zou, X.; Song, W. Optimization control strategy of driving torque for slope-crossing of pure electric vehicles. *J. Jiangsu Univ. (Nat. Sci. Ed.)* **2021**, *42*, 506–512.
30. Fan, L.; Ma, Z. Fuzzy comprehensive evaluation method for symmetry degree of mechanical structure symmetry. *Trans.-Can. Soc. Mech. Eng.* **2017**, *41*, 337–353. [CrossRef]
31. Liang, H.; Yue, M.; Yu, W.; Zhi, J.; Peng, Y. Research on Torque Optimization Allocation Strategy about Multi-wheel Vehicles. In *Innovative Techniques and Applications of Modelling, Identification and Control*; Springer: Singapore, 2018; pp. 63–92. [CrossRef]
32. Wu, X.; Zheng, D.; Du, J.; Liu, Z.; Zhao, X. Torque Optimal Allocation Strategy of All-wheel Drive Electric Vehicle. *Energies* **2019**, *12*, 1122. [CrossRef]
33. Chang, H.H.; Chia, W.C. Permanent Magnetic Brushless DC Motor Magnetism Performance depends on Different Intelligent Controller Response. *Trans. Can. Soc. Mech. Eng.* **2020**, *45*, 287–296. [CrossRef]
34. Li, J.; He, R. Optimization design and performance analysis of dual-rotor in-wheel motor based on parameter sensitivity. *J. Jiangsu Univ. (Nat. Sci. Ed.)* **2020**, *41*, 640–647.
35. Wu, Z.; Wu, Y.; He, S.; Xiao, X. Hierarchical fuzzy control based on spatial posture for a support-tracked type in-pipe robot. *Trans. Can. Soc. Mech. Eng.* **2020**, *44*, 133–147. [CrossRef]

World Electric Vehicle Journal

MDPI

Article

Nonlinear MPC-Based Acceleration Slip Regulation for Distributed Electric Vehicles

Wentong Shi [1], Yuyao Jiang [2,*], Zuying Shen [3], Zhongjing Yu [3], Hongqing Chu [1] and Dengcheng Liu [4]

1 School of Automotive Studies, Tongji University, Shanghai 201804, China
2 School of Customs and Public Administration, Shanghai Customs College, Shanghai 201204, China
3 Jiangxi Jiangling Group Electric Vehicle Co., Ltd., Nanchang 330013, China
4 Nanchang Automotive Institute of Intelligence and New Energy, Tongji University, Nanchang 330200, China
* Correspondence: jiangyuyao@shcc.edu.cn; Tel.: +86-021-2899-1551

Abstract: To address the problem in which wheel longitudinal slip rate directly affects the dynamics and handling stability of a vehicle under driving conditions, front and rear dual-motor four-wheel drive electric vehicles (4WD EVs) were selected as the research object in this study. An acceleration slip regulation (ASR) control strategy based on nonlinear model predictive control (NMPC) is proposed. First, the vehicle dynamics model and the Simulink/CarSim co-simulation platform were built. Second, an ASR controller with intervention and exit mechanisms was designed with the control objective of tracking reference speed or optimal slip rate. Then, considering the problem that the left and right wheels could not freely distribute torque under the condition of a split road surface, the motor output torque was determined in accordance with the wheel with the larger slip rate to enhance passibility. Finally, on the basis of the built Simulink/CarSim co-simulation platform, slip rate control simulation experiments were performed on a snow-covered road, a wet asphalt road, a docking road, and a split road. The designed controller can better track target slip rate and it exhibits better dynamic performance and stability than the method with PID control under different road conditions, especially under low speed and low adhesion road conditions, and its robustness can also meet the requirements.

Keywords: nonlinear model predictive control; four-wheel drive; acceleration slip regulation; intervention and exit mechanisms

Citation: Shi, W.; Jiang, Y.; Shen, Z.; Yu, Z.; Chu, H.; Liu, D. Nonlinear MPC-Based Acceleration Slip Regulation for Distributed Electric Vehicles. *World Electr. Veh. J.* **2022**, *13*, 200. https://doi.org/10.3390/wevj13110200

Academic Editors: Yong Li, Xing Xu, Lin Zhang, Yechen Qin and Yang Lu

Received: 30 September 2022
Accepted: 24 October 2022
Published: 27 October 2022

1. Introduction

Pure electric vehicles (EVs) are new energy vehicles driven by motors with batteries as a power source. Compared with traditional vehicles, they exhibit the advantages of not consuming fossil fuels and producing zero emissions during operation; thus, EVs are highly valued by society as an important tool for future development of the automotive industry [1]. EVs, with four-wheel drive (4WD EVs), will provide a higher degree of initiative and electrification to the chassis, improving the performance of the entire vehicle, making the overall structure more compact, and achieving higher transmission efficiency and freer control of the vehicle [2]. In contrast with centralized-drive EVs, front-and-rear axle independent-drive EVs can control the front and rear motors independently, and thus, they exhibit unique advantages in vehicle stability control, vehicle maneuverability enhancement, and motor efficiency improvement [3–5]. Front-and-rear axle dual-motor pure EVs have become a popular research topic [6–9].

The acceleration slip regulation (ASR) system has been widely used in vehicles driven by an internal combustion engine (ICE). Compared with torque output by ICE, torque output via motor presents the characteristics of quick response and large quantity, making EVs more prone to slipping [10–12]. This phenomenon will result in the speed of 4WD EVs being uncontrollable, whereafter stability and dynamics cannot be guaranteed. Therefore, the ASR control strategy of 4WD EVs is important.

To solve the aforementioned problems, scholars have conducted numerous related studies. MPC has received wide attention due to its advantages in solving constrained, nonlinear problems [13]. Yuan et al. [14] presented a nonlinear model predictive control (NMPC) for an antilock braking system (ABS) and a traction control system (TCS). Reference tracking was not used because slip rate was solely controlled through the constraints of NMPC formulation. Davide et al. [15] presented a TCS for EVs with in-wheel motors; this TCS is based on explicit NMPC. The proposed controller achieved good results in the simulation and real vehicle verification. Chen et al. [16] proposed an ASR strategy by using fuzzy control. The results of their study indicated that the proposed strategy is an effective technique for improving the dynamic performance and stability of EVs. They focused on controlling the slip rate to remain below the optimal slip rate. In this manner, vehicle stability is guaranteed, but dynamic performance is sacrificed to a certain extent. Guo et al. [17] presented an ASR control strategy based on the classification of road conditions and the calculation of the optimal slip rate for road conditions. In accordance with the real-time slip rate, an appropriate motor torque can be calculated by the control algorithm to track the reference value. These researchers focused on tracking the optimal slip rate, which can maximize the power of a vehicle. However, when driver acceleration demand is small, the actual slip rate is less than the optimal slip rate, and the vehicle is in a stable state. At this moment, turning on the controller will increase computational burden. Therefore, the problem of a controller's intervention and exit must be solved.

In the current study, an ASR control strategy for EVs based on NMPC with intervention and exit mechanisms is proposed. The algorithm is divided into three modules: the speed control module based on a proportional–integral–derivative (PID) controller, the slip rate control module based on an NMPC controller, and the torque selection module. The PID controller outputs the corresponding torque in accordance with the difference between actual and target vehicle speeds. The NMPC controller outputs the corresponding torque in accordance with the relationship between the actual and optimal slip rates. The torque selection module is responsible for coordinating the torque output of the two controllers, ensuring that the vehicle obtains the most suitable torque input in a certain state. The overall control block diagram is shown in Figure 1. In order to highlight the advantages of the proposed algorithm, we compared it with ASR-PID (contains PID speed controller, PID slip rate controller, intervention and exit mechanisms) and WASR (without acceleration slip regulation). The simulation results show that the proposed algorithm can achieve better results, especially under low speed and low adhesion road conditions.

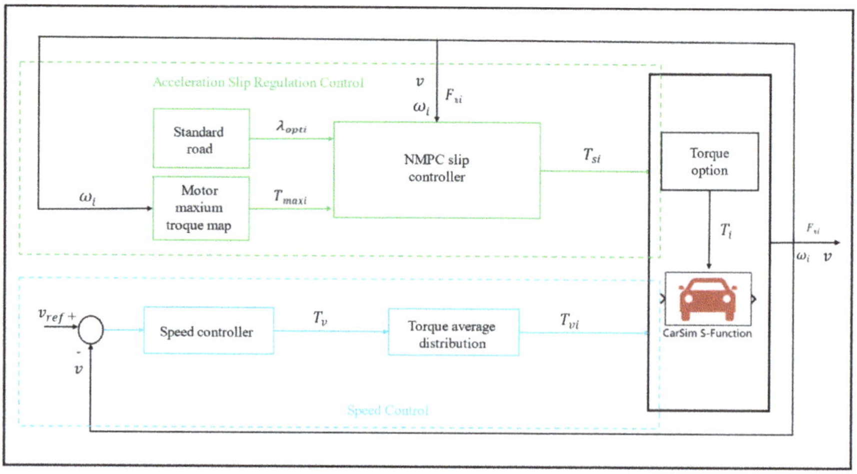

Figure 1. ASR control block diagram.

The structure of this paper is as follows. In Section 2, we establish the vehicle dynamics model, the motion model of a single wheel, the tire longitudinal force calculation model, and the longitudinal slip model. In Section 3, the speed control strategy is first introduced; then, the NMPC and torque selection mechanism are described. Finally, the overall control block diagram is provided. In Section 4, the control strategy is verified through the co-simulation of Simulink/CarSim.

2. System Model Establishment

2.1. Vehicle Longitudinal Movement

In accordance with Newton's law, the motion equations of a vehicle are shown as follows [18]. Figure 2 illustrates the longitudinal dynamic model of a vehicle.

$$m\dot{v} = F_{xfl} + F_{xfr} + F_{xrl} + F_{xrr} - F_w - F_f \tag{1}$$

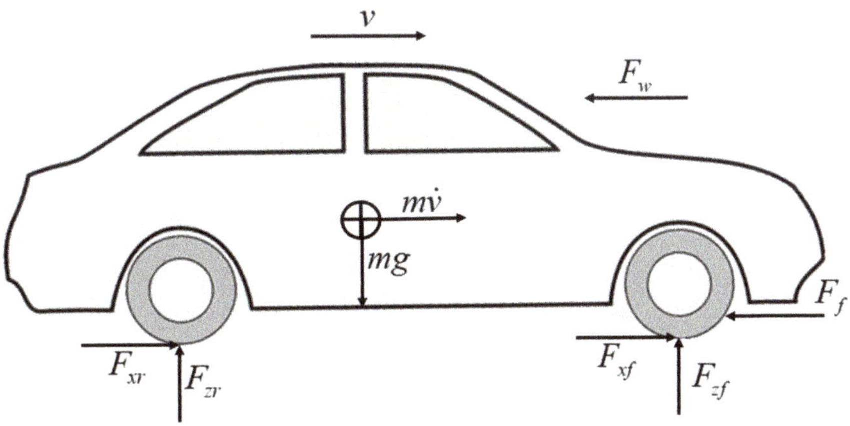

Figure 2. The longitudinal dynamic model of a vehicle.

where m is the mass of the vehicle; $\dot{v}$ is the acceleration of the vehicle; F_{xfl}, F_{xfr}, F_{xrl}, and F_{xrr} are the longitudinal forces of the front left, front right, rear left, and rear right tires, respectively; and F_w and F_f are the air resistance and rolling resistance, respectively.

2.2. Four Wheels' Rotation Movement

The wheels' rotation movement describes the relationship between the applied torque and the longitudinal force generated, which is shown in Figure 3.

$$J_i\dot{\omega}_i = T_{di} - F_{xi}R, \quad i = fl, fr, rl, rr \tag{2}$$

where J_i is the wheel moment of inertia, $\dot{\omega}_l$ is the wheel angular acceleration, T_{di} is the torque acting on the wheel, and R is wheel rolling radius.

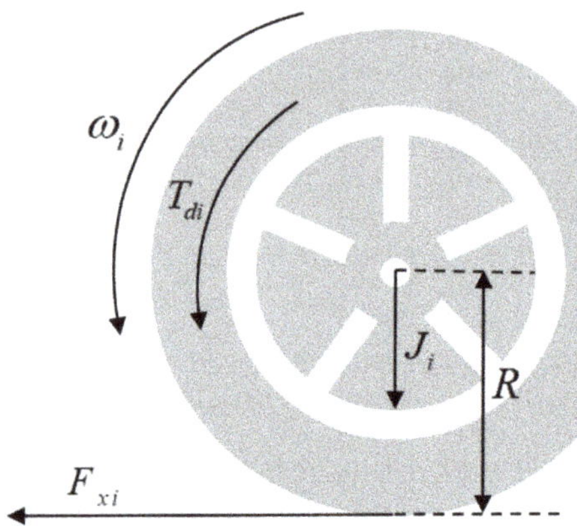

Figure 3. The wheel dynamics model.

2.3. Tire Longitudinal Force Calculation Model

The calculation of the longitudinal force of the tire is crucial for the longitudinal dynamics of a vehicle. Researchers typically establish a tire model to obtain the longitudinal force of a tire. Here, the μ–s curve of the standard road proposed by Burckhardt is used to replace the tire model [19], which is shown in Figure 4.

$$\mu(\lambda) = C_1\left(1 - e^{-C_2\lambda}\right) - C_3\lambda \tag{3}$$

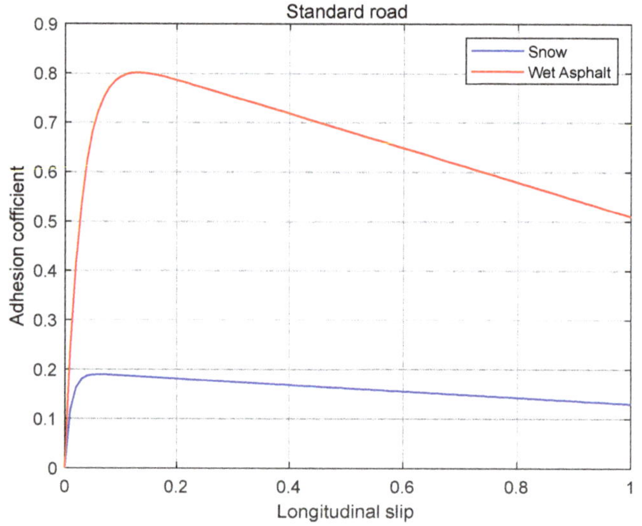

Figure 4. Adhesion coefficient vs. slip.

where C_1, C_2, and C_3 are the coefficients related to the road surface. Their values are provided in Table 1. λ is the slip rate.

Table 1. The relevant parameters of a standard road.

Road Surface	C_1	C_2	C_3	λ_{opt}	$\mu_{\max}$
Snow	0.1964	94.129	0.0646	0.06	0.19
Wet asphalt	0.8570	33.822	0.3470	0.13	0.80

In accordance with Formula (3), the following conclusion can be drawn:

$$\frac{d\mu(\lambda)}{d\lambda} = C_1C_2e^{-C_2\lambda} - C_3 \tag{4}$$

Let $\frac{d\mu(\lambda)}{d\lambda} = 0$. We obtain

$$\begin{cases} \lambda_{opt} = \frac{1}{C_2}\ln\frac{C_1C_2}{C_3} \\ \mu_{\max} = C_1 - \frac{C_3}{C_2}(1+\ln\frac{C_1C_2}{C_3}) \end{cases} \tag{5}$$

where λ_{opt} is the optimum slip rate, and $\mu_{\max}$ is the maximum road adhesion coefficient.

The calculation method for tire longitudinal force is

$$F_{xi} = \mu(\lambda)F_{zi}, \quad i = fl, fr, rl, rr \tag{6}$$

The vertical force of each tire can be calculated using the following formula:

$$\begin{cases} F_{zfl} = m\left(\frac{b}{a+b}g - \frac{h_g}{a+b}a_x\right)\left(\frac{1}{2} - \frac{h_g a_y}{t_f g}\right) \\ F_{zfr} = m\left(\frac{b}{a+b}g - \frac{h_g}{a+b}a_x\right)\left(\frac{1}{2} + \frac{h_g a_y}{t_f g}\right) \\ F_{zrl} = m\left(\frac{a}{a+b}g + \frac{h_g}{a+b}a_x\right)\left(\frac{1}{2} - \frac{h_g a_y}{t_f g}\right) \\ F_{zrr} = m\left(\frac{a}{a+b}g + \frac{h_g}{a+b}a_x\right)\left(\frac{1}{2} + \frac{h_g a_y}{t_f g}\right) \end{cases} \tag{7}$$

In the current study, the longitudinal motion of a vehicle is considered, and the influence of lateral acceleration is disregarded. The above formula can be simplified as

$$\begin{cases} F_{zfl} = \frac{1}{2}m\left(\frac{b}{a+b}g - \frac{h_g}{a+b}a_x\right) \\ F_{zfr} = \frac{1}{2}m\left(\frac{b}{a+b}g - \frac{h_g}{a+b}a_x\right) \\ F_{zrl} = \frac{1}{2}m\left(\frac{a}{a+b}g + \frac{h_g}{a+b}a_x\right) \\ F_{zrr} = \frac{1}{2}m\left(\frac{a}{a+b}g + \frac{h_g}{a+b}a_x\right) \end{cases} \tag{8}$$

where F_{zfl}, F_{zfl}, F_{zfl}, and F_{zfl} are the vertical loads of the four wheels; a and b are the distance from the center of mass to the front and rear axles, respectively; g is the gravitational acceleration; h_g is the height from the center of mass to the ground; and a_x is the longitudinal acceleration of a vehicle.

2.4. Longitudinal Slip Calculation Model

The longitudinal slip of the tire is the key to calculating its longitudinal force. The relationship of longitudinal slip with vehicle speed and wheel angular speed can be described as

$$\lambda_i = \frac{\omega_i R - v}{\omega_i R}, \quad i = fl, fr, rl, rr \tag{9}$$

3. Control Strategy

3.1. PID-Based Speed Controller Design

The speed control module is specifically established to simulate the driver's expected torque output. It consists of a PID controller and an average torque distribution. The total control torque can be expressed by the following formula:

$$\begin{cases} T_v = K_p e + K_i\int e dt + K_d\frac{de}{dt} \\ e = v_{ref} - v \end{cases} \tag{10}$$

where T_v is the total control torque; e is the difference between reference and actual speeds; and K_p, K_i, and K_d are the proportional, integral, and differential parameters of the PID controller, respectively.

The torque adopts the principle of equal distribution, and the torque of each wheel can be expressed as

$$T_{vi} = \frac{1}{4} T_v \tag{11}$$

where T_{vi} is the torque of each wheel.

3.2. NMPC-Based Slip Controller Design

Wheel slip rate is set as the direct control variable by the general slip rate control. When the vehicle uses slip rate as the control variable in the low-speed starting stage, errors and disturbances in vehicle speed estimation will exert a relatively greater effect, and the lag of the motor torque will cause larger buffeting. The primary reason for this phenomenon is that disturbances and delays are passed directly to the control variable. Considering this deficiency, a slip rate control algorithm with wheel speed as the control variable is proposed to compensate for the defect of the controller at low vehicle speed. From the definition of slip rate, the tracking reference wheel slip rate signal is equivalent to the tracking reference wheel speed signal.

$$\omega_o = \frac{v}{r(1-\lambda_o)} \tag{12}$$

A minimum vehicle speed strategy is designed to prevent the effects of vehicle speed errors and disturbances on the controller when starting at a low speed. That is, when the estimated vehicle speed is less than a certain value, the fixed value vehicle speed $v_{\min}$ is used in calculating reference wheel speed.

$$\omega_o = \frac{\max(v, v_{\min})}{r(1-\lambda_o)} \tag{13}$$

From Formula (2), the following can be determined:

$$\dot{\omega}_i = \frac{T_{di} - F_{xi}R}{J_i}, \quad i = f, r \tag{14}$$

From Formula (1),

$$\dot{v} = \frac{F_{xfl} + F_{xfr} + F_{xrl} + F_{xrr} - F_w - F_f}{m} \tag{15}$$

3.2.1. Model Discretization

The state-space equation of slip control can be described as

$$\begin{cases} \dot{\omega}_f = \frac{T_{df} - F_{xf}R}{J_f} \\ \dot{\omega}_r = \frac{T_{dr} - F_{xr}R}{J_r} \\ \dot{v} = \frac{F_{xfl} + F_{xfr} + F_{xrl} + F_{xrr} - F_w - F_f}{m} \end{cases} \tag{16}$$

The state-space equations of the system are discretized using Euler's method. Δt is defined as the sampling step size. By applying it to the state-space model at sampling time k, the preceding formula can be discretized as

$$\begin{cases} x(k+1) = f^k(x(k), u(k))\Delta t + x(k) \\ y(k) = C_y x(k) \end{cases} \tag{17}$$

where $x = \left[\omega_f, \omega_r, v\right]^T$ is the state variable, $u = \left[T_{df}, T_{dr}\right]^T$ is the input for system control, $y = \left[\omega_f, \omega_r, v\right]^T$ is the output for system control, f^k represents the gradient of the system state change at time k, and the output matrix is $C_y = \text{diag}(1;1;1)$.

N_C is defined as the control time domain, N_p is the prediction time domain, and $N_p \geq N_c \geq 1$. At sampling time k, the control sequence U_k and the output sequence of the system Y_k are expressed as follows:

$$U_k = \begin{bmatrix} u(k|k) \\ u(k+1|k) \\ \vdots \\ u(k+N_c-1|k) \end{bmatrix} \text{ and } Y_k = \begin{bmatrix} y(k+1|k) \\ y(k+2|k) \\ \vdots \\ y(k+N_p|k) \end{bmatrix}$$

Here, $R(k)$ is the output reference sequence that represents the optimal angular velocity sequence ω_0. At sampling time k, $r(k) = [\omega_{o1}(k), \omega_{o2}(k)]^T$ is the sequence of changes in the control input. It is calculated such that $\Delta u(k) = u(k) - u(k-1)$, and its value is 0 when k exceeds the control time domain.

$$R_k = \begin{bmatrix} r(k) \\ r(k) \\ \vdots \\ r(k) \end{bmatrix}, \ \Delta U_k = \begin{bmatrix} \Delta u(k|k) \\ \Delta u(k+1|k) \\ \vdots \\ \Delta u(k+N_c-1|k) \end{bmatrix}$$

At sampling time k, the predicted state and output are

$$\begin{gathered} x(k+1|k) = f^k(x(k|k), u(k|k))\Delta t + x(k|k) \\ x(k+2|k) = f^k(x(k+1|k), u(k+1|k))\Delta t + x(k+1|k) \\ \vdots \\ x(k+N_c|k) = f^k(x(k+N_c-1|k), u(k+N_c-1|k))\Delta t + x(k+N_c-1|k) \end{gathered} \tag{18}$$

$$\begin{gathered} y(k+1|k) = C_y\left(f^k(x(k|k), u(k|k))\right)\Delta t + x(k|k) \\ y(k+2|k) = C_y\left(f^k(x(k+1|k), u(k+1|k))\right)\Delta t + x(k+1|k) \\ \vdots \\ y(k+N_p \mid k) = C_y\left(f^k(x(k+N_p-1|k), u(k+N_p-1|k))\right)\Delta t + x(k+N_p-1|k) \end{gathered} \tag{19}$$

3.2.2. Description of Optimization Problem

The primary objective of the control requirement is to track the optimal slip rate and ensure that the vehicle can obtain a large longitudinal force to fully utilize the road adhesion coefficient and improve possibility. Therefore, a cost function J_1 is added to meet the requirements for tracking the optimal angular speed.

$$J_1 = \sum_{i=1}^{N_p} \|(y(k+i|k) - r(k+i)|k)\| \tag{20}$$

To preserve driver comfort, the change rate of the control action should not be excessively large, because such a condition may lead to serious torque ripple. That is, the change rate of the control variable should be sufficiently small to ensure comfort during acceleration. Thus, a cost function J_2 is added to limit the change rate of the control variable.

$$J_2 = \sum_{j=1}^{N_{c-1}} \| \Delta u(k+j-1)|k \| \tag{21}$$

By combining the aforementioned control objectives, the following cost function is proposed:

$$\min J = J_1 + J_2 = \sum_{i=1}^{N_p} \|(y(k+i|k) - r(k+i)|k)\|_Q^2 + \sum_{j=1}^{N_{c-1}} \|\Delta u(k+j-1)|k\|_R^2 \tag{22}$$

$$\text{s.t}\begin{cases} -T_{\text{fmax}} \le T_{df}(k+j|k) \le T_{\text{fmax}}, & j = 0,1,\ldots,N_c - 1 \\ -T_{\text{rmax}} \le T_{dr}(k+j|k) \le T_{\text{rmax}}, & j = 0,1,\ldots,N_c - 1 \\ x(k+1) = f^k(x(k),u(k))\Delta t + x(k) \\ y(k) = C_y x(k) \end{cases} \tag{23}$$

3.3. *Design of Intervention and Exit Mechanisms*

In practical applications, being fully controlled by the controller is unnecessary, and driver's demand torque must also be comprehensively considered. Therefore, for the selection of driver's demand torque and controller's control torque, the current study designs intervention and exit mechanisms for the control algorithm to ensure that the motor output torque conforms to the current state of the vehicle.

To avoid the failure of the intervention and exit mechanisms due to the chattering of the slip rate at low speed, the NMPC controller is always turned on when $v < v_{\min}$. The specific flowchart is shown in Figure 5:

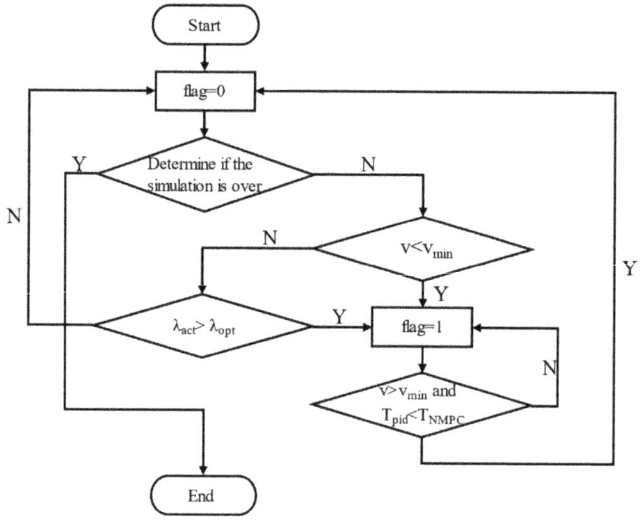

Figure 5. The process of the torque option.

where λ_{act} is the actual slip rate, λ_{opt} is the optimal slip rate, T_{PID} is driver's demand torque, and T_{NMPC} is the output torque of the NMPC controller.

The controller is turned on when *flag* = 1, and the driver takes over when *flag* = 0.

4. Simulation and Analysis

MATLAB/CarSim co-simulation was conducted to verify the effectiveness of the proposed algorithm. The optimization toolbox fmincon in MATLAB is suitable for solving constrained and nonlinear problems; therefore, it could be used to solve the optimization problem in this study.

The simulations were performed under four different conditions: low adhesion road, high adhesion road, docking road, and split road. In addition, every condition included two aspects: vehicle starting and vehicle acceleration, the former initial velocity was 0 km/h, the latter initial velocity was 30 km/h. The road surface is assumed to be flat and smooth, and the vehicle will drive straight on these roads except on the split road. These test maneuvers are typical maneuvers developed for evaluating the performance characteristics of a vehicle [20].

In order to ensure that the controller can adapt to various single road surfaces, the simulation tests of low adhesion and high adhesion road surfaces are carried out. When the vehicle enters the low adhesion road from the high adhesion road, the wheels will rotate violently if the driving torque cannot be reduced rapidly, which will greatly reduce the stability; from the low adhesion road to the high adhesion road, the vehicle's dynamic performance will be limited to a large extent if the torque cannot increase rapidly. In order to ensure that the controller can respond quickly to this, the simulation test of the docking road is carried out.

Driving on a split road is a very critical test maneuver, since the vehicle will experience severe instability if the driver does not react immediately to correct the course of the vehicle. During this test, due to the asymmetric driving forces generated on the left and right tires, the vehicle will be pushed to the side of the road that has a lower coefficient of friction. Therefore, it is necessary to a conduct simulation test on a split road.

The major parameters of the class A 4WD EV are listed in Table 2. To improve the realism of the simulation and the validity of the results, the white noise is added to the simulation process. Finally, the robustness of the whole control system is analyzed. The simulation results are as follows:

Table 2. Major parameters of the EV.

Parameter	Symbol	Value
Vehicle Mass	m	1710 kg
Height of the vehicle c.g.	h_g	0.552 m
Distance from c.g. to front axle	a	1.216 m
Distance from c.g. to Rear axle	b	1.613 m
Rolling radius of the tyre	R	0.32 m
Front area	A	2.3157 m^2
Front motor peak power	$P_{f\max}$	130 kW
Front motor peak torque	$T_{f\max}$	225 Nm
Rear motor peak power	$P_{r\max}$	60 kW
Rear motor peak torque	$T_{r\max}$	170 Nm

4.1. Low Adhesion Road

The simulation condition is a snow-covered road. The road adhesion coefficient is 0.19. The optimal slip rate is 0.06. The reference speed is 0–15 km/h and 30–45 km/h. The target speed is reached within 2 s, and this speed is maintained until the end of the simulation The simulation results are shown in Figures 6 and 7.

As shown in the figure, the designed controller achieves ideal results compared with the WASR (without acceleration slip regulation), which only contains the PID speed controller. ASR-MPC is the simulation result of the designed algorithm, ASR-PID contains PID speed and a PID slip rate controller, which also contains intervention and exit mechanisms. "Ref" and "Opt" are the reference vehicle speed and the optimal wheel slip rate, respectively. In order to clearly show the role of the intervention and exit mechanisms, only the simulation results of ASR-MPC are shown in the second and fourth lines of the above simulation figures. In vehicle starts condition of ASR-MPC, at 0–2.2 s, the slip rate controller is always on. At 0–0.8 s, vehicle velocity is less than 5 km/h, the actual slip rate cannot track the optimal slip rate because of the reference angular speed is a fixed value and the motor output torque is limited. An actual slip rate of 0.8–2.2 s can track the optimal slip rate well, and acceleration remains at 1.83 m/s^2. After 2.2 s, the NMPC is turned off, and acceleration rapidly drops close to 0, because the reference speed has been tracked at this time, and thus, continuing to accelerate becomes unnecessary. The simulation results of

ASR-PID are similar to those of ASR-MPC. The difference is that the first arrival reference speed of ASR-PID is slightly slower than that of ASR-MPC, because the latter has better slip rate control. In contrast with WASR, acceleration is maintained at 1.65 m/s^2 most of the time, and the slip rate is close to 1. After reaching the reference speed, acceleration cannot be reduced in time, and the simulation reports an error at 4.6 s. This working condition is extremely dangerous and should be avoided. In the torque diagram, MPC and PID are the slip rate control output torque and driver output torque, respectively, in ASR control. ASR is the final output torque of the front and rear axle motors. Under the action of the intervention and exit mechanisms, the front and rear axle motors can always select a more suitable torque as the output.

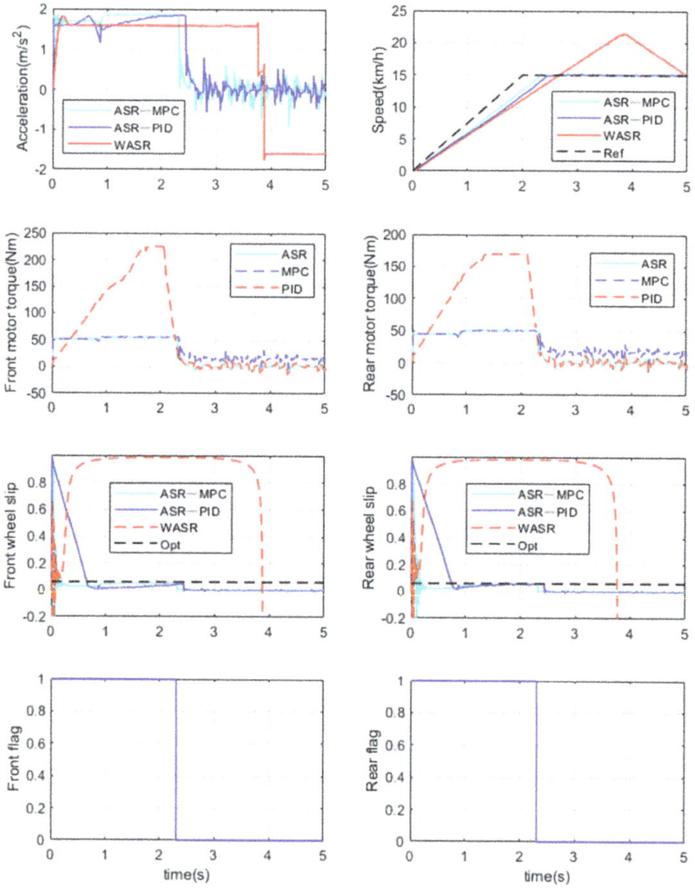

Figure 6. Vehicle starts under a low adhesion coefficient road.

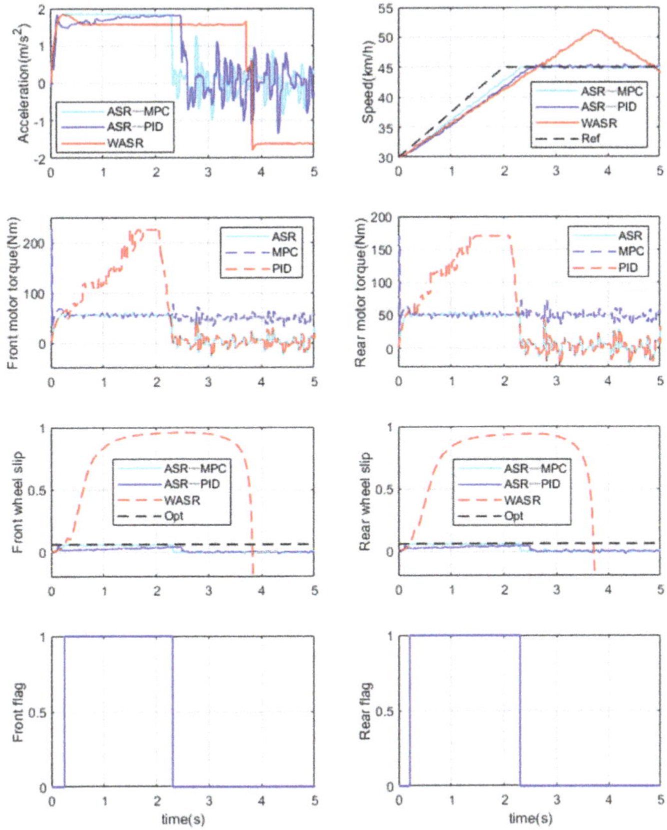

Figure 7. Vehicle accelerates under a low adhesion coefficient road.

During vehicle acceleration, the simulation results are similar to those of a vehicle starting. The difference is that the former slip rate changes smoothly and the NMPC is closed in the beginning of the simulation, while the latter fluctuates evidently and the NMPC is turned on. The reason for this phenomenon is that the latter starts at a speed of 0, and a small fluctuation in wheel angle can cause a considerable change in slip rate. To suppress this phenomenon, the NMPC is turned on when the vehicle speed is less than the minimum vehicle speed v_{min}.

4.2. High Adhesion Road

The simulation condition is a wet asphalt road. The road adhesion coefficient is 0.8, and the optimal slip rate is 0.13. The reference speed is 0–55 km/h and 30–85 km/h. The target speed is reached within 2 s, and this speed is maintained until the end of the simulation. The simulation results are shown in Figures 8 and 9.

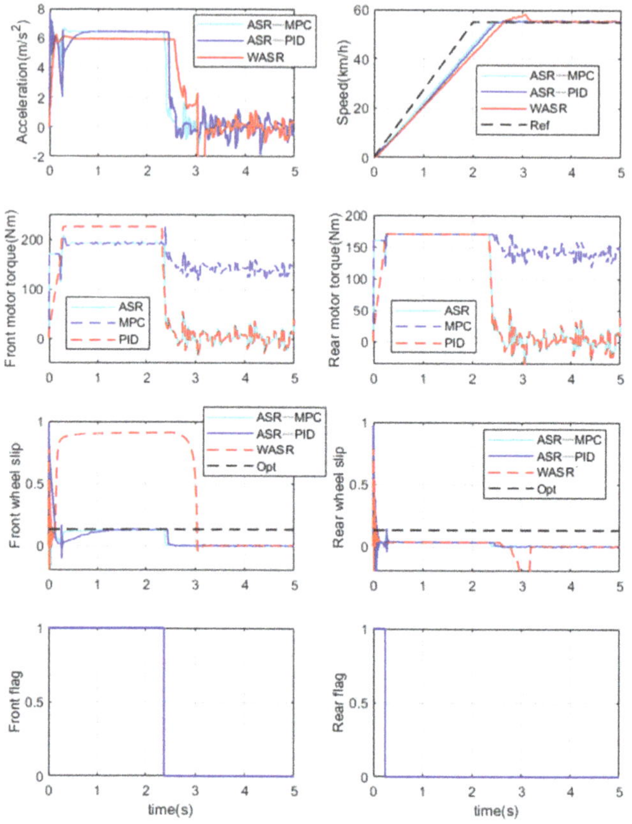

Figure 8. Vehicle starts under a high adhesion coefficient road.

As shown by the results, ASR-MPC and ASR-PID are more dynamic than WASR and has a faster speed response. In ASR-MPC control, the NMPC is turned on when vehicle speed is greater than v_{min}, and the front wheel slip rate can effectively track the optimal slip rate. However, the rear wheel slip rate cannot track the optimal slip rate, because the load is transferred backward during the acceleration process, and the rear axle wheels require a larger motor torque to generate the same slip rate. The slip rate cannot continue increasing due to the limitation of the motor torque. The control effect of ASR-PID is basically consistent with that of ASR-MPC. In WASR control, the wheel slip rate is large at 0.2–3 s. The handling stability of the vehicle is considerably reduced at this time. This condition is more dangerous, and the rear wheel slip rate is always kept at a small value due to load transfer. When speed is greater than v_{min}, the slip rate can track the optimal slip rate if *flag* = 1, and the slip rate remains below the optimal slip rate if *flag* = 0. Therefore, the designed algorithm can also play an important role in a road surface with a high adhesion coefficient. The expected results can be achieved regardless of the starting and acceleration.

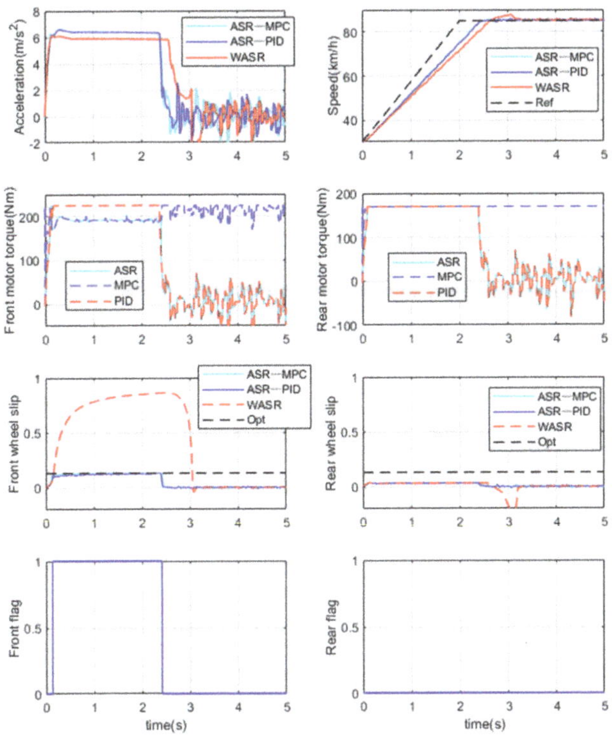

Figure 9. Vehicle accelerates under a high adhesion coefficient road.

4.3. Docking Road

The road adhesion coefficient is 0.8 in the first 5 m, 0.19 in 5–15 m, and 0.8 after 15 m. The reference speed is 0–50 km/h and 30–80 km/h. The target speed is reached within 2 s, and this speed is maintained until the end of the simulation. The target slip rates are 0.13 and 0.06 when the adhesion coefficients are 0.8 and 0.19, respectively. The simulation results are shown in Figures 10 and 11.

In accordance with the preceding simulation results, the vehicle starting process is as expected. However, the vehicle acceleration results have two unusual aspects. First, the slip rate of the rear wheel can keep up with the optimal slip rate immediately after entering the high adhesion road from the low adhesion road, because the reference slip rate is obtained on the basis of the front wheels. When the front wheels have just entered the high adhesion road from the low adhesion road, the rear wheels are still in the low adhesion road, and thus, their slip rate increases rapidly as the rear motor torque increases. Second, WASR acceleration is greater than ASR-MPC and ASR-PID acceleration in the second half of the acceleration process. The reason for this phenomenon is that wheel angular speed is excessively high, such that it can maintain a large slip rate after entering the high adhesion road from the low adhesion road in WASR. Meanwhile, in ASR-MPC and ASR-PID, the rear wheels' slip rate decreases rapidly after the vehicle enters the high adhesion road. The adhesion–longitudinal slip curve can determine that the longitudinal force generated by the rear wheels of WASR is greater than that of ASR-MPC and ASR-PID, and the total longitudinal force is also related in this manner.

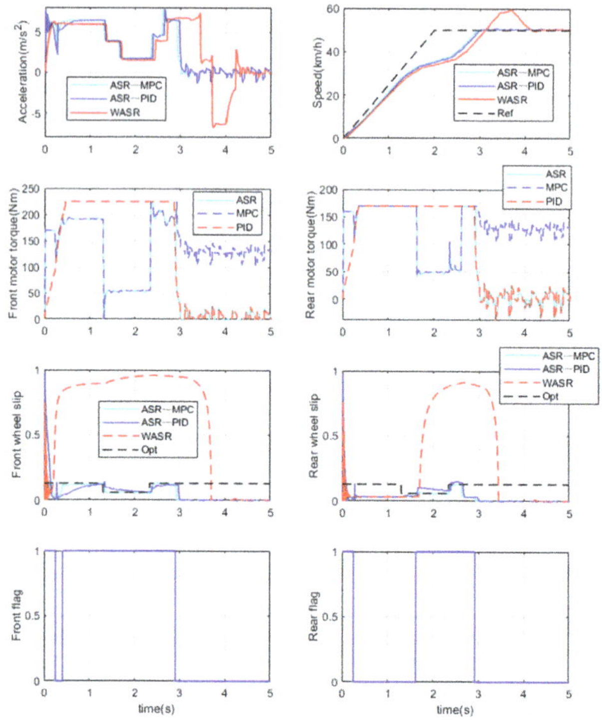

Figure 10. Vehicle starts on a docking road.

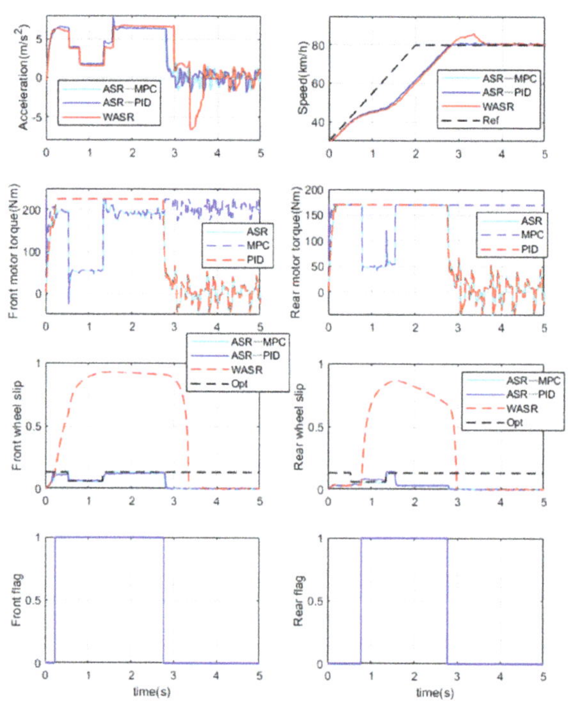

Figure 11. Vehicle accelerates on a docking road.

4.4. Split Road

The road adhesion coefficient of the left wheels is 0.8 in the first 5 m, 0.19 in 5–15 m, and 0.8 after 15 m. Meanwhile, that of the right wheels is 0.8. The reference speed is 0–5 km/h and 30–80 km/h. The target speed is reached within 2 s, and it is maintained until the end of the simulation. The simulation results are shown in Figures 12 and 13.

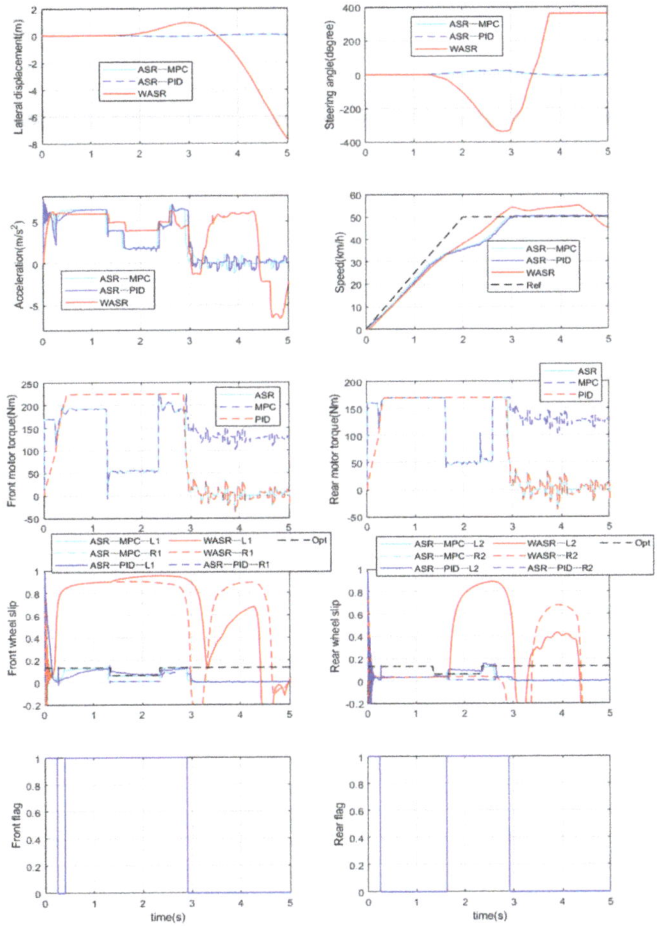

Figure 12. Vehicle starts on a split road.

When entering the split road, the output torque of the motor is determined by the wheel with the smaller optimum slip rate. At 5–15 m, the optimal slip rate of the left wheels is 0.06. The vehicle has dual motors on the front and rear axles; hence, the torque of the left and right wheels on the same axis should be equal. When the left wheels keep up with the optimal slip rate, the output torque of the motor makes the right wheels' actual slip rate less than 0.06, while the optimal slip rate of the right wheels is 0.13. If the actual slip rate is less than the optimal slip rate, then the generated longitudinal force increases with an increase in slip rate. In the simulation process, the longitudinal force generated by the right wheels is smaller than that of the left wheels. When the vehicle is traveling, differential steering is generated due to the difference in longitudinal force of the left and right wheels, and the vehicle is slightly deflected to the right. Accordingly, the PID controller is added to ensure that the vehicle drives on the split road within the 5–15 m interval.

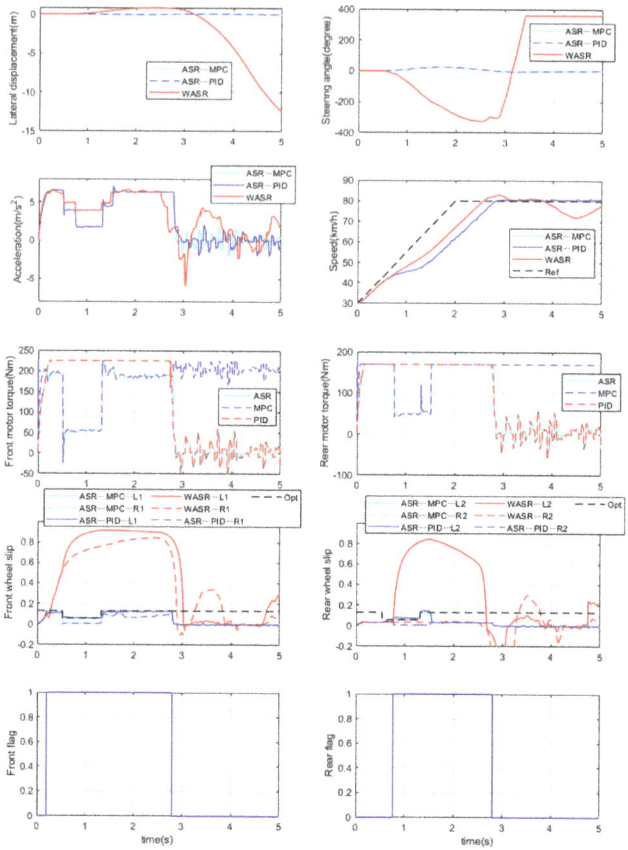

Figure 13. Vehicle accelerates on a split road.

As shown in the figure, ASR-MPC and ASR-PID have a smaller steering wheel angle and a smaller lateral displacement than WASR. The former is kept near the centerline of the lane. The maximum steering wheel angle is around 25°, and the maximum deviation does not exceed 0.08 m. The simulation runs smoothly throughout the whole process. Meanwhile, the latter's maximum steering wheel angle reaches the limit value of 360°, and the maximum deviation reaches 15 m. Safety is particularly important when driving on the split road; hence, power is sacrificed to a certain extent to ensure the smooth running of the vehicle.

A, B, C, D, and E in Table 3 represent the time taken to reach the reference speed for the first time, the time taken to stabilize at the reference speed, whether speed has an overshoot, the maximum steering wheel angle, and the maximum lateral displacement, respectively. "In" in B indicates that the steady speed is not reached at the end of the simulation. Different road surfaces and conditions have varying reference speeds. The difference between the maximum and minimum reference speeds of the same road surface under different conditions is the same.

Table 3. A comparison of the simulation results.

Road Surface	Condition	Algorithm	A	B	C	D	E
Low adhesion road	Starting	ASR-MPC	2.2 s	2.2 s	N	0	0
		ASR-PID	2.3 s	2.3 s	N	0	0
		WASR	2.6 s	In	Y	0	0
	Acceleration	ASR-MPC	2.2 s	2.2 s	N	0	0
		ASR-PID	2.4 s	2.4 s	N	0	0
		WASR	2.6 s	In	Y	0	0
High adhesion road	Starting	ASR-MPC	2.3 s	2.3 s	N	0	0
		ASR-PID	2.35 s	2.35 s	N	0	0
		WASR	2.6 s	3.2 s	Y	0	0
	Acceleration	ASR-MPC	2.3 s	2.3 s	N	0	0
		ASR-PID	2.3 s	2.3 s	N	0	0
		WASR	2.6 s	3.2 s	Y	0	0
Docking road	Starting	ASR-MPC	2.9 s	2.9 s	N	0	0
		ASR-PID	2.9 s	2.9 s	N	0	0
		WASR	3.1 s	4.2 s	Y	0	0
	Acceleration	ASR-MPC	2.8 s	2.8 s	N	0	0
		ASR-PID	2.8 s	2.8 s	N	0	0
		WASR	2.8 s	3.6 s	Y	0	0
Split road	Starting	ASR-MPC	2.9 s	2.9 s	N	24.8°	0.08 m
		ASR-PID	2.9 s	2.9 s	N	26.1°	0.08 m
		WASR	2.7 s	In	Y	360°	7.3 m
	Acceleration	ASR-MPC	2.8 s	2.8 s	N	25.2°	0.07 m
		ASR-PID	2.8 s	2.8 s	N	26.2°	0.07 m
		WASR	2.6 s	In	Y	360°	5.6 m

Table 3 clearly shows that, except for the docking road, the acceleration time of the same road surface under different conditions tends to be the same. The reason for the difference in the docking road is that under acceleration condition, when the low adhesion road enters the high adhesion road, the angular speed of the rear wheels of WASR is excessively high, such that the slip rate can still be maintained at a large value immediately after entering the high adhesion road. Meanwhile, the angular speed of the rear wheels of ASR-MPC and ASR-PID is low, and the slip rate of the rear wheels decreases rapidly after entering the high adhesion road. The longitudinal adhesion coefficient–slip rate relationship curve can determine that the longitudinal force generated by WASR is greater than that generated by ASR-MPC and ASR-PID. Thus, the acceleration of WASR in the second half is greater than that of ASR-MPC and ASR-PID.

To keep the vehicle running on the split road, a PID controller is added to offset the differential steering caused by the difference in longitudinal force of the left and right wheels. ASR-MPC and ASR-PID always achieve a smaller lateral displacement with a smaller steering wheel angle. The control effect of ASR-MPC at low speed and low adhesion is obviously better than that of ASR-PID, the simulation results of the two at other working conditions are close. The effectiveness of the proposed algorithm is fully verified through the preceding simulation tests.

4.5. Robustness Analysis

In order to verify the robustness of the control system, the vehicle body mass in CarSim is set to 0.8, 1 and 1.2 times the original body mass, and working condition is set to vehicle starts under a low adhesion coefficient road. The simulation results are shown in Figure 14.

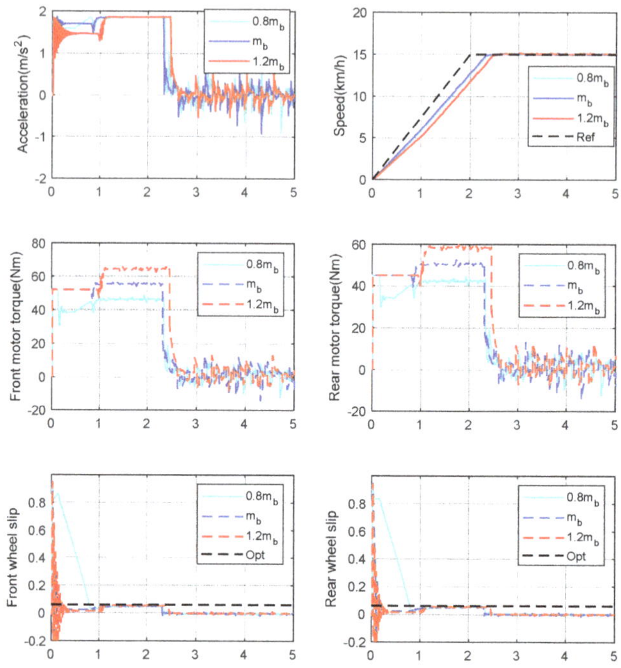

Figure 14. The robustness verification.

where m_b refers to the original body mass, $0.8m_b$ and $1.2m_b$ represent 0.8 times and 1.2 times the body mass, respectively.

From the above simulation results, it can be seen that the simulation result is the best when the body mass is m_b, because the body mass in the control algorithm is the same; when the body mass is $0.8m_b$, the slip ratio decreases slowly from a large value. The reason for this is that the optimal wheel angular speed is a fixed value if the vehicle speed is less than v_{min}. It takes a short time to track the optimal wheel speed for the reason that the vehicle weight is small. With the increase of vehicle speed, the slip ratio decreases gradually; when the body mass is $1.2m_b$, the reference speed tracking time of the vehicle speed is greater than the $0.8m_b$ and m_b. This phenomenon occurs because the maximum output torque of the motor is limited when the vehicle speed is less than v_{min}, which results in the wheel slip rate being kept at a small value, and therefore the acceleration is small. In general, under these three body masses, the simulation can achieve the desired results, and the robustness of the controller has been verified.

5. Conclusions

This study proposes an ASR control system for 4WD EVs based on NMPC with intervention and exit mechanisms to solve the problem of wheel slipping and uncontrollable speed under acceleration conditions. Considering dynamic performance and comfort problems caused by severe torque fluctuation, a cost function for tracking the optimal slip rate and limiting torque change rate is established. An optimal solution is then found using the optimization toolbox fmincon in MATLAB. Intervention and exit mechanisms are designed for the controller to solve the problem of switching between slip rate tracking and speed tracking. The effectiveness of the proposed controller is evaluated on a low adhesion road, a high adhesion road, a docking road, and a split road. The simulation results show that the proposed controller can accurately control the longitudinal slip of the four wheels in accordance with the optimal slip rate and reference speed, and the robustness can also meet the requirements.

Author Contributions: Conceptualization, Y.J. and Z.S.; methodology, W.S., Y.J., Z.S. and H.C.; software, W.S. and Y.J.; validation, W.S. and Z.Y.; formal analysis, H.C., Z.Y. and D.L.; investigation, W.S. and Z.Y.; writing—original draft preparation, W.S. and Y.J.; writing—review & editing, H.C. and Z.S.; visualization, D.L.; supervision, Y.J. and H.C.; project administration, H.C.; funding acquisition, H.C. All authors have read and agreed to the published version of the manuscript.

Funding: This research was funded in part by the Perspective Study Funding of Nanchang Automotive Institute of Intelligence and New Energy, Tongji University, grant number [TPD-TC202110-09], and in part by the Starting Research Fund from the Shanghai Customs College, grant number [kyqd202204].

Conflicts of Interest: The authors declare no conflict of interest.

References

1. Liu, Z.; Hao, H.; Cheng, X.; Zhao, F. Critical issues of energy efficient and new energy vehicles development in China. *Energy Policy* **2018**, *115*, 92–97. [CrossRef]
2. Emadi, A. *Advanced Electric Drive Vehicles*; CRC Press: Boca Raton, FL, USA, 2014.
3. Nguyen, C.T.; Walker, P.D.; Zhang, N. Optimization and coordinated control of gear shift and mode transition for a dual-motor electric vehicle. *Mech. Syst. Signal Process.* **2021**, *158*, 107731. [CrossRef]
4. Yang, Y.; He, Q.; Chen, Y.; Fu, C. Efficiency optimization and control strategy of regenerative braking system with dual motor. *Energies* **2020**, *13*, 711. [CrossRef]
5. Hu, X.; Li, Y.; Lv, C.; Liu, Y. Optimal energy management and sizing of a dual motor-driven electric powertrain. *IEEE Trans. Power Electron.* **2018**, *34*, 7489–7501. [CrossRef]
6. Interrupt-Free Operation of Dual-Motor Four-Wheel Drive Electric Vehicle under Inverter Failure. Available online: https://ieeexplore.ieee.org/document/9103127 (accessed on 30 September 2022).
7. Xiong, H.; Zhu, X.; Zhang, R. Energy recovery strategy numerical simulation for dual axle drive pure electric vehicle based on motor loss model and big data calculation. *Complexity* **2018**, *2018*, 4071743. [CrossRef]
8. Driving-Cycle-Oriented Design Optimization of A Permanent Magnet Hub Motor Drive System for A Four-Wheel-Drive Electric Vehicle. Available online: https://ieeexplore.ieee.org/abstract/document/9141333 (accessed on 30 September 2022).
9. Heydari, S.; Fajri, P.; Shadmand, M.; Sabzehgar, R. Maximizing harvested energy through regenerative braking process in dual-motor all-wheel drive electric vehicles. In Proceedings of the 2020 IEEE Transportation Electrification Conference & Expo (ITEC), Chicago, IL, USA, 23–26 June 2020; pp. 1246–1250.
10. He, H.; Yu, X.; Sun, F.; Zhang, C. Study on power performance of traction motor system for electric vehicle. *Proc. -Chin. Soc. Electr. Eng.* **2006**, *26*, 136.
11. Ding, X.; Wang, Z.; Zhang, L. Hybrid Control-Based Acceleration Slip Regulation for Four-Wheel-Independent-Actuated Electric Vehicles. *IEEE Trans. Transp. Electrif.* **2021**, *7*, 1976–1989. [CrossRef]
12. Hori, Y. Future vehicle driven by electricity and control-research on four wheel motored "UOT Electric March II". In Proceedings of the 7th International Workshop on Advanced Motion Control (Cat. No.02TH8623), Maribor, Slovenia, 3–5 July 2002; pp. 1–14. [CrossRef]
13. Wang, F.; Chen, H.; Guo, L.; Hu, Y. Predictive safety control for road vehicles after a tire blowout. *Sci. China Inf. Sci.* **2018**, *61*, 1–3. [CrossRef]
14. Yuan, L.; Zhao, H.; Chen, H.; Ren, B. Nonlinear MPC-based slip control for electric vehicles with vehicle safety constraints. *Mechatronics* **2016**, *38*, 1–15. [CrossRef]
15. Tavernini, D.; Metzler, M.; Gruber, P.; Sorniotti, A. Explicit Nonlinear Model Predictive Control for Electric Vehicle Traction Control. *IEEE Trans. Control Syst. Technol.* **2019**, *27*, 1438–1451. [CrossRef]
16. Chen, Q.; Kang, S.; Chen, H.; Liu, Y.; Bai, J. Acceleration slip regulation of distributed driving electric vehicle based on road identification. *IEEE Access* **2020**, *8*, 144585–144591. [CrossRef]
17. Guo, L.; Xu, H.; Zou, J. Acceleration slip regulation control strategy for four-wheel independent drive electric vehicles. *IEEJ Trans. Electr. Electron. Eng.* **2019**, *14*, 630–639. [CrossRef]
18. Peng, J.; He, H.; Feng, N. Simulation research on an electric vehicle chassis system based on a collaborative control system. *Energies* **2013**, *6*, 312–328. [CrossRef]
19. Burckhardt, M. *Fahrwerktechnik: Radschlupf-Regelsysteme: Reifenverhalten, Zeitablaeufe, Messung des Drehzustands der Raeder, Anti-Blockier-System (ABS), Theorie Hydraulikkreislaeufe, Antriebs-Schlupf-Regelung (ASR), Theorie Hydraulikkreislaeufe, elektronische Regeleinheiten, Leistungsgrenzen, ausgefuehrte Anti-Blockier-Systeme und Antriebs-Schlupf-Regelungen*; Vogel Business Media: Würzburg, Germany, 1993; Available online: https://www.zvab.com/9783802304774/Fahrwerktechnik-Radschlupf-Regelsysteme-Burckhardt-Manfred-3802304772/plp (accessed on 29 September 2022).
20. Jalali, K. Stability control of electric vehicles with in-wheel motors. *Mech. Syst. Signal Process.* **2010**, *118*, 340–359.

World Electric Vehicle Journal

MDPI

Review

A Review of Position Sensorless Compound Control for PMSM Drives

Yong Li [1,*], Han Hu [1] and Peicheng Shi [2,*]

1 Automotive Engineering Research Institute, Jiangsu University, Zhenjiang 212013, China
2 Automotive New Technique of Anhui Province Engineering Technology Research Center, Anhui Polytechnic University, Wuhu 241000, China
* Correspondence: liyong@ujs.edu.cn (Y.L.); shipeicheng@ahpu.edu.cn (P.S.)

Abstract: As position sensorless control technology can avoid many disadvantages caused by mechanical position sensors, improve the reliability of the motor, reduce costs and other advantages, a large number of researchers have conducted research on compound control technology in order to achieve position sensorless control technology in a wide speed range. In this article, the position sensorless compound control technology of a permanent magnet synchronous motor is reviewed, and the compound control technology of a permanent magnet synchronous motor without a position sensor is elaborated. Finally, the existing problems and development trend of sensorless compound control technology are summarized and prospected.

Keywords: permanent magnet synchronous machine (PMSM); position sensorless compound control; high frequency (HF) signal injection method; I/F control; model-based techniques

Citation: Li, Y.; Hu, H.; Shi, P. A Review of Position Sensorless Compound Control for PMSM Drives. *World Electr. Veh. J.* **2023**, *14*, 34. https://doi.org/10.3390/wevj14020034

Academic Editors: Joeri Van Mierlo and Walter Lhomme

Received: 16 December 2022
Revised: 4 January 2023
Accepted: 19 January 2023
Published: 30 January 2023

1. Introduction

In recent years, with the shortage of energy and the deterioration of the environment, energy conservation and environmental protection have become the main theme of the development of automobiles in the future [1–5]. Permanent magnet synchronous motors (PMSMs) have been continuously improved in the field of automotive motor applications due to their high-power density, large torque inertia ratio and fast dynamic response speed [6]. To achieve high performance control of the motors, the current vector and the rotor position must be synchronized. Therefore, the exact position of the rotor needs to be obtained in real time [7]. Many mechanical detection devices, such as an optical encoder and a rotary encoder, are usually installed on the motor to detect the rotor position [8]. However, the traditional motor rotor position observer occupies a certain size, which is not conducive to the installation of the motor and can also cause the motor to go out of control due to sensor faults working in harsh operating environments [9,10]. Thus, position sensorless control technology plays a key role in the field of PMSM research [11–13].

The research on position sensorless control technology for PMSM is gradually maturing and is currently focused on two main aspects of position sensorless control technology for zero- and low-speed and for medium- and high-speed operation [14,15]. As shown in Figure 1 and Table 1, the former relies on the saliency of the motor, including high-frequency (HF) signal injection [16–23], etc.; the latter generally relies on the mathematical model of the motor and is usually divided into two categories: open-loop algorithms, including direct calculation [24–27], the back electromotive force (BEMF) integration method [28–30], flux estimation [31,32], etc., and closed-loop algorithms, including the model reference adaptation system (MRAS) [33–38], sliding mode observer (SMO) [39–42], etc.

While the current position sensorless control technology at zero–low speed and medium–high speed is now well established, it is still a weak point in the full-speed domain, which appeals to numerous researchers for improvement. In the zero–low-speed range, saliency-based position sensorless control techniques are able to estimate the motor speed and position signal very well; similarly, in the medium- and high-speed range, model-based control algorithms have very good performance. As a result, most of the current position sensorless control techniques in the full-speed range are compound control technology, which combine control techniques at zero–low speed with control techniques at medium and high speed, and use switching algorithms to achieve a smooth switch between the two algorithms in the over-range. The block diagram of the compound control structure in the full-speed domain is generally shown in Figure 2.

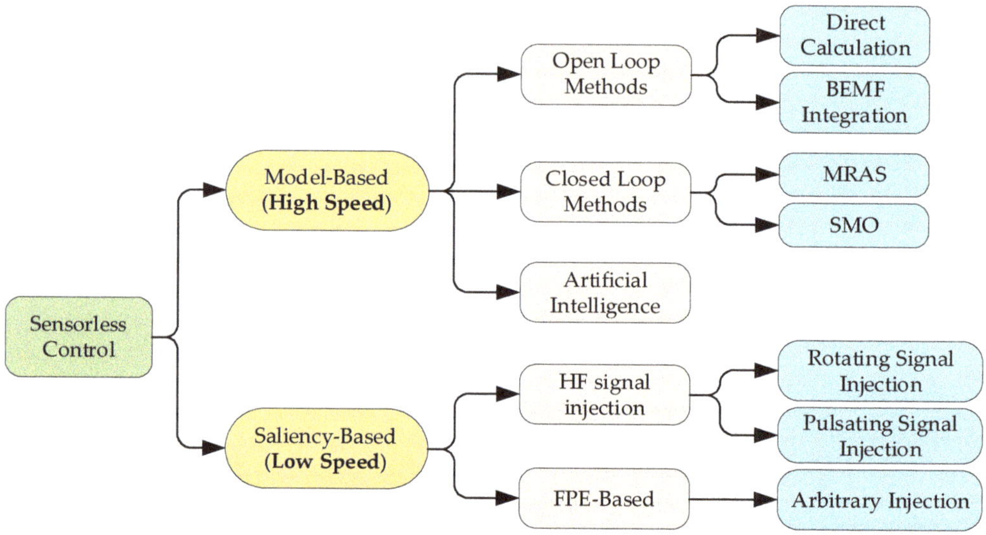

Figure 1. Categories of PMSM sensorless control methods.

Table 1. Position sensorless control algorithms.

Algorithms	Reference	Description
high frequency (HF) injection	[16–23]	The HF injection method is not reliant on the spatial protrusion of the tracking rotor rather than the mathematical equation of the motor, which addresses the sensitivity to the change in motor parameters and leads to a strong robustness. Yet the filter is needed, which has the defects of low signal-to-noise ratio and large phase lag in signal processing.
including direct calculation	[24–27]	This method does not depend on the speed of the motor, but it needs to increase the integral circuit and increase the hardware complexity and may bring additional integral error.
back-electromotive force (EMF)	[28–30]	The realization is simple, but the back EMF signal is small when the motor is low speed or static. The back EMF needs to be filtered, which will cause phase shift of the signal.
flux estimation	[31,32]	The rotor flux of the motor cannot be detected directly. It is necessary to measure the phase voltage and current of the motor, and to establish the function equation, which is directly related to the rotor flux without relying on the rotor speed. The calculation is large.
model reference adaptation system (MRAS)	[33–38]	The position observation is based on the accuracy of the reference model, and the accuracy of the parameters of the reference model itself directly affect the effectiveness of the identification.
sliding mode observer (SMO)	[39–42]	It can solve the problem that the motor is difficult to control at high speed and heavy load, and has strong robustness, but it needs a large amount of operation.

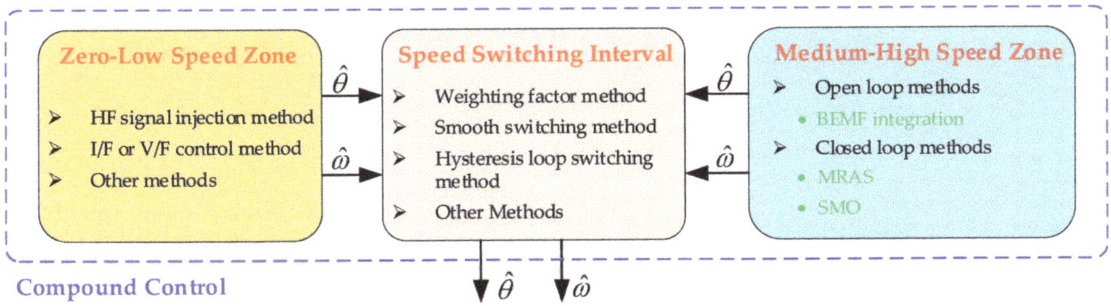

Figure 2. The block diagram of the compound control structure in full speed.

In addition to the conventional compound control algorithms above, there are also many special control algorithms in the full-speed range, such as compound control based on three-stage control [43–45] and compound control based on arbitrary injection [46,47].

The contribution of this paper is a classification of state-of-the-art compound position estimation methods and advances in sensorless control. The paper provides a complete overview of the advantages and disadvantages of HF-based, as well as IF-based, compound estimation techniques, which provides an effective guide for researchers working in this field.

The rest of this review paper is organized as follows. Section 2 introduces sensorless position estimate strategy for compound control based on the high-frequency injection method and Section 3 for compound control based on I/F control. Section 4 introduces the switching strategy in the full-speed range. Conclusions and future trends are drawn in Section 5.

2. Compound Control Based on High-Frequency Injection Method

High-frequency signal injection methods are generally divided into two categories. One is the high-frequency voltage injection method, and the other is the high-frequency current injection method [17]. As the high-frequency current injection method is difficult to control due to the high requirements of the current regulator used, the high-frequency voltage signal injection method is more commonly used. According to the different injection signals, high-frequency voltage injection methods are divided into two categories. One is the high-frequency rotary voltage injection, the other is the high-frequency pulse voltage injection. The former enables sensorless control of saliency-pole motors, and the latter is suitable for sensorless control of both saliency-pole and hidden-pole motors. A large number of researchers have made many improvements to traditional signal injection [18–23]. Therefore, the current system of high-frequency signal injection methods has been formed. Since this paper focuses on compound control, instead of giving a detailed account of each high-frequency signal injection method, two common high-frequency voltage signal injection methods have been selected for introduction.

The principle of the high-frequency voltage signal injection method is that when the motor is at standstill or running at low speed, the injection of a high-frequency voltage signal into the winding, due to the saliency of the motor, causes the feedback high-frequency signal to carry rotor position information.

1. High-frequency rotating injection method:

In general terms, the basic principle of this method is to add a high-frequency voltage excitation signal to the static coordinate system of the motor and to use the saliency of motors' own structural convex pole or saturated convex pole effect to produce a high-frequency current response. This current response contains both positive- and negative-phase sequence components, and only the phase of the negative-phase sequence high-frequency current component contains rotor position information. Therefore, appropriate signal processing techniques are required to extract rotor speed and position information by constructing suitable observer methods or the phase-locked loop method. The block diagram of the system structure is shown in Figure 3. The injected high-frequency voltage signal can be expressed as [8]

$$\begin{bmatrix} u_{\alpha h} \\ u_{\beta h} \end{bmatrix} = \begin{bmatrix} -U_{\mathrm{h}} \sin(\omega_{\mathrm{h}} t) \\ U_{\mathrm{h}} \cos(\omega_{\mathrm{h}} t) \end{bmatrix} \quad (1)$$

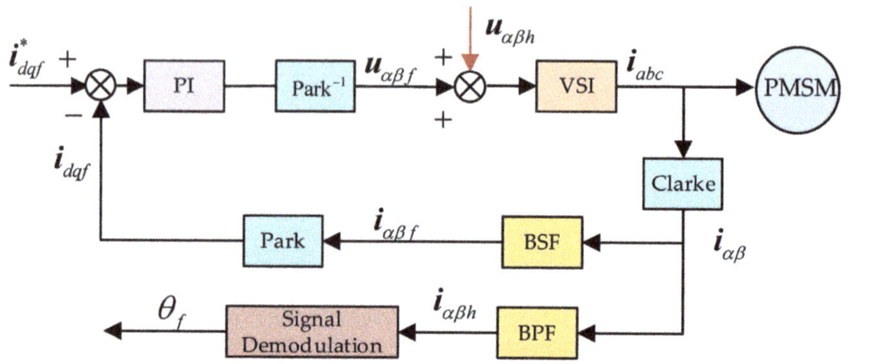

Figure 3. The control block diagram of the HF rotating injection method.

The high-frequency rotating injection method is insensitive to the uncertainty of motor parameters. However, its signal demodulation process is relatively complicated, the use of multiple filters deteriorates the dynamics of the system, and it is susceptible to multiple uncertainties leading to poor position-estimation accuracy.

2. High-frequency pulsating injection method:

The basic principle of this method is similar to that of the rotating high-frequency voltage signal method, except that the pulsating high-frequency voltage injection method only injects a high-frequency sinusoidal voltage signal into the d-axis of the estimated synchronous rotating coordinate system and produces a pulsating voltage vector in space. Figure 4a establishes the relationship between the estimated rotor and the actual rotor synchronous rotation coordinate system. In addition, in terms of the signal processing method for the feedback current, the pulsating high-frequency voltage injection method needs to multiply the high-frequency current component with the high-frequency sinusoidal signal for amplitude modulation, and then perform filtering and rotor position observation. The block diagram of the system structure is shown in Figure 4b. The injected high-frequency voltage signal can be expressed as [8].

$$\begin{bmatrix} \hat{u}_{\mathrm{dh}} \\ \hat{u}_{\mathrm{qh}} \end{bmatrix} = \begin{bmatrix} U_{\mathrm{h}} \cos(\omega_{\mathrm{h}} t) \\ 0 \end{bmatrix} \quad (2)$$

Compared with the high-frequency rotating injection method, the high-frequency pulsating signal injection method does not require the salient polarity of the motor and has the advantages of reliability, with inverter non-linearity and high accuracy in position identification. However, it has the problems of a long convergence time, poor dynamic performance and small stability range [19,20].

This section then provides an in-depth analysis of compound control based on the two widely used high-frequency signal injection methods mentioned above.

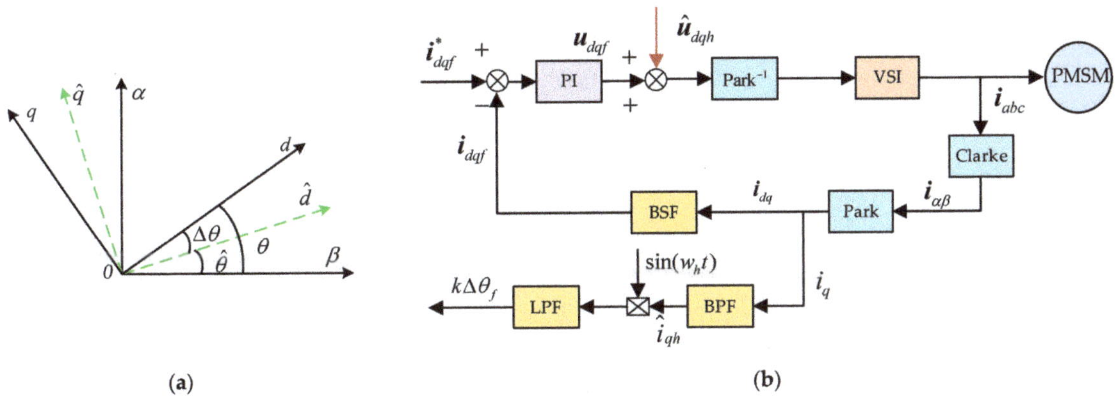

Figure 4. (**a**) The relationship between the estimated rotor and the actual rotor synchronous rotation coordinate system; (**b**) The control block diagram of the HF pulsating injection method.

2.1. HF Compounded with Model Reference Adaptive Method

The model reference adaptive system (MRAS) is a common method for estimating rotor positions based on the fundamental wave model method [33,34]. The basic principle of MRAS is to use the mathematical equations containing the parameters to be estimated as the adjustable model and the PMSM itself, which does not contain the unknown parameters, as the reference model. The output error of the two models is used to achieve the tracking of the adjustable model to the reference model by designing a suitable adaptive law to achieve the estimation of the rotor position and speed [35–37]. The basic structure is shown in Figure 5.

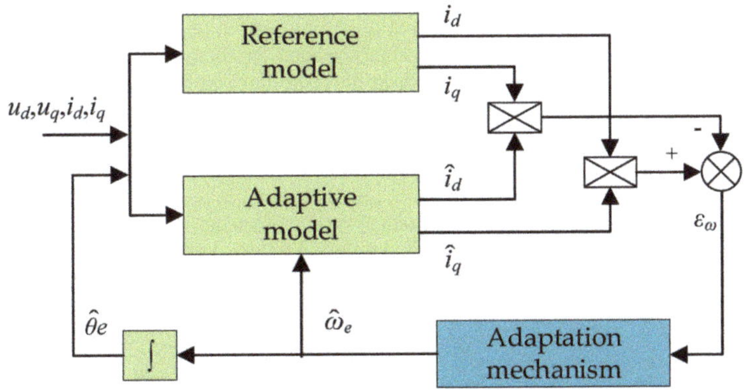

Figure 5. Block diagram of parallel structure MRAS.

In [48,49], Qin and Xu combined HF signal injection with MRAS in order to obtain rotor position estimates in the full-speed range. Rotor position and speed signals are obtained by the HF signal injection method at zero–low speeds, and by the model reference adaption method at medium and high speeds. The switching of the two algorithms is achieved by a switching method with weighting factors, which is described in detail separately in Section 4. In [50], the error term constructed using the HF pulse injection method was applied to correct the model reference adaptive observer, and the rapid dynamic response performance of the adaptive observer was combined with the steady-state accuracy of the HF injection method. The framework diagram of the compound system is shown in Figure 6.

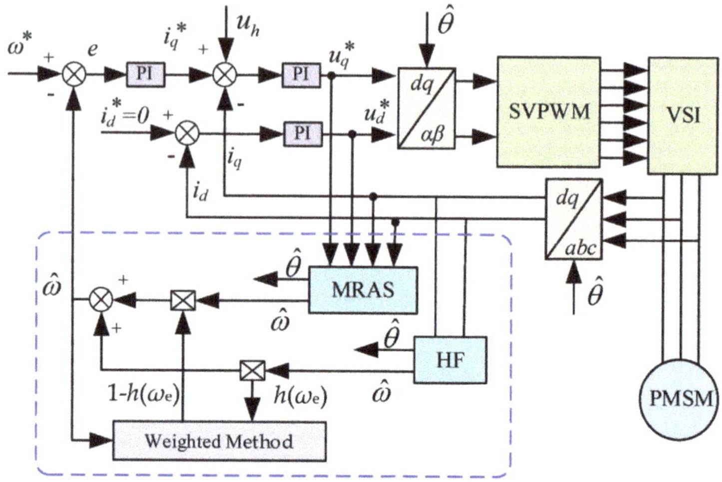

Figure 6. Block diagram of a compound control structure based on MRAS and HF.

The MRAS has the advantages of parameter adaption, structural simplicity and good steady-state performance [51]. As a result, it is often used in combination with HF for position estimation in the full-speed range. However, the performance of this algorithm depends on the selection of the reference model and the design of the adaptive law, which directly affect the stability and robustness of the algorithm, as well as the accuracy of the estimation. Therefore, the design of the adaptive law has been a problem that needs to be studied in depth in this method in order to obtain superior performance in the full-speed range [35]. In order to improve the robustness of this algorithm, the literature [36] proposes to combine the sliding mode algorithm with MRAS to improve the robustness of the system. In [52], Li introduced ADRC into the adaptive law design of MRAS to improve the estimation accuracy and increase the robustness of the system.

2.2. HF Compounded with Sliding Mode Observer

The Sliding Mode Observer (SMO) algorithm uses a switching characteristic that makes the structure of the system change over time, allowing the control to take on a discontinuous character [53,54]. The basic principle of the method is to build the SMO from a mathematical model of a permanent magnet synchronous motor and to design the sliding surface from the estimated error between the observed and actual currents [55,56]. By measuring the estimation error of the current, the back electromotive force is reconstructed and the rotor position and speed information is estimated using the back electromotive force. The basic structure is shown in Figure 7.

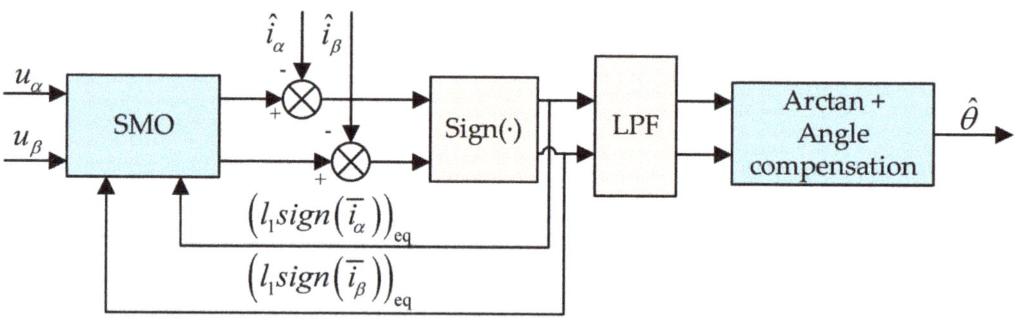

Figure 7. Block diagram of SMO.

In [57,58], the HF signal injection method was combined with SMO to determine the rotor position. The driver not only adopted a nonlinear adaptive SMO for the estimation of the rotor speed, but also conducted Lyapunov stability analysis to improve the low-speed and stationary performance of the drive. In [58], the running speed range of the motor was split into three sections, including a low-speed zone, transition zone and medium high-speed zone, so as to realize the smooth switching of the control method of the whole speed range of the motor, while verifying the sudden addition load and the discharge load of the transition area. In [59], a nonlinear sliding mode speed regulation scheme was proposed for IPMSM to be combined with the maximum torque amperometry trajectory. The global asymptotic stability of the controller and observer was ensured by Lyapunov stability analysis. As for IPMSM, Wang proposed a hybrid observation method based on a combination of the location error information of high-frequency signal injection and the anti-electric potential model method. The HF voltage signal of pulse vibration was injected during low-speed operation, and the medium–high-speed operation obtained the information on the position error through the anti-electric potential model SMO, thus normalizing the position error signal captured by the two methods and merging the information in a weighted way [60]. The scheme of the hybrid methods adopted in these papers is shown in Figure 8. Aiming at the compound control method based on rotational speed or position information fusion, it shows the disadvantages of high operation complexity and the difficulty in realization.

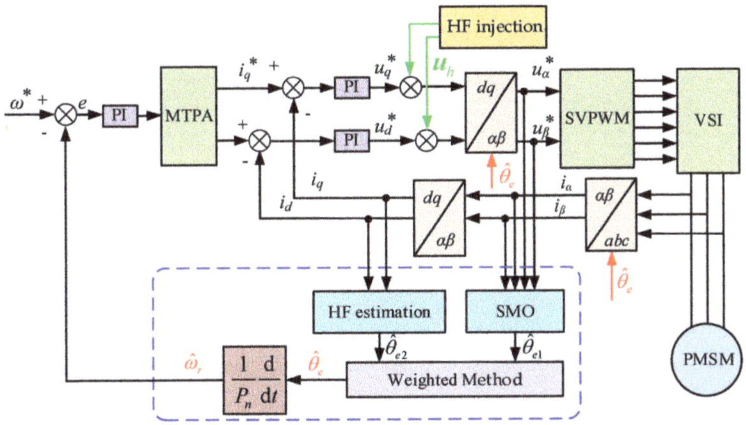

Figure 8. The block diagram of the rotor position hybrid observer based on HF and SMO.

2.3. HF Compounded with BEMF Integration Method or Flux Estimation

The BEMF integration method, or flux estimation, is an open-loop algorithm, which is different from the two-position sensorless algorithms already described above at medium and high speeds [61,62]. The key point of this algorithm is to obtain an accurate rotor magnetic chain vector and to use it to derive rotor position information. The stator voltage equation and the stator-rotor magnetic chain relationship in the α-β coordinate system of a permanent magnet synchronous motor are used for integration and inverse trigonometric operations to find the rotor position angle and speed [63].

$$\psi_f = \boldsymbol{\psi}_s - \boldsymbol{L}_{\alpha\beta}\boldsymbol{I}_s = \int(\boldsymbol{U}_s - R_s\boldsymbol{I}_s)\mathrm{d}t - \boldsymbol{L}_{\alpha\beta}\boldsymbol{I}_s \tag{3}$$

$$\theta_e = \arctan\frac{\boldsymbol{\psi}_{f\alpha}}{\psi_{f\beta}} \tag{4}$$

where $\boldsymbol{\psi}_f$ is the permanent magnet chain vector; $\boldsymbol{\psi}_s$ is the stator synthetic magnetic chain vector; $\boldsymbol{U}_s$ is the stator voltage vector; $\boldsymbol{I}_s$ is the stator current vector; $\boldsymbol{L}_{\alpha\beta}$ is the inductance matrix of the motor in the α-β coordinate system; and $\boldsymbol{\psi}_{f\alpha}$ and $\boldsymbol{\psi}_{f\beta}$ are the components of the rotor magnetic chain on the α-β axis, respectively. The basic structure is shown in Figure 9.

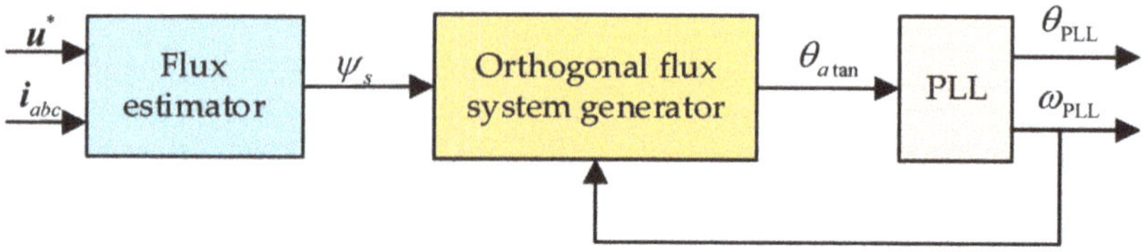

Figure 9. The block diagram of Flux Estimation.

In [64,65], a hybrid structure integrating flux observer and signal injection technology was proposed to make the rotor position signal independent of motor parameters at low and zero speed. In [66], a single Luenberger position observer compound control method based on the integration of standardized position error information was suggested. The standard position error signals were captured by square-wave voltage injection and the back-EMF model, respectively, at different speeds, and the information on the weighted fusion of the standardized position error signals in the transition region was collected through velocity information. In [67], when the motor remained up and running at low speeds, the real-time three-phase inductance of the motor was obtained by the rotating HF injection method, and the stator flux was obtained by combining the phase current. When the motor was running at high speeds, the back-EMF filter was applied to estimate the stator flux, according to which the DTC method based on the torque angle was combined, the motor operation was controlled, and the hysteresis method was used to ensure the smooth transition. In [68], through a combination of the EEMF and the HF signal injection method, the amplitude of the signal current was adjusted to maintain sufficient EEMF amplitude for mitigating interference. As the lower bound of the EEMF could be adjusted according to the degree of interference, signal setting was made easier. The scheme of the hybrid methods adopted in these papers is similar to Figure 8.

3. Compound Control Algorithm Based on I/F Control

I/F control is a frequency conversion speed control process to keep the current stable through the current closed loop, avoiding too much or too little current []. I/F control can directly control the torque current, which improves the ability to match the motor output torque with the load torque and can avoid low-frequency oscillation during motor operation. V/F control is similar to I/F control in that it is also a variable frequency speed control strategy, but in contrast to I/F control, V/F control is open loop for both speed and current, whereas I/F control is open loop for speed and closed loop for current. The schematic diagram for starting and running a permanent magnet synchronous motor via I/F control is shown in Figure 10.

At the initial start, the virtual synchronous coordinate system lags behind the synchronous coordinate system by a $\pi/2$ electrical angle. As the virtual synchronous coordinate system rotates, the motor rotor also starts to rotate with the virtual synchronous coordinate system. The motor output electromagnetic torque is determined by the phase difference between the two coordinate systems, and the motor output electromagnetic torque is shown as

$$T_e = \frac{3}{2} n_p \psi_f i_q^* \cos\theta_L \tag{5}$$

The equation for the torque balance of the motor during acceleration is

$$T_e - T_L = \frac{J}{n_p} \cdot \frac{d\omega_{re}}{dt} = \frac{J}{n_p} \cdot \frac{d^2\theta_{re}}{dt} \tag{6}$$

where i_q* is the given current; J is the rotational inertia; T_L is the load torque; and ω_{re} is the rotor electric angular velocity.

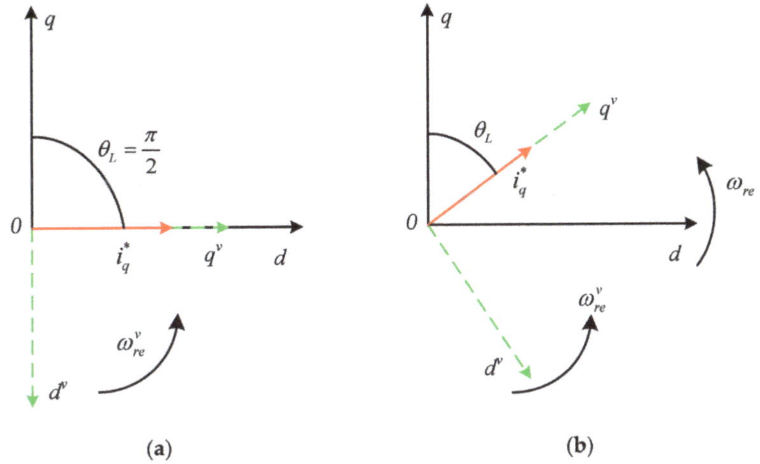

Figure 10. The phase relation of virtual synchronous frame and rotor synchronous frame in IF control. (**a**) Start state; (**b**) Running state.

3.1. I/F Compounded with Sliding Mode Observer

In [69,70], the I/F control method combined with a SMO composite control strategy was presented. In order to analyze the phase relationship of the virtual synchronous coordinate system and the rotor synchronous coordinate system used in the I/F control, the I/F control strategy of rotational speed open loop and current closed loop was adopted in the low-speed region of the motor. The adaptive SMO estimated the rotor flux by introducing the electric angular velocity. Based on I/F open-loop control theory, single-current closed-loop I/F control was achieved based on instantaneous reactive power, thus improving the current utilization [71]. The block diagram of the rotor position compound control methods based on I/F and SMO is shown in Figure 11.

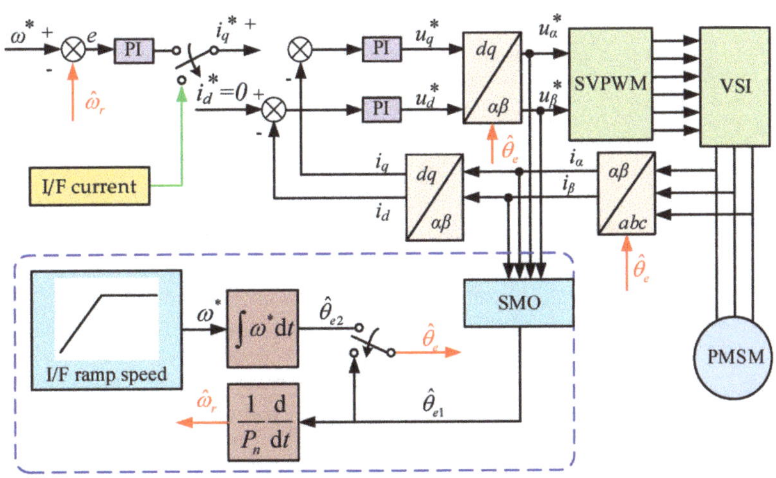

Figure 11. The block diagram of the rotor position compound control methods based on I/F and SMO.

3.2. I/F Compounded with Flux Obsever

In [72], a hybrid sensorless control strategy integrated with an improved flux linkage observer and the current-frequency (I-F) starting method was proposed. The I/F control method was utilized for the stable startup and strong antijamming capability, and the improved flux linkage observer based on the sliding-mode compensator is designed for the closed-loop sensorless operation. An adaptive transition algorithm was designed in order to achieve smooth operation between the two different control schemes.

4. Switching Methods

From the above sections, it can be found that the current compound control methods mainly use one or more zero–low-speed control methods combined with one or more methods of position sensorless control at medium and high speeds to achieve the full-speed range of the compound control technology, so how to achieve a smooth transition and switching between the two speed detection methods is an extremely important part of whether the compound control can achieve reliable operation [73]. The switching control algorithm between the two control methods is also the focus of this paper.

4.1. Weighting Factor Method

The basic principle of the weighted coefficient switching method is that when the rotor speed is above the upper limit of the switching interval, the medium–high speed control method is used for control; when the speed is below the lower limit of the switching interval, the position sensorless control method at zero–low speed is used for control; and when the estimated speed is within the switching interval, the weighted value of the results of the two algorithms is used to ensure smooth switching of the two methods. At the same time, the lower speed limit in the switching zone should be higher than the minimum speed at which the control algorithm can operate at medium and high speeds, while the upper speed limit in the switching zone should be lower than the maximum speed at which the control algorithm can start itself at zero–low speed. To ensure that there are no jumps in position and speed signals in the switching zone, the two methods are required to have essentially the same speed and position errors in the switching zone [5,60]. The block diagram of the system structure is shown in Figure 12. The estimation equation for the rotor speed is

$$\hat{n} = W_{\mathrm{h}}\hat{n}_{\mathrm{h}} + (1 - W_{\mathrm{h}})\hat{n}_{\mathrm{m}} \tag{7}$$

where $\hat{n}$ is the estimated speed of the compound; are the lower and upper limits of the speed switching interval respectively; and a and b are estimated speeds at zero–low speed and medium–high speed, respectively. The weighting factor W_{h} is

$$W_{\mathrm{h}} = \begin{cases} 1, & \hat{n} \leqslant n_1 \\ \frac{n-n_2}{n_1-n_2}, & \hat{n}_1 < n < n_2, \\ 0, & \hat{n} \geqslant n_2. \end{cases} \tag{8}$$

Although this weighted coefficient method has a simple structure and is easy to implement, in practical applications, in order to achieve reliable switching during speed changes, both algorithms need to run all the time, resulting in a large amount of wasted hardware and software resources, and also causing an impact on the system control performance due to the constant injection of high-frequency signals. In [74], Zhao proposed an improvement scheme: under the premise of ensuring the upper and lower limits of the switching interval remain unchanged, based on the basis of the variable weighted switching algorithm, the working interval of the two algorithms is reduced, while providing sufficient margin for the convergence of the algorithms, and the speed and rotor position identified by the other algorithm in real time when each algorithm starts working is used as the initial value for the identification of the new working algorithm in order to enable the algorithm that suddenly switches to the working state to quickly converge to a stable value. The lower limit of the switching interval for both algorithms is set to 100 r/min, and the upper limit is set to 200 r/min. On top of this, the speed of the MRAS method is additionally set to 50 r/min at the start of the speed rise phase, and the speed of the pulse vibration injection method is set to 300 r/min at the start of the operation. The operation of the improved switching algorithm in each speed interval is shown in Figure 13.

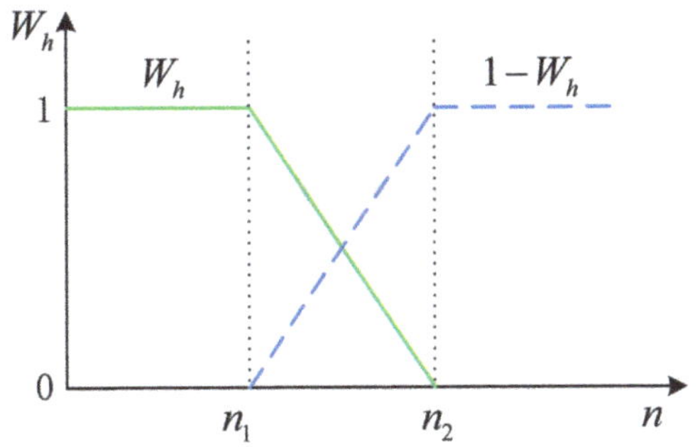

Figure 12. The block diagram of the system structure for weighting factor method.

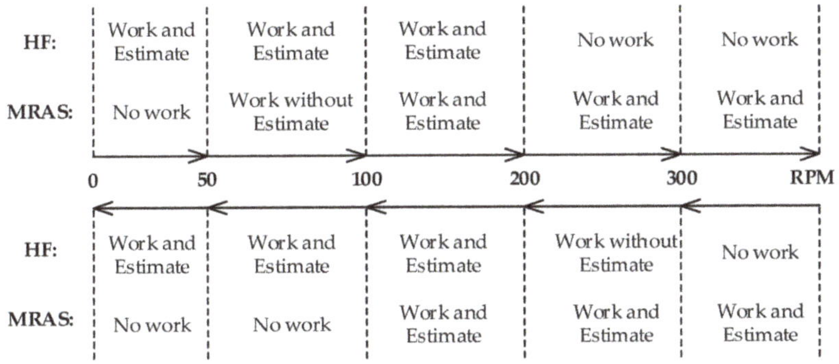

Figure 13. Operations of high-frequency pulse injection method and MRAS in each speed interval after improvement.

4.2. Smooth Switching Method

The smooth switching algorithm is mostly used in the control strategy of I/F start-up. The I/F control uses a virtual synchronous coordinate system, which has a phase difference, θ_L, between it and the rotor synchronous coordinate system [69,70]. The process of smooth switching control strategy is actually the process of adjusting the phase difference θ_L from an acute angle to close to zero. The phase angle difference θ_L during I/F control preparation for switching cannot be equal to zero, otherwise it will cause the motor to miss-step [71,72,75]. The ideal control strategy is shown in Figure 14.

The traditional smooth switching strategy is actually a strategy of smooth switching by gradually adjusting the given angle of the virtual synchronous coordinate system in the IF control. However, due to the "torque-work angle self-balancing" principle, the virtual synchronous coordinate system can never "catch up" with the rotor synchronous coordinate system. In order to solve this problem, in [8], Liu proposed a smooth switching strategy by adjusting the amplitude of the virtual q^v-axis current given indirectly to adjust the phase difference θ between the two coordinate systems at heavy load, and by adjusting the virtual q^v-axis and virtual d-axis current given simultaneously at no load or light load. This improved switching strategy has a wider range of application, and at the same time, in this improved switching strategy, the synthetic current vectors of the q^v and d^v axes have a larger phase difference with the q-axis of the rotor synchronous coordinate system, such that this switching strategy is more resistant to disturbances and has a lower risk of motor miss-steps. The improved switching strategy is shown in Figure 15.

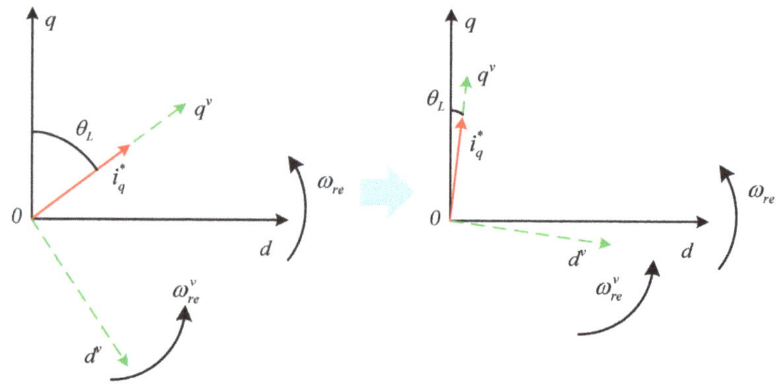

Figure 14. The block diagram of the ideal control strategy for smooth switching method.

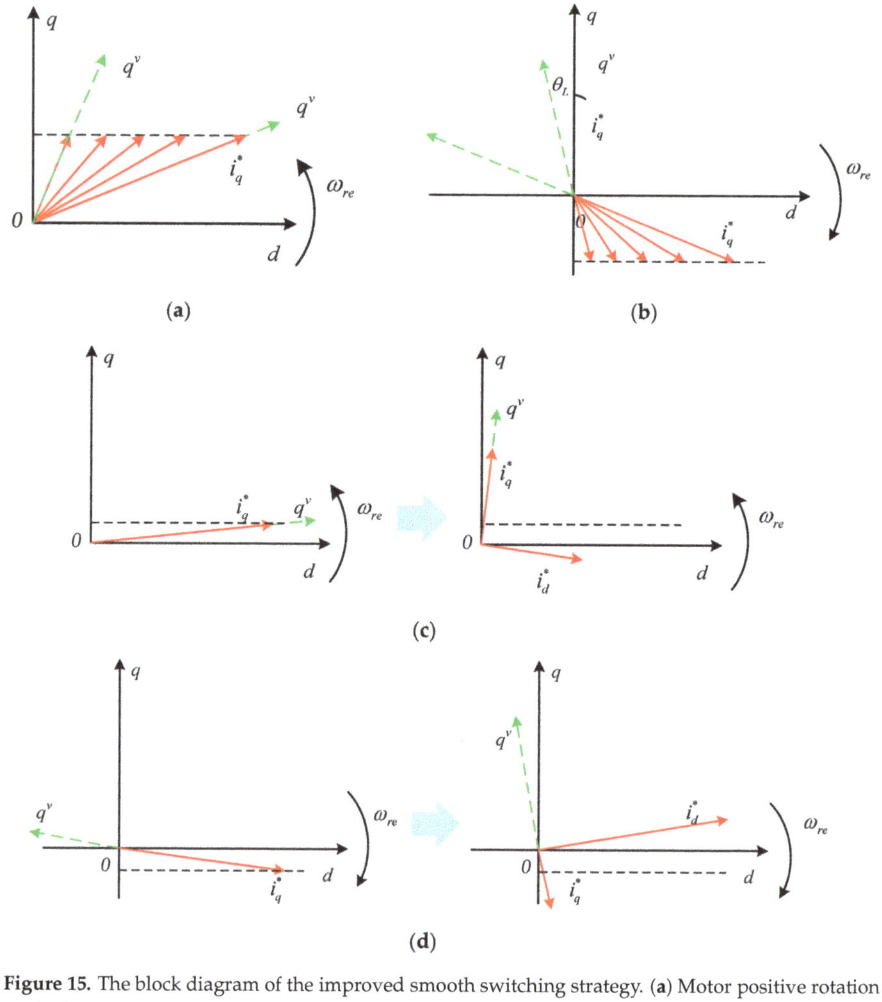

Figure 15. The block diagram of the improved smooth switching strategy. (**a**) Motor positive rotation under heavy load; (**b**) Motor negative rotation under heavy load; (**c**) Motor positive rotation under light load; (**d**) Motor negative rotation under light load.

4.3. Hysteresis Loop Switching Method

The hysteresis loop switching strategy is shown in Figure 16. When the speed rises to $|\omega_{eH}|$, the estimated position is quickly switched from the position sensorless estimation algorithm at zero–low speed to the position sensorless estimation algorithm at medium and high speed within a certain time T_{sw}. The switching process uses a smooth transition strategy as

$$\hat{\theta}_e = g_h\hat{\theta}_h + (1 - g_h)\hat{\theta}_m \tag{9}$$

$$\hat{\omega}_e = g_h\hat{\omega}_h + (1 - g_h)\hat{\omega}_m \tag{10}$$

where $\hat{\theta}_{hf}$, $\hat{\omega}_{hf}$ are the estimated position and speed obtained by the zero–low-speed sensorless estimation algorithm, respectively; and $g_{hf} = t/T_{sw}$, where $0 \le t \le T_{sw}$. The zero–low-speed sensorless estimation algorithm can be removed when all positions estimated by the medium- and high-speed sensorless estimation algorithm are used in order to reduce losses, noise, etc. To avoid the problem of switching back and forth between the two methods due to speed fluctuations, a hysteresis loop strategy is designed so that the position-free estimation algorithm at medium and high speeds is switched to the position sensorless estimation algorithm at zero–low speed, only when the speed is below $|\omega_{eL}|$. As $|\omega_{eL}|<|\omega_{eH}|$, it can avoid switching back and forth between the two methods at two switching points and improve the smoothness of speed operation at the switching point [76,77].

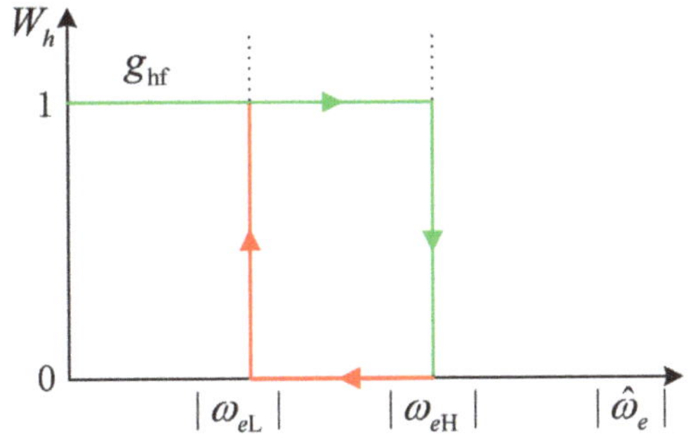

Figure 16. The block diagram of Hysteresis loop switching method.

5. Conclusions

This article reviews the two main types of state-of-the-art compound control based on high-frequency signal injection, and compound control based on IF algorithms without position sensors and their switching algorithms. The advantages and disadvantages of compound control combined with MRAS, SMO, BEMF, et al. are presented under compound control based on the HF signal injection method and I/F control, and the position estimation process is illustrated with examples. In addition, the advantages and disadvantages of the various switching algorithms in the compound control algorithm are presented. The following points can be drawn from the content of the sections set out above:

1. The high-frequency signal injection method has many advantages, but there are also problems of long convergence time, poor dynamic performance and small stability range. For compound control, the advantages of the high-frequency signal injection method inherit many disadvantages, which can have more or less impact on the estimation accuracy and other aspects in the compound control process. Therefore more advanced high-frequency signal injection methods should be studied to improve the reliability of the dynamic range of motor compound control estimation.
2. I/F control is simple in structure, easy to implement, and has the advantages of smooth start-up and no current overcharge. It is the current start-up strategy for most position sensor-free control at medium and high speeds. However, the basic

I/f control strategy is an open-loop scheme with disadvantages, such as the current amplitude and frequency cannot be automatically adjusted, and there is a tendency to lose steps, and the speed is easily disturbed, so its improvement will largely improve the performance of the compound control.

3. There is a wide variety of control strategies for position sensorless control at high speeds, each with its own advantages, but also some shortcomings. So, the improvement of such algorithms will greatly improve the performance of the current compound control.
4. Switching algorithms play a pivotal role in compound control. In the current research field of compound control, whether based on HF control or I/F control, switching strategies are required. Most of the switching algorithms commonly used today have the advantages of, for example, simpler methods and relatively stable algorithm switching, but their switching smoothness is not ideal, which means that the existing switching algorithms should be improved to achieve smoother and more stable switching, or smoother and more stable switching algorithms should be developed to meet the current high demand for motor position estimation.

Author Contributions: Conceptualization, Y.L. and H.H.; methodology, Y.L. and P.S.; investigation, H.H.; data curation, H.H.; writing—original draft preparation, Y.L. and H.H.; writing—review and editing, Y.L. and P.S.; supervision, Y.L.; project administration, Y.L.; funding acquisition, Y.L. and P.S. All authors have read and agreed to the published version of the manuscript.

Funding: This work is supported by the National Natural Science Foundation of China (Grant No. 51705213), supported by the Opening Project of Automotive New Technique of Anhui Province Engineering Technology Research Center (Grant No. QCKJ202201A) and supported by the Opening Project of National Engineering Laboratory of Energy-saving Motor & Control Technique of Anhui University (Grant No. KFKT202215).

Data Availability Statement: Not applicable.

Conflicts of Interest: The authors declare no conflict of interest.

References

1. Li, Y.; Wu, H.; Xu, X.; Sun, X.; Zhao, J. Rotor Position Estimation Approaches for Sensorless Control of Permanent Magnet Traction Motor in Electric Vehicles: A Review. *World Electr. Veh. J.* **2021**, *12*, 9. [CrossRef]
2. Jiang, H.B.; Li, A.X.; Ma, S.D.; Chen, L. Design and performance analysis of airflow energy recovery device of electric vehicle. *J. Jiangsu Univ.* **2017**, *38*, 125–132.
3. Liu, C.H.; Luo, Y.X. Overview of Advanced Control Strategies for Electric Machines. *Chin. J. Electr. Eng.* **2017**, *3*, 53–61.
4. Li, S.Q.; Ding, X.M.; Yu, B. Optimal control strategy of efficiency for dual motor coupling drive system of pure electric vehicle. *J. Jiangsu Univ.* **2022**, *43*, 1–7.
5. Zhang, H.; Liu, W.G.; Chen, Z.; Mao, S.; Meng, T.; Peng, J.C.; Jiao, N.F. A Time-Delay Compensation Method for IPMSM Hybrid Sensorless Drives in Rail Transit Applications. *IEEE Trans. Ind. Electron.* **2019**, *66*, 6715–6726. [CrossRef]
6. Li, Y.; Wu, H.; Zhang, B.H. Frontier techniques and prospect of in-wheel motor for electric vehicle. *J. Jiangsu Univ.* **2019**, *40*, 261–268.
7. Xue, H.T.; Zhou, Y.; Wang, M.; Li, Z.X. Fault diagnosis method for in-wheel motor based on wolf pack algorithm. *J. Jiangsu Univ.* **2019**, *40*, 579–584.
8. Liu, J.L.; Xiao, F.; Shen, Y.; Mai, Z.Q.; Li, C.R. Position-Sensorless Control Technology of Permanent-Magnet Synchronous Motor-a Review. *Trans. China Electrotech. Soc.* **2017**, *32*, 76–88.
9. Li, J.M.; He, R. Optimization design and performance analysis of dual-rotor in-wheel motor based on parameter sensitivity. *J. Jiangsu Univ.* **2020**, *41*, 640–647.
10. Sun, X.D.; Cao, J.H.; Lei, G.; Guo, Y.G.; Zhu, J.G. Speed Sensorless Control for Permanent Magnet Synchronous Motors Based on Finite Position Set. *IEEE Trans. Ind. Electr.* **2020**, *67*, 6089–6100. [CrossRef]
11. Wu, D.R.; Tian, S.P. New control strategy of motor for pure electric vehicle based on TLGI technology. *J. Jiangsu Univ.* **2021**, *42*, 9–14.
12. Guo, J.G.; Dog, H.X.; Sheng, W.H.; Tu, C. Optimum control strategy of regenerative braking energy for electric vehicle. *J. Jiangsu Univ.* **2018**, *39*, 132–138.
13. Zhang, L.; Zhu, X.Y.; Gao, J.; Mao, Y. Design and Analysis of New Five-Phase Flux-Intensifying Fault-Tolerant Interior-Permanent-Magnet Motor for Sensorless Operation. *IEEE Trans. Ind. Electron.* **2020**, *67*, 6055–6065. [CrossRef]

14. Wang, G.; Valla, M.; Solsona, J. Position sensorless permanent magnet synchronous machine drives—A review. *IEEE Trans. Ind. Electron.* **2019**, *67*, 5830–5842. [CrossRef]
15. Mohan, H.; Pathak, M.K.; Dwivedi, S.K. Sensorless control of electric drives–a technological review. *IETE Tech. Rev.* **2020**, *37*, 504–528. [CrossRef]
16. Lin, T.C.; Zhu, Z.Q. Sensorless operation capability of surface-mounted permanent-magnet machine based on high-frequency signal injection methods. *IEEE Trans. Ind. Appl.* **2015**, *51*, 2161–2171. [CrossRef]
17. Liu, Y.; Zhou, B.; Li, S.; Feng, Y. Initial Rotor Position Detection of Surface Mounted Permanent Magnet Synchronous Motor. *Proc. CSEE* **2011**, *31*, 48–54.
18. Liu, J.M.; Zhu, Z.Q. Sensorless control strategy by square aveform high-frequency pulsating signal injection into stationary reference frame. *IEEE J. Emerg. Sel. Top. Power Electron.* **2014**, *2*, 171–180. [CrossRef]
19. Zhou, T.; Jiang, Q. Overview of Sensorless Control Technology for Full Speed Range Permanent Magnet Synchronous Motors. *Electron. Sci. Technol.* **2021**, *34*, 59–69.
20. Xu, X.; Mi, J.; Wang, F.; Ma, S.D.; Tao, T. Design of differential braking control system of travel trailer based on multi-objective PID. *J. Jiangsu Univ.* **2020**, *41*, 172–180.
21. Yoon, Y.D.; Sul, S.K.; Morimoto, S.; Ide, K. High-Bandwidth Sensorless Algorithm for AC Machines Based on Square-Wave-Type Voltage Injection. *IEEE Trans. Ind. Appl.* **2011**, *47*, 1361–1370. [CrossRef]
22. Lu, J.D.; Liu, J.L.; Wei, L.C. Estimation of the Initial Rotor Position for Permanent Magnet Synchronous Motors. *Trans. China Electrotech. Soc.* **2015**, *30*, 105–111.
23. Zhou, Y.J.; Cai, M.F. Initial rotor position inspection of PMSM based on rotating high frequency voltage signal injection. *Electr. Mach. Control.* **2010**, *14*, 68–72.
24. Chen, S.H.; Liu, G.; Zhu, L.Q. Sensorless Start-Up Control Strategy for a 315 kW High-Speed Magnetic Suspension BLDC Motor with Small Inductance and Non-ideal Back-EMF. *IEEE Trans. Ind. Electron.* **2018**, *66*, 1703–1714. [CrossRef]
25. Damodharan, P.; Vasudevan, K. Sensorless Brushless DC Motor Drive Based on the Zero-Crossing Detection of Back Electromotive Force (EMF) From the Line Voltage Difference. *IEEE Trans. Energy Convers.* **2010**, *25*, 661–668. [CrossRef]
26. Zhang, G.Q.; Wang, G.L.; Xu, D.G.; Fu, Y.; Ni, R.G. Adaptive Notch Filter Based Rotor Position Estimation for Interior Permanent Magnet. *Proc. CSEE.* **2016**, *36*, 2521–2527.
27. Wang, D.F.; Zhu, Y.Q.; Jin, Y.; Liu, Z.Q. A novel research on detecting position of brushless DC motors. *Trans. China Electrotech. Soc.* **2013**, *28*, 139–144.
28. Shen, J.X.; Iwasaki, S. Sensorless control of ultrahigh-speed PM brushless motor using PLL and third-harmonic back EMF. *IEEE Trans. Ind. Electron.* **2006**, *53*, 421–428. [CrossRef]
29. Wang, D.F.; Zhu, Y.Q.; Jin, Y.; Zhao, G.Y. Tentative Strategy of Starting Sensorless BLDCM with the Method of Integrating the Back EMF. *Trans. China Electrotech. Soc.* **2012**, *27*, 178–184.
30. Chen, Z.Q.; Tomita, M.; Doki, S. An extended electromotive force model for sensorless control of interior permanent- magnet synchronous motors. *IEEE Trans. Ind. Electron.* **2003**, *50*, 288–295. [CrossRef]
31. Lee, K.G.; Lee, J.S.; Lee, K.B. Wide-range sensorless control for SPMSM using an improved full-order flux observer. *J. Power Electron.* **2015**, *15*, 721–729. [CrossRef]
32. Qiu, T.F.; Wen, X.H.; Zaho, F.; Wang, Y.X. Design Strategy of Permanent Magnet Flux Linkage Adaptive Observer for Permanent Magnet Synchronous Motor. *Proc. CSEE* **2015**, *35*, 2287–2294.
33. Zhu, Y.; Cheng, M.; Hua, W.; Zhang, B.F.; Wang, W. Sensorless control for electrical variable transmission based on sliding mode model reference adaptive system. *Trans. China Electrotech. Soc.* **2015**, *30*, 64–72.
34. Rai, R.; Shukla, S.; Singh, B. Electromagnetic Torque-Based Model Reference Adaptive System Speed Estimator for Sensorless Surface Mount Permanent Magnet Synchronous Motor Drive. *IEEE Trans. Ind. Inform.* **2020**, *16*, 4714–4725. [CrossRef]
35. Hu, W.H.; Wang, Y.; Li, M.X.; Li, M.; Wang, Z.A. Research on sensorless control strategy of direct drive multiphase PMSG wind power generation system based on MRAS. *Power Syst. Prot. Control.* **2014**, *42*, 118–124.
36. Li, Y.H.; Zhao, Y.Q. Path tracking of 4WIS autonomous vehicle based on double-layer control strategy. *J. Jiangsu Univ.* **2022**, *43*, 386–393.
37. Zhang, H.S.; Wang, P.; Han, B.C. Rotor Position Measurement for High-speed Permanent Magnet Synchronous Motors Based on Fuzzy PI MRAS. *Proc. CSEE* **2014**, *34*, 1889–1896.
38. Yao, X.L.; Ding, D.D.; Yang, Y.; Ge, W.J.; Wu, M.Y. Failure compensation control design of adaptive actuator for vehicle robot driver. *J. Jiangsu Univ.* **2021**, *42*, 642–647.
39. Xu, B.; Shen, X.K.; Ji, W.; Shi, G.D. Adaptive Nonsingular Terminal Sliding Model Control for Permanent Magnet Synchronous Motor Based on Disturbance Observer. *IEEE Access* **2018**, *6*, 48913–48920. [CrossRef]
40. Zhang, X.; Guo, L.L.; Yang, S.Y.; Cao, R.X. Speed Sensorless Control of Permanent Magnet Synchronous Generators. *Proc. CSEE* **2014**, *34*, 3440–3447.
41. Su, J.Y.; Li, T.C.; Yang, G.J. PMSM sensorless control based on four-order hybrid sliding mode observer. *Proc. CSEE* **2009**, *29*, 98–103.
42. Zhao, Y.; Qiao, W.; Wu, L. Improved Rotor Position and Speed Estimators for Sensorless Control of Interior Permanent- Magnet Synchronous Machines. *IEEE J. Emerg. Sel. Top. Power Electron.* **2014**, *2*, 627–639. [CrossRef]

43. Zhang, Z.Y.; Lin, M.Y.; Zhou, G.Q. Anti-reverse rotation startup and smoothly switching of sensorless brushless DC motor. *Trans. China Electrotech. Soc.* **2009**, *24*, 26–32.
44. Wu, C.; Qi, R.; Li, B.Q.; Ma, D.L. Whole Speed Range Sensorless Control of Permanent Magnet Synchronous Motor Considering Saturation Effect. *Trans. China Electrotech. Soc.* **2017**, *32*, 171–179.
45. Chen, S.Y.; Pi, Y.G. Position Sensorless Control for Permanent Magnet Synchronous Motor Based on Sliding Mode Observer and Sliding Mode Controller. *Trans. China Electrotech. Soc.* **2016**, *31*, 108–117.
46. Wang, Z.; Lu, K.; Blaabjerg, F. A simple startup strategy based on current regulation for back-EMF-based sensorless control of PMSM. *IEEE Trans. Power Electron.* **2012**, *27*, 3817–3825. [CrossRef]
47. Barnard, F.J.; Villet, W.T.; Kamper, M.J. Hybrid active-flux and arbitrary injection position sensorless control of reluctance synchronous machines. *IEEE Trans. Ind. Appl.* **2015**, *51*, 3899–3906. [CrossRef]
48. Qin, F.; He, Y.K.; Jia, H.P. Investigationn of the Sensorless Control for PMSM Based on a Hybird Rotor Position Self-Sensing Approach. *Proc. CSEE* **2007**, *27*, 12–17.
49. Xu, H.Z.; Xie, S.Y.; Zhang, L.S.; Wang, S.L. Investigation of hybrid sensorless control approach for dualrotor PMSM. *Electr. Mach. Control.* **2012**, *16*, 12–16.
50. Antti, P.; Marko, H. Adaptation of motor parameters in sensorless PMSM drives. *IEEE Trans. Ind. Appl.* **2009**, *45*, 203–212.
51. Jiang, H.B.; Zhu, C.; Tang, B.; Yin, C.H.; Hua, Y.F.; Xie, J. Lateral stability control of ECHBPS commercial vehicle based on extension active disturbance rejection. *J. Jiangsu Univ.* **2021**, *42*, 166–172.
52. Li, Y.; Hu, H.; Qin, Z.C.; Wu, H. Sensorless control of permanent magnet in-wheel motor based on improved model reference adaptive strategy. *J. Automot. Saf. Energy* **2022**, *13*, 560–570.
53. Yang, M.H.; Li, S.Q.; Li, Z.; Feng, B.; Li, J. Active disturbance rejection controller of direct torque for permanent magnet synchronous motor based on super-twisting sliding mode. *J. Jiangsu Univ.* **2022**, *43*, 680–684, 696.
54. Jiang, H.B.; Cao, F.G.; Zhu, W.W. Control method of intelligent vehicles cluster motion based on SMC. *J. Jiangsu Univ.* **2018**, *39*, 385–390.
55. Chen, M.S.; Hwang, Y.R.; Tomizuka, M. A state-dependent boundary layer design for sliding mode control. *IEEE Trans. Autom. Control.* **2010**, *47*, 1677–1681. [CrossRef]
56. Zhang, J.C.; Wang, L.M.; Zou, X.J.; Song, W. Optimization control strategy of driving torque for slope-crossing of pure electric vehicles. *J. Jiangsu Univ.* **2021**, *42*, 506–512.
57. Wang, G.L.; Li, Z.M.; Zhang, G.Q.; Yu, Y. Quadrature PLL-Based High-Order Sliding-Mode Observer for IPMSM Sensorless Control with Online MTPA Control Strategy. *IEEE Trans. Energy Convers.* **2013**, *28*, 214–224. [CrossRef]
58. Li, R. *Research on the Sensorless Control. Technique of Permanent Magnet Synchronous Motor*; Zhejiang University: Hangzhou, China, 2012.
59. Gilbert, F.; Rahman, M.F. Sensorless sliding-mode MTPA control of an IPM synchronous motor drive using a sliding-mode observer and HF signal injection. *IEEE Trans. Ind. Electron.* **2010**, *57*, 1270–1278.
60. Wang, G.L.; Yang, R.F.; Xu, D.G. DSP-Based Control of Sensorless IPMSM Drives for Wide-Speed-Range Operation. *IEEE Trans. Ind. Electron.* **2013**, *60*, 720–727. [CrossRef]
61. Zhou, L.Y.; Wang, S.J. Torque control of BLDCM based on predictive current control. *J. Jiangsu Univ.* **2015**, *36*, 79–86.
62. Gupta, N.; Pandey, D.A.K. *A Review: Sensorless Control of Brushless DC Motor*; Esrsa Publications: Gandhinagar, India, 2012.
63. Boldea, I.; Paicu, M.C.; Andreescu, G.D.; Blaabjerg, F. Active flux" DTFC-SVM sensorless control of IPMSM. *IEEE Trans. Energy Convers.* **2009**, *24*, 314–322. [CrossRef]
64. Silva, C.; Asher, G.M.; Sumner, M. Hybrid rotor position observer for wide speed-range sensorless PM motor drives including zero speed. *IEEE Trans. Ind. Electron.* **2006**, *53*, 373–378. [CrossRef]
65. Andreescu, G.D.; Pitic, C.I.; Blaabjerg, F.; Boldea, I. Combined Flux Observer with Signal Injection Enhancement for Wide Speed Range Sensorless Direct Torque Control of IPMSM Drives. *IEEE Trans. Energy Convers.* **2008**, *23*, 393–402. [CrossRef]
66. Zhang, G.Q.; Wang, G.L.; Xu, D.G.; Yu, Y. Hybrid Sensorless Control Based on Single Position Observer Using Error Combination for Interior Permanent Magnet Synchronous Machine Drives. *Proc. CSEE* **2017**, *37*, 6077–6082.
67. Qiu, X.; Huang, W.X.; Bu, F.F. Sensorless direct torque control of interior permanent magnet synchronous machines over wide speed range. *Trans. China Electrotech. Soc.* **2014**, *29*, 92–99.
68. Ohnuma, T.; Makaino, Y.; Saitoh, R. Adaptive signal injection method combined with EEMF based position sensorless control of IPMSM drives. *IEEJ J. Ind. Appl.* **2014**, *4*, 914–918.
69. Song, G.Y.; Li, J.L. A Sensorless Control for PMSM with combined IF and Improved Sliding Mode Observer. *Electr. Mach. Control.* **2020**, *24*, 63–72.
70. Liu, J.L.; Xiao, F.; Mai, Z.Q.; Gao, S.; Yu, X.W. Hybrid Position-Sensorless Control Scheme for PMSM Based on Combination of IF Control and Sliding Mode Observer. *Trans. China Electrotech. Soc.* **2018**, *33*, 919–929.
71. Wang, M.; Yang, J.Q.; Zhang, X.; Zhu, C.S. An I/f Control Method with Closed-loop Regulation of Current Vector for Surface Permanent Magnet Synchronous Motor Drives. *Proc. CSEE* **2015**, *35*, 2513–2521.
72. Tang, Q.P.; Chen, D.X.; He, X.N. Integration of Improved Flux Linkage Observer and I-f Starting Method for Wide-Speed-Range Sensorless SPMSM Drives. *IEEE Trans. Power Electron.* **2020**, *35*, 8374–8383. [CrossRef]
73. Tao, W.; Liu, Z.Q. Seat suspension control of wheel loader based on damping state switching. *J. Jiangsu Univ.* **2019**, *40*, 504–510.
74. Zhao, Q.J.; Liao, Z.L.; Zhang, Y.Y.; Cai, L.C. Research on Position Sensorless Control of Hub Motor in Full Speed Range. *Acta Armamentarii* **2019**, *40*, 915–926.

75. Wu, D.D.; Ge, Q.; Xu, X.F.; Qiu, B.Y. Grid-connected control strategy of virtual synchronous generator based on virtual power. *J. Jiangsu Univ.* **2022**, *43*, 458–463.
76. Panin, S.V.; Bogdanov, A.A.; Eremin, A.V.; Buslovich, D.G.; Alexenko, V.O. Estimating Low- and High-Cyclic Fatigue of Polyimide-CF-PTFE Composite through Variation of Mechanical Hysteresis Loops. *Materials* **2022**, *15*, 4656. [CrossRef]
77. Wu, C.; Fu, Z.J.; Sun, M.X.; Liu, Z. Sensorless Control of PMSM in All Speed Range Based on Extended State Observer for Load Toque Compensation. *Trans. China Electrotech. Soc.* **2020**, *35* (Suppl. S1), 172–181.

www.ingramcontent.com/pod-product-compliance
Lightning Source LLC
Chambersburg PA
CBHW061838210926
51850CB00006B/919
* 9 7 8 3 0 3 6 5 8 0 5 6 2 *

MDPI
St. Alban-Anlage 66
4052 Basel
Switzerland
Tel. +41 61 683 77 34
Fax +41 61 302 89 18
www.mdpi.com

World Electric Vehicle Journal Editorial Office
E-mail: wevj@mdpi.com
www.mdpi.com/journal/wevj

MDPI